Nanophase Materials

Nanophase Materials

Proceedings of the
2nd National Meeting on Nanophase Materials
Rome, Italy, May 1995

Editors:

E. Bonetti
University of Bologna

D. Fiorani
ICMAT-CNR, Rome

TRANS TECH PUBLICATIONS
Switzerland • Germany • UK • USA

ISBN 0-87849-709-9

Volume 195 of
Materials Science Forum
ISSN 0255-5476

***Distributed** in the Americas by*

Trans Tech Publications Ltd
Whitman Distribution Center
10 Water Street, Room 310
Lebanon, NH 03766
USA
Phone: (603) 448 0317
Fax: (603) 448 2576

and worldwide by

Trans Tech Publications Ltd
Brandrain 6
CH-8707 Zurich-Uetikon
Switzerland
Fax: (++41) 62 74 10 58
E-Mail: ttp@transtech.ch
http://www.transtech.ch/ttp

Printed in the United Kingdom
by Hobbs the Printers Ltd,
Totton, Hampshire SO40 3YS

C.N.R.

National Research Council

Progetto Finalizzato "Special materials for advanced technologies"
and
GNSM, National Condensed Matter Group

II National Meeting on

NANOPHASE MATERIALS

Rome, Italy May 23-24, 1995

Organizers: E. Bonetti (University of Bologna)
D. Fiorani (ICMAT-CNR)

Scientific Committee: P. Allia (Polythecnic of Torino)
L. Battezzati (University of Torino)
E. Bonetti (University of Bologna)
S. Enzo (University of Sassari)
D. Fiorani (ICMAT-CNR, Roma)
P.G. Merli (LAMEL-CNR, Bologna)
G. Piccaluga (University of Cagliari)
F. Ronconi (University of Ferrara)
A. Stella (University of Pavia)
E. Tondello (University of Padova)
M. Vittori (ENEA, Roma)

Secretary: G. Ianni (GNSM-CNR, Roma)

Sponsors:
- GNSM Sections: "Magnetic properties of condensed matter" and "Metals and low teperatures"
- Institute of Materials Chemistry, CNR, Rome
- Physics Department, University of Bologna
- Rivoira
- Oxford Instruments

PREFACE

The II Italian Symposium on Nanophase Materials was held in Rome, May 22-23, 1995, under the auspices of the National Research Council (National Condensed Matter Group, GNSM, and the "Progetto Finalizzato: Special materials for advanced technologies").

The purpose of the Symposium was to provide an interdisciplinary forum for the national research community for discussing latest developments in the field of Nanophase Materials. The Symposium topics were indeed interdisciplinary, covering theory, materials preparation, structural characterization, thermodynamic aspects and mechanical, optical, electrical and magnetic properties.

We wish to thank all the partecipants in the Symposium for their stimulating presentations and contributions to these Proceedings. We also thank the members of the Scientific Committee and in particular the Symposium secretary Grazia Ianni.

The support coming from the GNSM Sections "Magnetic properties of condensed matter" and "Metals and low temperatures", from the CNR Institute of Materials Chemistry and from the Physics Department of the University of Bologna is acknowledged. Finally, thanks are due to Rivoira and Oxford Instruments for kindly supporting the Symposium.

E. Bonetti
Physics Department
University of Bologna

D. Fiorani
Materials Chemistry Institute
C.N.R., Rome

Table of Contents

Section II: Physical Properties

SECTION I

Synthesis and Structural Investigations

Materials Science Forum Vol. 195 (1995) pp. 1-6

Fabrication, Characterization and Application of Ni/SiO_2 Nanocomposite Materials Prepared by Sol-Gel

R. Monaci, A. Musinu, G. Piccaluga and G. Pinna

Dipartimento di Scienze Chimiche, Via Ospedale 72, I-09124 Cagliari, Italy

Keywords: Nanocomposites, NiO/SiO_2 Catalyst

Abstract.

Composites of nanoparticles of nickel dispersed in amorphous silica have been prepared by the classical sol-gel method. Mild conditions have been used whenever possible to prevent the formation of large particles of nickel; times and temperatures of the reduction step were varied in the treatment of samples with 7% nickel to verify their influence on the particle size. The samples with Ni particles of about 5 nm showed good catalytic activity in a standard hydrogenation reaction.

Introduction.

Materials composed by ultrafine particles dispersed in amorphous matrices to form nanocomposite materials have been studied extensively in the last few years because of their interesting properties and potential applications in catalysis [1], electric devices [2], in magnetic [3], optical [4] and electronic fields [5]. Among the physical and chemical methods devised for preparation of nanocomposite and nanosized materials, syntheses from atomic or molecular precursors have proven to give the better control over particle size and over omogeneity in particle distribution. Sol gel methods in particular are showing to be effective in producing crystalline particles of nanometric dimensions [6,7], and nanocomposites formed of a nanocrystalline phase dispersed in an amorphous (or ceramic) matrix [8-9].

Recently, we have started a research project aimed to the preparation and structural characterization of materials composed of nanocrystalline metals dispersed in an amorphous phase. In particular, we intend to investigate selected Me/SiO_2 systems with wide ranges of metal content; accurate evaluation of the preparation method and of the experimental conditions are therefore necessary to contain metal particles to the nanometer size. To this, the sol-gel method seems to be a good candidate in view of its versatility [10-14]; however, it is also extremely sensitive to all the possible variations introduced in anyone of the steps leading to the final product. So, the first stages of the preparation, usually carried out at temperatures not much higher than ambient, greatly influence the amorphous network and the omogeneity of the sample, while the following thermal treatments are the most delicate for the control of the crystalline particles. Therefore, as a first approach in our investigation we concentrated our attention on the high temperature stages, while keeping constant all the other preparation conditions.

We started our study preparing samples of nickel dispersed in silica. Samples with different amounts of nickel have been obtained and the evolution of particle size has been followed using X-ray Diffractometry (XRD) and Transmission Electron Microscopy (TEM). Besides, the samples with low metal content have been tested as catalysts. In fact, Ni on silica has been used for decades in hydrogenation reactions. As catalytic activity is related to the catalyst surface area, we wanted to verify the correlation between catalytic activity of our samples and the size of their metal phase.

Experimental.

Composites of crystalline nickel and amorphous silica were prepared by mixing alcoholic solutions of TEOS ($Si(OC_2H_5)_4$) and aqueous solutions of Ni nitrate ($Ni(NO_3)_2.6H_2O$). Several preparations were carried out, keeping the molar ratio ethyl alcohol to TEOS to water equal to 3.85:1:10.2 and the ratio Ni / ($Ni+SiO_2$) equal to 7%. Samples with greater Ni content were also obtained (14% and 21% respectively), but only the ones with less nickel were subjected to catalytic tests. A few drops of nitric acid were also added to maintain the solution under pH = 4. The mixtures were stirred for one hour, then poured in teflon beakers and allowed to gel in ambient atmosphere. After two days the gels were put in stove for a drying treatment, and kept there for one week at temperatures slowly raised to 100 °C. Thermogravimetric Analysis (TGA) carried out in ambient atmosphere, Fig. 1, showed that continuous modifications of the samples take place in a wide range of temperatures; however, the most dramatic changes occur above 300

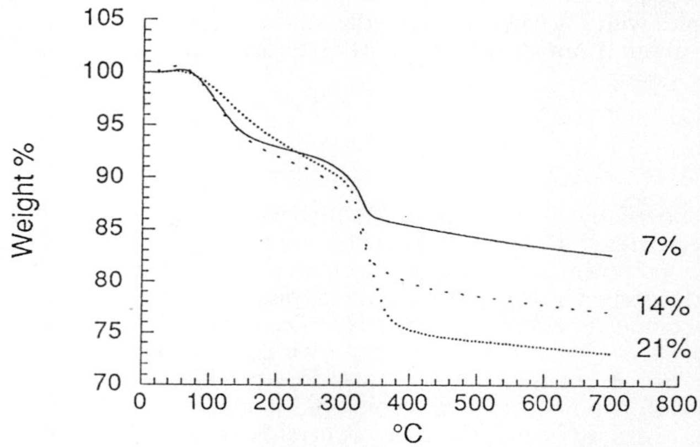

Fig. 1 - TGA measurements of dried gels composed of Ni/SiO_2 with Ni content equal to 7%, 14%, and 21% respectively.

°C, with heavy weight loss corresponding to the formation of crystallites of nickel oxide. Therefore, mild thermal treatments are better to be chosen at this stage if one wants to avoid the formation of large oxide particles, which in turn would need severe and prolonged reducing treatments to be transformed into metal. The dried gels were then crushed to powder, transferred into an oven and kept at 220 °C for 6 hours. After this treatment the samples had lost much of ethanol and water, but no crystalline peaks appeared in their XRD spectra. The following reduction step has to be carried out at high temperature in reducing atmosphere, so it needs to be carefully controlled. Several parameters might be important for the final dimensions of Ni particles: temperature of the treatment, reduction time, temperature gradient during the heating and cooling processes, concentration of H_2 in the hydrogen stream. We concentrated our attention on the influence of the first two parameters, reduction temperature and time, keeping the other conditions rigorously constant.

The structural modifications of the samples subjected to thermal treatment were followed by X-ray diffractometry using a conventional equipment (Seifert XRD 3000TT) and Cu K_α wavelength; nickel metal particles were observed and measured on electron micrographs obtained with a electron microscope (Jeol 200CX), TGA measurements were carried out using a Perkin-Elmer TGA7.

Results and Discussion.

Preliminary trial tests confirmed the effect of long thermal treatments in air or nytrogen: at temperatures around 500 °C the formation of a crystalline phase of nickel oxide was apparent in XRD spectra, while TEM images showed crystallites of sizes ranging from 50 to 150 nm; the same samples put in reducing atmosphere gave rise to composites with the nickel oxide only partially transformed into metallic nickel. The following procedure was therefore adopted: the pretreated gels were put into a reactor tube under a steady flow of pure hydrogen (40ml/min) and heated to the preset temperature with a rate of 15 °C/ min.

To verify the effect of the permanence time at high temperature, four samples were brought to 500 °C under the same conditions and kept in reducing atmosphere for different times, 10 min, 30 min, 60 min, 240 min, then allowed to cool. Fig. 2 shows their diffraction patterns. The spectra are very similar, with small humps emerging from the spectrum of the amorphous silica which has its maximum around θ=11.5 deg. The comparison with the lines at the bottom, which mark the positions of the three strongest reflections of metallic nickel (full lines) and nickel oxide (dashed lines), shows that the features emerging from the silica background are due to metallic nickel, clearly organized in small and disordered crystallites. A very weak evolution is perhaps detectable with increasing reduction time, but of so little importance that TEM micrographs seem not to record them; Fig. 3 in fact shows a TEM image of the sample kept at 500 °C for 4 hours, with Ni crystallites of dimension around 5 nm homogeneously dispersed in the silica matrix.

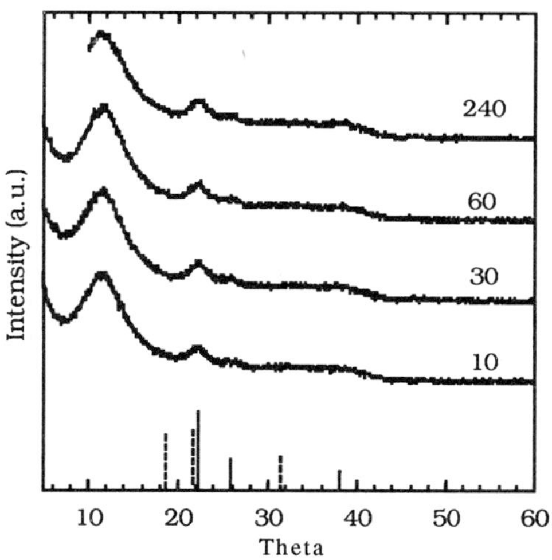

Fig. 2 - XRD spectra of Ni/SiO_2 composites obtained reducing the dried gels at 500 °C for different times: 10', 30', 60', 240' (from bottom). The vertical bars mark the positions of the strongest Ni (full) and NiO (dashed lines) reflections.

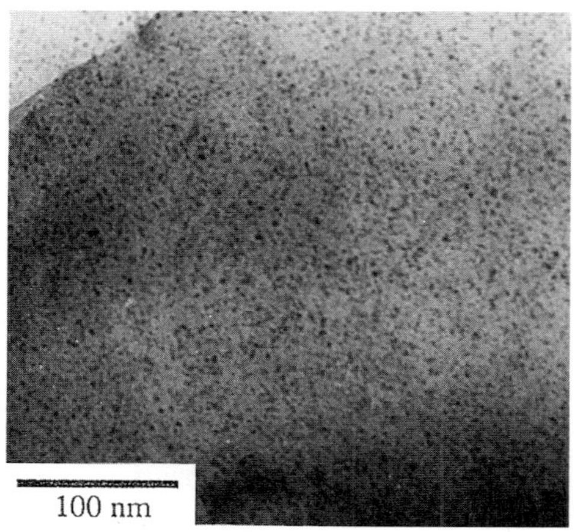

Fig. 3 - TEM image of the sample reduced at 500 °C for 240 minutes.

The effect of reduction temperature on particle size was also investigated: four samples were brought to 450 ^{o}C, 500 ^{o}C, 550 ^{o}C, and 600 ^{o}C, respectively, and kept at those temperatures for 30 minutes. The XRD spectra of these samples are reported in Fig. 4. Also in this case the spectra appear similar, with traces of the features due to nickel strongest reflections; only in the bottom spectrum an additional shoulder is visible at $\theta\sim$ 21.8 deg, due to traces of nickel oxide not completely reduced at 450 ^{o}C. The four spectra display a weak evolution of the halo at θ= 22.2 deg with increasing temperature, which should indicate a slight increase in the average size of nickel particles. This fact is confirmed by TEM images: all the samples appear very similar, with uniform distributions of nickel crystallites of dimensions around 5 nm. However, careful examination of the sample kept at 600 ^{o}C shows the sporadic appearance of regions with particles of about 10 nm; the frequency of these larger particles has been evaluated around 1%. An example is given in Fig. 5.

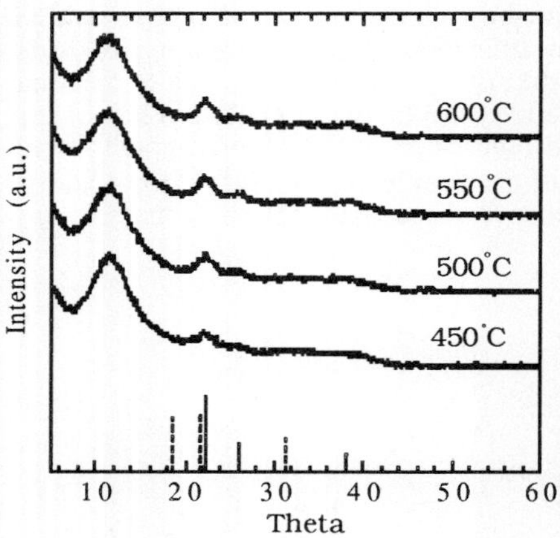

Fig. 4 - XRD spectra of the 7% Ni composite kept for 30' in H_2 atmosphere at the temperature reported.

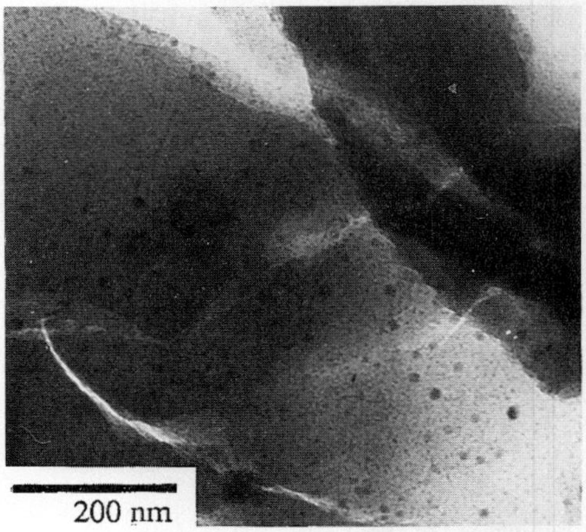

Fig. 5 - TEM image of a portion of the kept sample reduced at 600 ^{o}C, showing an unusual clustering of Ni particles larger than the average.

In summary, a mild effect of the temperature has been revealed in the 150 Celsius degrees explored, starting from a temperature (450 ^{o}C) barely sufficient for the reduction procedure to be effective; on the other hand, the effect of reduction time at 500 ^{o}C on particle dimension could not be visualized by TEM.

The catalytic properties of 7% Ni samples were tested in the hydrogenation reaction of toluene to methylcycloexane; the experiments were carried out in a microcatalytic flux reactor at atmospheric pressure and 200 ^{o}C. All the samples showed high catalytic activity, with conversion rates above 85% and practically constant for at least 6 hours. To test the catalyst stability, one portion of the same material was re-used after three months. Before the reaction, the powder was activated in H_2 stream at 500 ^{o}C for 10 minutes, then the temperature was lowered to 200 ^{o}C and the reagent introduced. The reaction was followed for 49 hours: the conversion rate was very high (about 90%) with no apparent decrease. From all the tests carried out, the proof of a strict correlation between catalytic activity and particle dimension emerged: high catalytic activity was shown by materials with small Ni particles (5-10 nm), medium activity in presence of medium particles (20-50 nm), low activity with large Ni nanoparticles (50-100 nm). Summing up, the samples were easy to handle, easy to prepare, showed high activity and good reproducibility.

Fig. 6 shows XRD spectra of samples with higher Ni content 14% (bottom), 21% (top). The materials were obtained following the procedure given above, except that the reduction treatment of the 21% Ni had to be carried out at 600 °C. The spectra display progressive modifications of the Ni features, which, at least in the top curve, may be called peaks; however, even in this case, they appear very large and not too prominent in comparison with the amorphous silica background, indicating that he metal is still organized in small and disordered crystallites. New preparations are now being carried out under slightly different experimental conditions, with the purpose to understand and hence better control the preparative procedures.

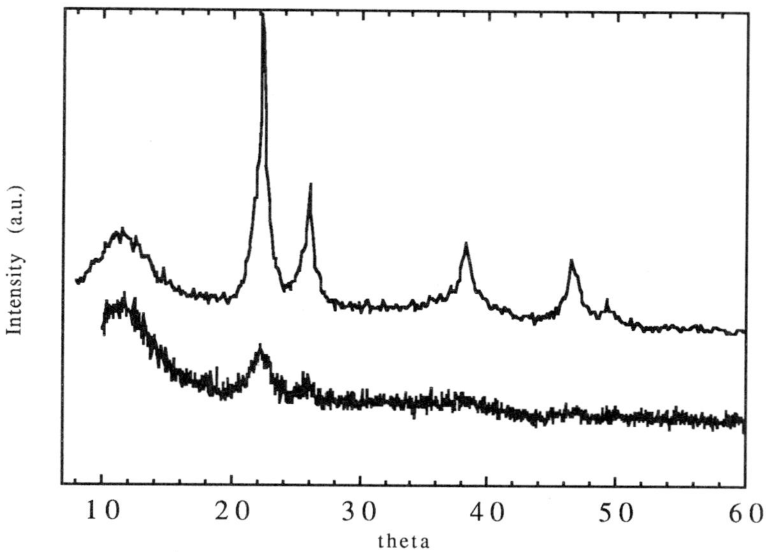

Fig. 6 - XRD spectra of Ni/SiO_2 composites with higher Ni content : 14% and 21% from bottom.

Conclusions.

Nanocomposites of nickel dispersed in an amorphous silica matrix have been prepared by sol-gel. The weak effect of the high temperature reduction step on the size of nickel particles has been investigated in the case of 7% Ni samples. These same samples have shown to be good catalysts in a trial hydrogenation reaction. Further investigation on the influence of the preparation parameters are needed in order to control the size of Ni crystallites with increasing Ni content.

Acknowledgements.

This work was supported by CNR under the national project "Progetto finalizzato per tecnologie avanzate" and MURST.

References.

[1] A. Ueno, H. Suzuki, and Y. Kotera, J. Chem. Soc., Faraday Trans. 1, 79, 1277 (1983).
[2] B. Roy, S. Roy, and D. Chakravorty, J. Mat. Res. 9, 2677 (1994).
[3] I.M.L. Billas, W.A. de Heer, and A. Chatelain, J. Non-Cryst. Solids 179,316 (1994).
[4] R. Dagani, Chem. Eng. News, 23 Nov. 1992, 18.
[5] P. Marquardt, G. Nimtz, and B. Mühlschlegel, Solid State Comm. 65, 539 (1988).
[6] H. Bönnemann, W. Brijoux, R. Brinkmann, E. Dinjus, T. Joußen, and B. Korall, Angew. Chem. Int. Ed. Engl. 30, 1312 (1991).
[7] S. Utamapanya, K.J. Klabunde, and J.R. Schlup, Chem. Mat. 3, 175 (1991).
[8] A. Chatterjee and D. Chakravorty, J. Phys. D: Appl. Phys. 22 1386 (1989).
[9] B. Breitscheidel, J. Zieder, and U. Schubert, Chem. Mat. 3, 559 (1991).
[10] C.J. Brinker and G.W. Scherer, "Sol-gel Science", Academic Press, San Diego, CA, USA (1990).
[11] "Aerogels", J. Fricke ed., Springer-Verlag, Berlin Heidelberg (1986).
[12] S.J. Teichner, G.A. Nicolaon, M.A. Vicarini, and G.E.E. Gardes, Adv. Coll. Interf. Sci, 5, 245 (1976).
[13] M.S. Gebert, R.W. Pekala, Chem. Mater. 6, 220 (1994).
[14] C.D.E. Lakeman and D. Payne, Mater. Chem. Phys. 38, 305 (1994).

Materials Science Forum Vol. 195 (1995) pp. 7-12

Preparation and Characterization of Sol-Gel Derived Pure and Doped $CdTiO_3$

M. Canali[1], A. Montenero[1], G. Gnappi[2], D. Bersani[3] and P.P. Lottici[3]

[1] General Chemistry Dept., University of Parma, Viale delle Scienze, I-43100 Parma, Italy

[2] CO.R.I.VE., Viale delle Scienze, I-43100 Parma, Italy

[3] Physics Dept., University of Parma, Viale delle Scienze, I-43100 Parma, Italy

Keywords: Cadmium Titanate, Sol-Gel, CdS Doping, CdSe Doping

$CdTiO_3$, pure and doped with CdS and CdSe, was prepared by the sol-gel technique in bulk and in thin film form. $CdTiO_3$ powders were treated at different temperatures: both the ilmenite-like phase and the orthorhombic perovskite were obtained. In particular, these phases were obtained at lower temperature and in a shorter time with respect to conventional preparation methods. A simple way to dope the titanate with CdS and CdSe is proposed. Bulk and thin films samples were obtained.

INTRODUCTION

Among ferroelectric materials with perovskite structure, those based on lead titanates and zirconates are so far the most widely studied, while very few attention has been paid to cadmium titanate $CdTiO_3$, probably because this material does not exhibit ferroelectricity at room temperature. Nevertheless, $CdTiO_3$ is an interesting system for the investigation of the nature of ferroelectric and structural phase transition; furthermore, specially when doped with CdS and CdSe, it might have interesting properties in the field of non-linear optics.

In this work we report about the preparation of cadmium titanate both in bulk and thin film form following the sol-gel route, also discussing the structural evolution of the material as a function of different thermal treatments. $CdTiO_3$ was not only prepared as a pure compound, but it was also doped with CdS and CdSe: the presence of microcrystals of CdS and CdSe was detected by means of diffraction methods and spectroscopic measurements.

To the author's knowledge there are no literature data about the sol-gel preparation of $CdTiO_3$ doped with CdS and CdSe. Processes are reported about the sol-gel preparation of silica-based glasses doped with CdS and CdSe. Most of the methods reported for CdS doped SiO_2 glasses use a silicon alkoxide and small quantities of a cadmium salt, which lead to a porous final product, usually containing about 95% SiO_2 and 5% CdO: the bulk is then treated in H_2S flow, after putting the system under vacuum, to make cadmium react [1,2,3]. Another approach implies the use of thiourea in the starting solution, which brings to the formation of a complex between thiourea and cadmium salt, which is then destroyed during heating and organics combustion, giving rise to CdS [4,5,6]. For CdSe doping, Tohge [6] prepared a SiO_2 film doped with CdO: the porous obtained product is immersed in a solution of selenourea, to let selenide ions enter the pores and react with cadmium; Nogami [7] added Se, dissolved in HNO_3, to silicon alkoxide and cadmium salt and treated the resulting gel in hydrogen and nitrogen atmosphere.

In this work a method is proposed for obtaining doped materials without using vacuum equipments or gas flows. The source of sulfide and selenide ions was directly introduced in the starting solution, so allowing to simplify the method and a better control of stoichiometry.

EXPERIMENTAL

The preparation of pure $CdTiO_3$ via the sol-gel route has been extensively reported elsewhere [8]. The same procedure was adopted for the preparation of the doped compounds, choosing thioacetamide (CH_3CSNH_2, Carlo Erba RPE) and selenourea (CH_4N_2Se, Acros 99%) as the sources of S^{2-} and Se^{2-} ions respectively. The molar ratios were chosen in such a way to have Cd : Ti : S^{2-} (or Se^{2-}) = 1.00 : 1.00 : 0.05. The final gels, dried al 120°C, yielded bulk samples exhibiting the yellow and reddish colours typical of CdS and CdSe.

Crystalline phases were identified by using a powder X-ray diffractometer Philips PW1050, using Cu Kα radiation (λ = 1.54178 Å). Raman spectra were collected with back-scattering geometry with a double monochromator SPEX 1403, and a Coherent 2020 Ar^+ laser (λ = 5145 Å) as the light source. Mono- and multi-layer films were prepared by using a dip coating apparatus, using different withdrawal speeds. UV spectra were recorded on a Perkin Elmer Lambda 2 spectrophotometer.

RESULTS AND DISCUSSION

Bulk samples were heat treated under air atmosphere from 300 to 800°C, with a dwell of 30 min every 100°C. At the end of each dwell, powders were sampled and the correspondent and XRD analyses were recorded.

Fig. 1 shows the thermal evolution of pure $CdTiO_3$ powders. At low temperature only CdO is present, whose contribution decreases as temperature increases, until it disappears at 800°C (see also [8]). At 500°C the ilmenite phase of $CdTiO_3$ starts forming. The opposite behavior is noticed for the hexagonal form of $CdTiO_3$. Traces of TiO_2, both in anatase and rutile forms, can be due to the small fraction of cadmium decomposed at high temperature. A treatment at 500°C for 24 hours gave rise to pure ilmenite phase. Orthorhombic $CdTiO_3$ is formed after a treatment at 1100°C, for 4 hours. Traces of rutile are present, whose contribution increases with the duration of the thermal treatment. This would confirm cadmium decomposition.

A different sample containing an excess of cadmium (molar ratio Cd:Ti = 1.3 : 1.0) underwent a treatment similar to the previous ones. At 1100°C the detected phases are CdO and orthorhombic $CdTiO_3$ (see Fig. 2): this further confirms that the formation of TiO_2 is due to the lack of cadmium due to decomposition. Data were confirmed by Raman spectroscopy measurements [8].

XRD analysis recorded on CdS gel dried at 120°C showed the amorphous state of the samples. This would confirm the formation of microcrystals dispersed in the amorphous matrix, whose dimensions are below the instrumental detection limit, and not of larger agglomerates.

After performing thermal treatments, the diffractograms recorded at 700 and 800°C (Fig. 3) exhibit, besides the phases already discussed for pure $CdTiO_3$, the presence of the peaks of CdS whose intensity is consistent with the growth of this phase, which is present in low concentration.

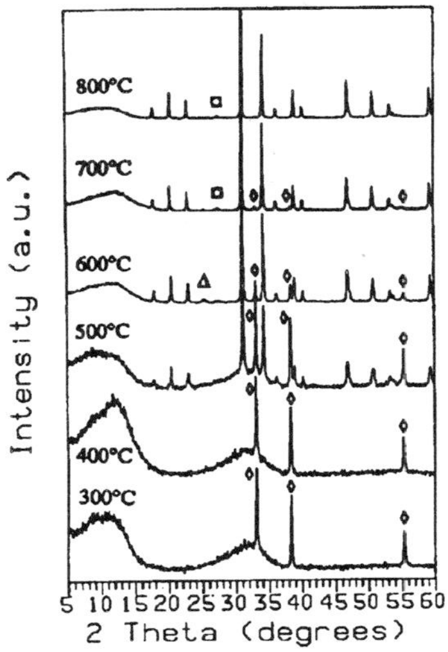

Fig. 1: Diffractograms recorded on heat-treated powders (◊ = CdO; Δ = TiO_2 A, ◻ = TiO_2 R; unmarked peaks correspond to ilmenite phase).

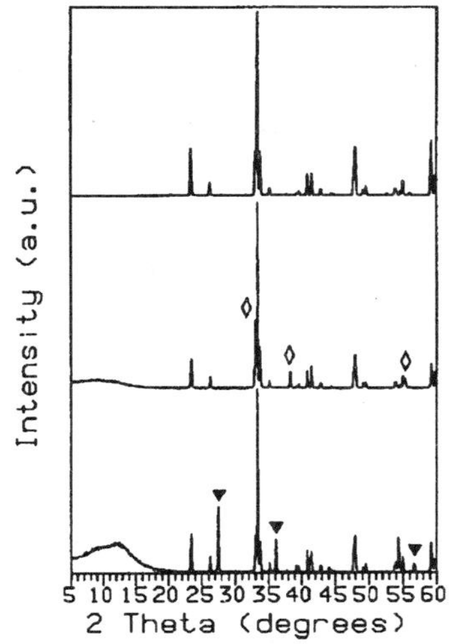

Fig. 2: From bottom: diffractogram of the stoichiometric sample (▼ = TiO_2 R), of that with excess cadmium (◊ = CdO), and calculated pattern of the perovskite phase. 1100°C

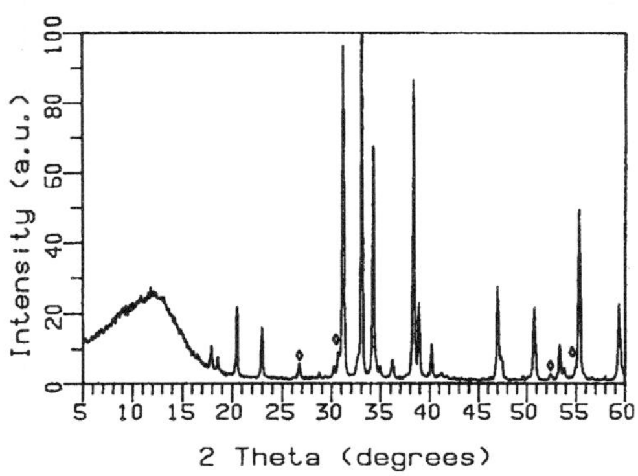

Fig. 3: Diffractogram of the CdS doped sample treated at 800°C (◊ = CdS)

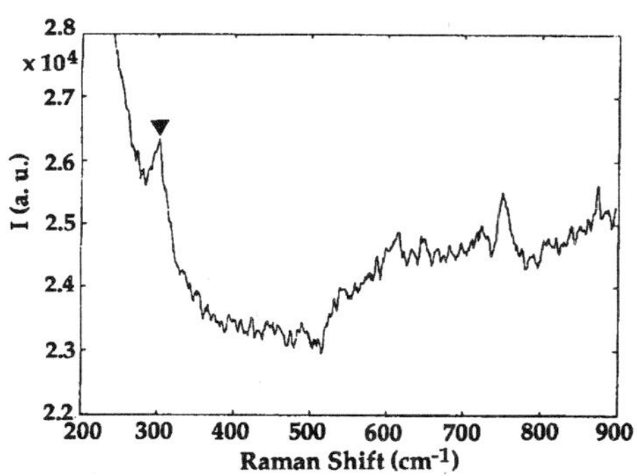

Fig. 4: Raman spectrum of the CdS doped sample treated at 70°C (▼ = CdS).

CdS-doped heat treated powders were also examined with Raman spectroscopy (Fig. 4): the spectrum recorded on the sample treated at 70°C shows a peak at about 300 cm^{-1}, which is due to the presence of CdS [9]. The spectra of the samples treated at higher temperatures are not particularly interesting, due to the disturbing effect of CdO.

Concerning the samples doped with CdSe, the diffractogram recorded on the dried gel confirms a predominantly amorphous state of the powders, plus the presence of small peaks, among which the three main signals of CdSe: their intensity is by the way extremely small.

XRD measurements on heat treated samples are totally similar to those obtained for pure $CdTiO_3$, and do not exhibit any signal attributable to CdSe. This result, which finds macroscopic evidence in the disappearance of the red colour, is also reported in literature [7], where the tendency of CdSe to decompose at temperatures above 300°C has been already evidenced.

Also with Raman results were not as clear as in the case of CdS doped samples. No peaks were found in the expected range, nevertheless fluorescence effects occurred, which might have covered weak signals.

The gel parent solutions were also used for depositing films with the dip coating method on glass microscope slides, after a dilution process.

For pure $CdTiO_3$ the as deposited films were first dried at 120°C: after cooling they were white and opaque. A further treatment at 450°C for 30 min makes the films transparent. Adhesion was successfully tested also on aluminum and steel.

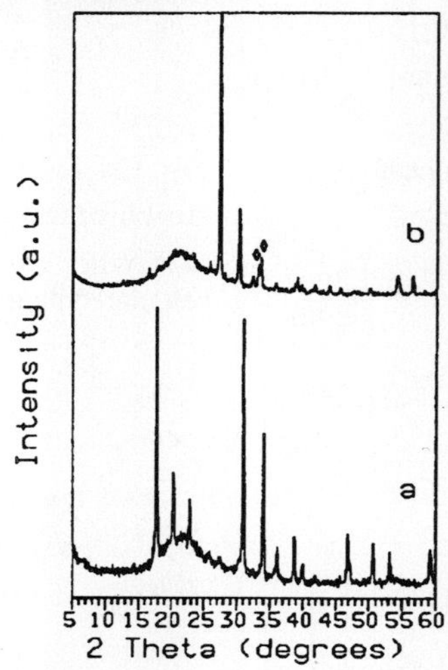

Fig. 5: Diffractograms of multilayer films treated at (a) 450°C (ilmenite $CdTiO_3$) and (b) 1100°C (◊ = perovskite $CdTiO_3$; unmarked peaks correspond to TiO_2).

Multilayer samples were prepared by making subsequent deposits on the same substrate, and treating every obtained film for 30 min at 450°C. A series of 19 thin layers was successfully obtained, with good homogeneity. Thanks to the increased thickness, it was possible to make XRD analysis on the samples: initially, the layers were amorphous. A further treatment at 450°C for 48 hours made the film opalescent and XRD measurements reveal the presence of rhombohedric $CdTiO_3$ (see Fig. 5a). 19-layer films were deposited on quartz and finally heated at 1100°C for 24 hours: X-ray diffraction pattern show the presence of TiO_2 (Rutile), which can be explained with cadmium evaporation (Fig. 5b) together with traces of perovskite $CdTiO_3$.

Films doped with CdS were dried at 60°C for 15 min after preparation, then thermal treatments were performed at 100, 150, 200, 250, 300, 350, 400 and 450°C for 1.5 hours. The coatings treated at 100÷250°C showed a yellow colour typical of CdS and a slight opalescence. The tendency of the sulfide to decompose and the darkening due to the formation of CdO make the deposits treated at higher temperatures similar to that of pure $CdTiO_3$. Such films were analyzed with a double-beam UV-Visible spectrophotometer, in the range 1100 ÷ 190 nm. Using an undoped slide as the reference, the spectra of the CdS doped samples were compared with those homologous (for treatment temperature and time, concentration, preparation condition) of pure $CdTiO_3$.

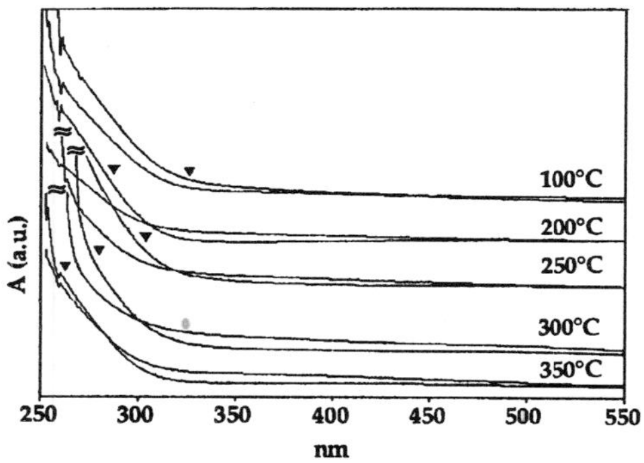

Fig. 6: Influence of treatment temperature on absorption spectra (▼ = CdS doped samples; unmarked lines = undoped samples).

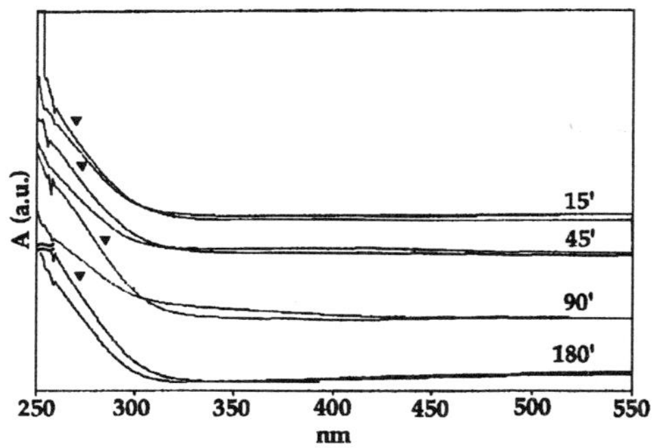

Fig. 7: Influence of treatment duration on absorption spectra (▼ = CdS doped samples; unmarked lines = undoped samples).

It is known [4,6] that products with bulk CdS or with big CdS crystals show a strong absorption at 510 nm, while microcrystalline dispersions absorb at lower wavelength. All the spectra reported in Fig. 6, relative to sample pairs treated at increasing temperatures for 90 minutes, show an absorption peak below 300 nm and the absence of signals around 500 nm: a similar trend was reported in literature [2,10]. Confirming the first impressions linked to the colour of the deposit, it is possible to note that the absorption difference between the doped and the undoped samples increases from 100°C to 250°C and decreases at higher temperatures (above 400°C the signals are practically superimposed). So, up to 250°C the supplied heat causes the formation and growth of microcrystals, while at higher temperatures CdS decomposes.

Also the duration of the treatments influences the absorption curves: CdS first grows, then decomposes (Fig. 7), relative to samples treated for different times at 200°C.

For what concerns CdSe-doped films, samples obtained at temperatures lower than 150°C exhibited a yellowish colour, which evolves to the reddish as time passes**. Those treated at higher temperatures resulted to be equal to the correspondent undoped, supporting the thesis of CdSe tendency to decompose. A series of samples was treated at 60, 100 and 150°C. The recorded spectra (Fig. 8) show that the absorption peak starts around 310 nm. Tohge [6] reports about a wavelength of 750 nm for bulk CdSe, while absorptions at lower wavelength are noticed as the particles diameter decreases. These experimental data would confirm the formation of microcrystals.

Furthermore, one can see as the temperature increase tends to uniform the two curves. These confirms the tendency of CdSe to form at low temperature and to decompose just above 150°C (as also reported in literature [7]). This is further confirmed by the analysis of three samples treated at 150°C for 10, 15 and 60 minutes respectively. The spectra reported in Fig. 9 show that the absorption decreases as treatment duration increases.

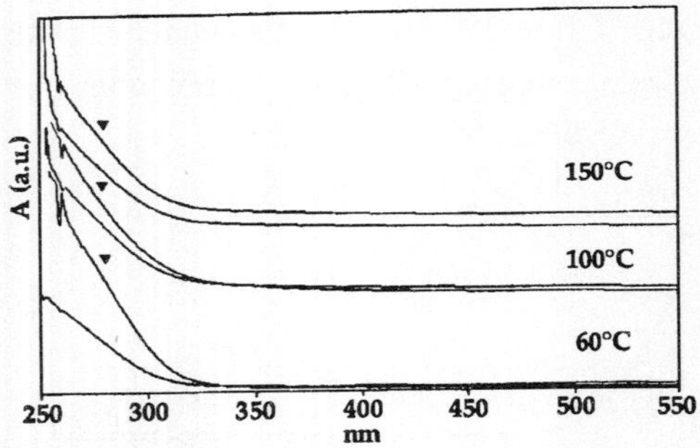

Fig. 8: Influence of treatment temperature on absorption spectra (▼ = CdSe doped samples; unmarked lines = undoped samples).

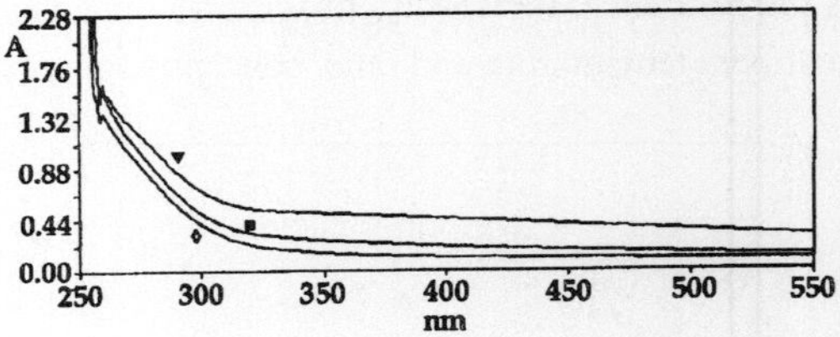

Fig. 9: Influence of treatment duration on absorption spectra of CdSe doped samples treated at 150°C (▼ = 10', ■ = 15'; ◊ = 60').

CONCLUSIONS

$CdTiO_3$ was prepared by the sol-gel method and characterized, in both its crystalline forms. XRD and Raman analyses confirmed the formation and the evolution of the ilmenite phase (not ferroelectric) into the perovskite one (ferroelectric). Due to its importance from the point of view of potential applications in electronics, the possibility of obtaining mono- and multi-layer films was studied. Homogeneous and transparent films were prepared, with different thickness and on various kind of substrates.

$CdTiO_3$ doped with CdS e CdSe were successfully obtained, both in bulk and in thin film form. Spectroscopic characterization confirmed the presence of microcrystallites of the dopants.

This work was partially supported by Progetto Strategico Materiali Innovativi of Italian National Research Council.

REFERENCES

[1] M. Nogami and K. Nagasaka: J. Non-Cryst. Solids 122 (1990) *101-106*.

[2] M. Nogami, K. Nagasaka and E. Kato: J. Am. Ceram. Soc. 73 (1990) *2097-99* .

[3] T. Takada, T. Yano, A. Yasumori, M. Yamane and J.D. Mackenzie: J. Non-Cryst. Solids 147&148 (1992) *631-635*.

[4] N. Tohge, M. Asuka and T. Minami: Chemistry Express, 5 (1990) *521-524*.

[5] N. Tohge, M. Asuka and T. Minami: SPIE Vol. 1328 Sol-Gel Optics (1990) *125-132*.

[6] N. Tohge, M. Asuka and T. Minami: "Preparation of Metal (II) Chalcogenide Semiconductor Doped Glasses by the Sol-Gel Process and Their Optical Properties", presented at "VI Int. Workshop on Glasses and Ceramics from Gels, Seville (Spain), 6-11 October 1991.

[7] M. Nogami, S. Suzuki and K. Nagasaka: J. Non-Cryst. Solids 135 (1991) *182-188*.

[8] M. Canali, A. Montenero, G. Gnappi, D. Bersani and P.P. Lottici, to be presented at ECerS - Symposium Electroceramics, 2-6 October 1995.

[9] D. Bersani and P.P. Lottici: Phys. Stat. Sol. (B) 170 (1992) *K5*.

[10] M. Nogami and A. Nakamura: Phys. Chem. Glasses 34 (1993) *109-113*.

Materials Science Forum Vol. 195 (1995) pp. 13-18

Electron Microscopy Characterization of Cu-Fe and Ag-Fe Alloys Obtained by Plastic Deformation

M. Angiolini[1], M. Krasnowski[1,2], G. Mazzone[1], A. Montone[1], M. Urchulutegui[1,3] and M. Vittori-Antisari[1]

[1] ENEA. CR Casaccia, INN-NUMA. C.P. 2400, I-00100 Roma, Italy

[2] Permanent Address: Institute of Atomic Energy, Swierk, Poland

[3] Permanent Address: Dpto. Fisica de Materiales, Fac. C. Fisicas, U.C.M. Madrid, Spain

Keywords: Nanostructures, Mechanical Alloying, Immiscible Systems

ABSTRACT

The behaviour of Cu-Fe and Ag-Fe powder blends plastically deformed in a high energy ball mill has been studied by transmission electron microscopy (TEM). Both systems do not show any intermediate phase or extended solid solution in the equilibrium phase diagram owing to their positive enthalpy of mixing. While Cu and Fe can be mixed in the liquid state, the enthalpy of mixing is so large in the Ag-Fe binary system that only a very limited solubility is observed even in the liquid state. Previous reported data show that upon high energy ball milling the former system forms a relatively homogeneous solid solution, while Ag and Fe do not show any appreciable solubility even after prolonged processing. The microstructural characterization of both systems, which show a different microstructure at the end of the milling process, and the comparison of their microstructural evolution during ball milling, allow to advance an hypothesis concerning the thermodynamic driving force giving rise to mechanical alloying in systems with a positive heat of mixing.

INTRODUCTION

The large plastic deformation imparted to the powders during high energy ball milling is one of the causes of the so called mechanical alloying, that is alloy formation by solid state reaction assisted by plastic deformation [1]. The process is carried out in a steel vial where some balls and the powder to be processed are sealed under inert atmosphere and subjected to mechanical shaking. The powder particles trapped between the colliding balls are subjected to impulsive plastic deformation experiencing both a strain and a strain rate well above those of the classical plastic deformation tests. When elemental powders blended at the desired average atomic ratio are processed, it is possible to observe how ball milling induces frequent cold welding and fracture among the different materials, giving rise to a particle morphology consisting of a convoluted lamellar structure whose spacing generally decreases with milling time. At the same time diffraction and calorimetric observations show that interdiffusion does occur and that intermediate phases or extended solid solutions begin to form. The reaction path depends critically on the thermodynamic properties of the processed elements as well as on the process parameters and on the presence of contaminants. The final products are often metastable phases, both crystalline or amorphous; the generally accepted interpretation of this behaviour is that the structure of the resulting phase or mixture of phases depends more on the kinetics of phase transformation than on the thermodynamic properties like the free energy difference between competing phases. In fact, the process is carried out at rather low temperatures such that the nucleation kinetics of the stable phases can be extremely sluggish. Generally, the formation of phases with a complicated crystal structure is hindered by this process and simpler phases are favoured even if less stable. In any case, most of the systems studied so far are characterized by a negative value of the enthalpy of mixing which has been considered responsible for the thermodynamic driving force of the reaction. Recently, also Cu-Fe and Cu-Co binary alloys have been synthesized with this method despite their positive heat of mixing[2], showing that high energy ball milling can affect not only the kinetics of the reaction but it can can also reverse the reaction direction expected on the basis of thermodynamic properties. However the ability of the process to this purpose appears limited to system with a positive enthalpy of mixing not exceeding some critical value, since Ag and Fe, whose positive enthalpy of mixing is so large that only a very limited mutual solubility is observed even in the liquid state, can not be mixed at an atomic level by mechanical alloying[3]. The hypotheses advanced to explain this effect take into account the modification of the free energy of the starting phases as a

consequence either of a high density of dislocations induced by the milling process [4], or of a high density of Cu-Fe interfaces produced by mechanical refinement of the multilayer down to nanometric level [5], or due to the presence of small particles or sharp tips produced by mechanical fragmentation [6], which cause the Gibbs-Thompson effect. Moreover the presence of oxygen contamination affecting the mixture composition and therefore the relevant thermodynamic parameters has been considered [7]. To this purpose it is necessary to remind that a not negligible oxygen contamination cannot be avoided with this material processing method, because of the high specific surface of the starting powders, where oxygen is present [8].
In a previous paper [9] we have shown that the microstructural features of ball milled Cu-Fe, as observed by TEM, do not appear to be coherent with any of the above models. In fact, we have observed the formation of the Cu(Fe) fcc solid solution with an Fe concentration above the solubility limit, in grains which have a size in the µm range, are relatively free of dislocations, and have flat interfaces with the neighbouring grains made of the pure starting elements. Moreover small voids, probably due to condensation of excess vacancies have been observed in the grain interior. Furthermore, we have observed the presence of the stable oxides expected on the basis of the relative phase diagrams, that is Cu_2O in the Cu rich grains and FeO in the Fe rich ones. These oxides are present as small particles precipitated in the matrix as evidenced by the fact that they have all the same crystallographic orientation relative to the matrix; this excludes the mechanism of mechanical dispersion of oxide particles already present. The change in the localization of the oxide, which is present as a surface layer in the starting powders and as precipitated particles in the interior after a certain time of processing, shows again that ball milling can affect, at least during some steps of the process the phase equilibrium. In fact, while both in the initial and in the final state the oxygen is in a stable compound, the precipitation of oxide particles in the matrix requires a supersaturated solid solution whose presence has to be hypothesized at some time, despite the negligible solubility of oxygen in these elements at low temperature. All the experimental observations reported in ref [9] show that high atomic mobility appears to be induced by the process, at least during some steps; the presence of intragranular voids seems to point the attention to the concentration of vacancies, which can be increased during plastic deformation by dislocation dynamics [10,11]. These considerations have allowed us to advance the hypothesis that the excess enthalpy associated with an abnormally high vacancy concentration can play a role on the phase equilibrium in this system. The purpose of this paper is to investigate further this point by microstructural observations of ball milled Cu-Fe and by a comparison with the behaviour of Ag-Fe powders processed in a similar way.

EXPERIMENTAL METHOD

The electron microscopy characterization of ball milled powders allows to study the crystallographic and chemical features of the material with high spatial resolution providing informations which cannot be generally obtained with bulk methods like X-ray Diffraction or Differential Scanning Calorimetry. This ability is particularly suited to describe the microstructural evolution of this kind of powders, considering their intrinsic inhomogeneity, at least in the intermediate processing steps. In fact, we are dealing with a solid state reaction, whose rate depends critically on those parameters like ball velocity at the impact, amount of powder trapped in the collision, collision frequency, which are experienced by the single powder particle within a statistical distribution. Different microstructures can then be present in different parts of the same sample, so that the bulk characterization methods providing results averaged on large volumes can hardly give information on the sequence of reaction steps leading to the final result. However reliable TEM analysis is restricted to a regime where the average grain size is larger than the typical sample thickness, which, for conventional TEM imaging and X-ray microanalysis, can be of the order of 100 nm. In fact owing to the fact that TEM observations project a tridimensional structure on a plane and that the X-ray microanalysis measures the average composition through the sample thickness, the overlapping of grains within the sample thickness, will negatively affect the reliability of the observations. On the other hand ball milling promotes a grain refinement, so that the average grain size of the final product can be as small as a few nm. In this work we have then studied powders milled for 1 hour to meet the above requirements on the average grain size, and also because our purpose is to study the first steps of the mechanical alloying process. All our results have then to be interpreted considering the structural inhomogeneity of the sample, where powder particles which have experienced a completely different thermo-mechanical history owing to the stochastic nature of the ball milling process, can be present in the same TEM sample. We have therefore to identify some microstructural mark which can be representative of the processing stage of the small volume of matter under examination by TEM. In the case of the Cu-Fe system the final product is a

rather homogeneous solid solution, so that to this purpose we have used the local chemical composition, while for Ag-Fe we have prepared also samples milled for longer times in order to study the microstructural evolution upon milling.
For both systems the powders were blended at the equiatomic average composition and milled in a Spex 8000 mixer-mill, under inert atmosphere.
Specimens for TEM observations have been prepared by embedding the powders in an Al matrix with the following procedure. A small amount of powder was placed between two Al foils about 0.5 mm thick and this composite was plastically deformed in a hydraulic press up to cold-welding of the Al foils. Since this method can in principle affect the powder microstructure, even if this effect is not expected to have an important role, owing to the severe plastic deformation experienced by the powders during the milling process and to the considerable difference in mechanical properties between the powder itself and the Al matrix, some specimens have also been prepared by embedding the powders in an epoxy resin. For both kinds of samples 3 mm discs could be punched after suitable mechanical grinding. Further sample preparation has been performed by dimpling, while final thinning was performed by Xe ion beam milling on a liquid N2 cooled stage.
Observations have been made with a Jeol 4000 FX Transmission Electron Microscope, operated up to 400 kV and equipped with a double tilt specimen holder and with an Energy Dispersive Microanalysis System.
Complementary information have been obtained by Scanning Electron Microscopy (SEM) with a Cambridge 250 MKIII instrument on samples prepared by embedding the powders in epoxy resin followed by conventional metallographic preparation.

EXPERIMENTAL RESULTS AND DISCUSSION

As far as the Cu-Fe system is concerned, most of the results have been previously reported [9] and they will be shortly summarized here, together with some new experimental findings.
As it can be expected, the sample microstructure after one hour of milling is strongly inhomogeneous and large volumes made only of bcc Fe or fcc Cu are present in the sample, in the interior of which some alloyed grains can be found. In agreement with previous considerations we have limited our analysis to those crystal grains crossing the sample from top to bottom and so showing a single crystal microdiffraction pattern. In fact only in this case the presence of both atomic species, revealed by X-ray microanalysis is indicative of a truly alloyed grain and not of overlapping grains of the two metals.
As stated before, most of the microstructural features appear to indicate that considerable atomic mobility is induced in the system by rapid plastic deformation. In fact the crystal grains, both the alloyed ones and those made of the pure metals, have an equiaxic shape, relatively flat grain boundaries and a low density of dislocations, indicating that dynamic recovery or dynamic recrystallization are caused by the process, probably because of the extremely high strain rate imparted to the powders by the colliding balls. Moreover, in the alloyed grains the chemical composition, measured with a spatial resolution of a few tens of nm, appears to be quite uniform without any gradient at the heterogenous interface between the two metals. It is important to notice that we did not observe, inside these grains, any fragment of the pure starting elements, suggesting that the alloying process proceeds through atomic diffusion at the heterogeneous boundaries. If this hypothesis is accepted, the absence of any measurable gradient at the interface is again indicative of an abnormally high diffusion coefficient[12]. Further indication of high mobility is given by the oxide precipitation discussed above. Further TEM analyses have revealed that in some alloyed grains with fcc structure and Fe concentration higher than about 30 at%, a modulated structure can be observed in particular conditions. Fig.1 shows the bright field image of one of such zones; the corresponding electron microdiffraction pattern is shown in the inset keeping relative orientation. The diffraction pattern, which can be indexed as corresponding to the [110] zone axis of a fcc crystal, indicates that the whole grain has the same fcc structure. The contrast modulation is parallel to one of the two sets of (111) planes as it can be seen from a comparison of the image with the diffraction pattern. To a first approximation the contrast remains the same if the reflections observed in the [110] diffraction pattern are excited one by one in the two beam diffraction condition. However, the contrast vanishes if the sample is rotated around any direction but that normal to the contrast modulation. This happens also if the sample is rotated without affecting the two beam diffraction condition, that is around the diffracting reciprocal lattice vector, thus showing that the nature of the contrast is not just related to the diffraction conditions. This behaviour can be interpreted considering a lamellar structure of the grain, which, while keeping the same crystal structure, shows a modulation of the chemical composition with an associated lattice strain affecting the diffracted intensity. In this case, in fact, the contrast observed

depends both on the diffraction conditions and on the lamellae orientation relative to the electron beam. If the wavevector of the incoming electron beam is not contained into the interface plane, the contrast, averaged through the sample thickness, does not show any definite periodical modulation even if the diffraction conditions able to reveal the contrast between different lamellae are satisfied. We have to notice that the observed periodicity is of the order of a few nm while the sample thickness is typically in the hundred nm range; a small tilt angle is sufficient to average the structure when projected into the beam direction. All these features suggest that this structure is due to spinodal decomposition of the supersaturated Cu(Fe) solid solution from which composition modulation waves arise. While the crystal structure remains fcc, the difference in the atomic scattering factor as well as the development of strain due to the different lattice parameters of the Cu-rich and the Fe-rich solid solutions can be at the origin of the observed contrast. This hypothesis is supported by the fact that some fcc iron is always observed by Mössbauer spectroscopy in this kind of alloy even without any subsequent thermal treatment [13]. Moreover the modulation periodicity is in good agreement with that computed for similar structures[14].

As far as the Ag-Fe system is concerned, no intermediate phases have been observed even after prolonged milling. The major microstructural evolution is from a convoluted lamellar structure, observed after a short time of milling, to an equiaxed structure where Fe and Ag grains with a size of a few tens of nm are randomly mixed. Fig.2 shows a SEM micrograph, obtained with backscattered electrons, of a sample milled for 1 hour. It is possible to observe some spots with uniform contrast and with a size in the µm range, embedded in a matrix made of alternating layers of Ag and Fe . A similar feature is shown with a better resolution in the TEM micrograph of Fig.3, where it is possible to observe the equiaxed grain structure of the crystals made of the pure metals. It is interesting to notice that, on the basis of what is shown in Fig.2, the transformation from lamellar to equiaxed structure does not appear to proceed by continuous refinement of the original multilayer but it seems to arise suddenly in small localized volumes. This is confirmed by TEM observations where it is possible to observe some lamellae a few nm thick without any tranformation of the lamellar structure into the granular one (Fig.4).

A possible explanation for the change in the microstructure can be related also in this case to the formation of a solid solution of the elements during the plastic deformation transient, followed by a fast decomposition driven by the large positive enthalpy of mixing, as soon as the deformation stops. The decomposition probably occurs by nucleation and growth owing to the large atomic mismatch between the atomic species. In the case of the strongly immiscible Ag-Fe system we can make the hypothesis that only a particularly energetic collision between the balls is able to induce the reaction, so that the transformation from the lamellar to the granular structure proceeds by localized spots which after some processing time cover the whole sample volume.The change in the thermodynamic properties of a system during rapid plastic deformation can be related to the large amount of energy stored as crystal defects during plastic flow, and subsequently released as heat by defect annihilation. In fact, it has been experimentally observed that the ball collision is almost completely inelastic when some powder is trapped in between [15].On the other hand it has been shown by low temperature deformation experiments that a fraction of the energy required to plastically deform a crystal is stored into crystal defects and released when some atomic mobility is allowed by increasing the temperature [10,11]. It has been deduced from the annealing behaviour that point defects can play a dominant role in this process. Our observations of small voids seem to support this hypothesis. In the present case the deformation is carried out at effective temperatures somewhat higher than room temperature, such that some annihilation mechanism for point defects can be operative mainly in defective structures and at large supersaturation. The actual concentration of point defects during powder deformation is then expected to result from a balance between generation and annihilation kinetics; in this frame the high strain rate connected to the kind of deformation we are dealing with may play an important role in defining the defect supersaturation.

The presence of a high point defect supersaturation affects the system energy owing to their large energy of formation. As an example, the formation energy of a vacancy in Cu is about 1.5 eV, while the maximum value of the chemical potential of Fe in Cu is about 0.2 eV/atom[6]. Even if a detailed thermodynamic analysis is beyond the purpose of the present paper, from the comparison of these figures it is possible to notice that the system free energy can be lowered when vacancy annihilation occurs by atomic jumps of foreign atoms across the Cu-Fe interface.

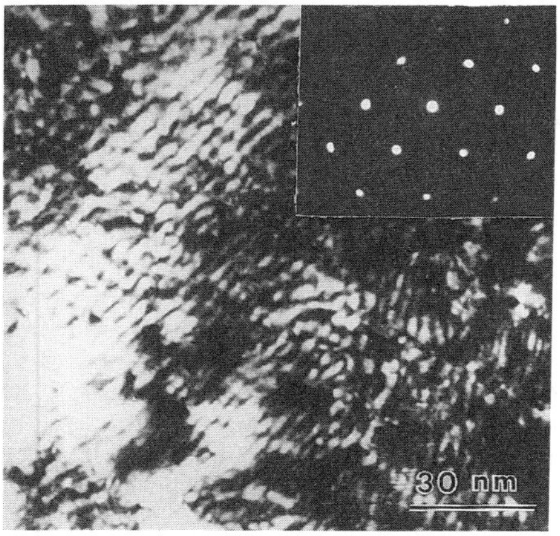

Fig.1. Bright field image of spinodal structure in a Cu(Fe) crystal. The diffraction pattern is reported in the inset.

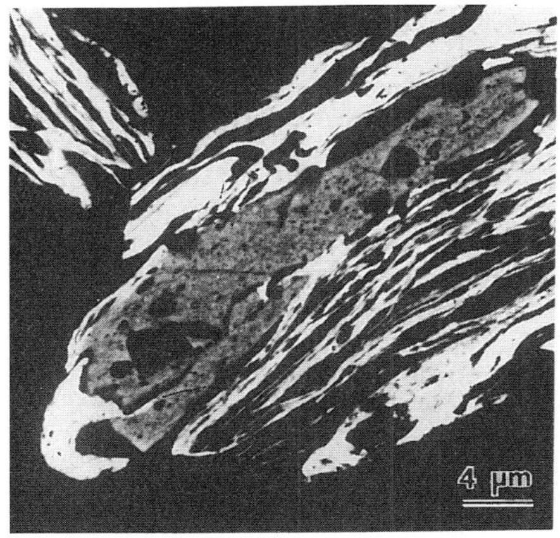

Fig.2. SEM image of the Ag-Fe blend milled for 1 hour.

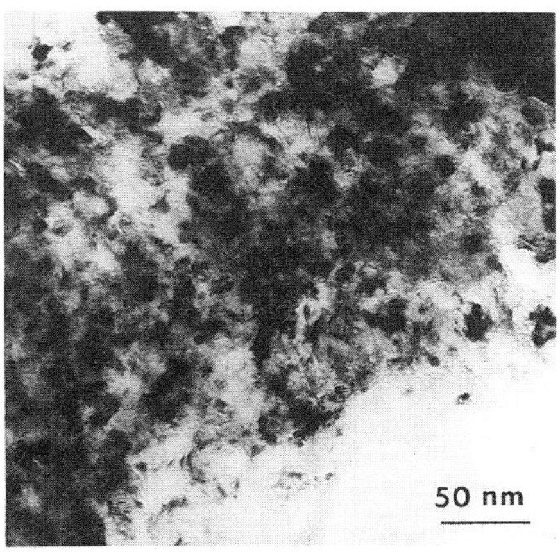

Fig.3 TEM image of the Ag-Fe blend milled for 1 hour.

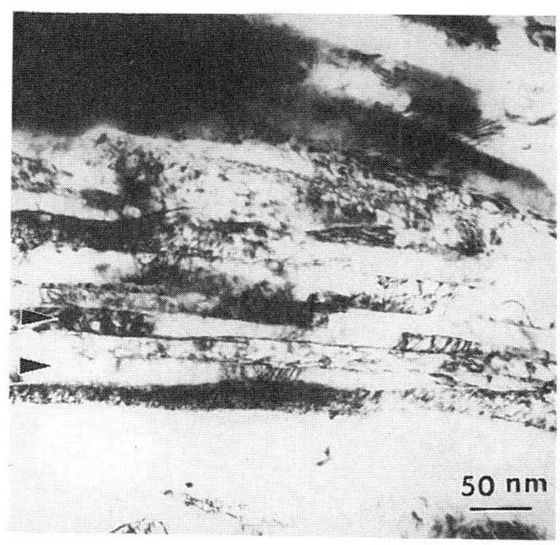

Fig.4. TEM image of the Ag-Fe blend showing thin Ag lamellae (arrowed).

CONCLUSIONS

The microstructural characterization of ball milled powders made of immiscible elements shows that the change in the thermodynamic properties required to explain the alloying reaction cannot be caused by the microstructural modifications permanentely present after milling, like small grain size or high dislocation density. This suggests that only during the plastic deformation transient the defect supersaturation can be high enough to affect significantly the free energy. According to this hypothesis, alloying occurs during the plastic flow and is assisted by an excess concentration of point defects generated by the dislocation dynamics. The high strain rate related to the impulsive deformation imparted by the ball impact is then expected to be necessary for high vacancy supersaturation. When plastic flow stops, defect annihilation lowers to a near equilibrium value the defect concentration. Probably during this step the system uses the residual atomic mobility to reduce the free energy of the solid solution by decomposition processes whose features are related to both the operative driving forces and to the actual atomic mobility. We can so observe oxygen precipitation as stable oxides, decomposition by nucleation and growth in AgFe and spinodal decomposition in CuFe. The last reaction is not complete probably because of a greater metastability related to the lower value of the positive enthalpy of mixing.

ACKNOWLEDGMENTS

We want to acknowlege F.Cardellini and R.Marazzi for their help in operating the ball mill.

REFERENCES

[1] R.Bormann, in "Materials by powder technology,PTM 93", F. Aldinger Ed, DGM Informationsghesellschaft, Oberursel, p 247 (1993).
[2] R.Uenishi, R.Kobayashi, S.Naso, H.Hatano, K.N.Ishihra, P.H.Shingu, Z.Metallk. 83, 132 (1992).
[3] J.Kuyama, H.Inui, S.Imaoka, S.Nasu, K.N.Ishihara, P.Shingu, Jap. J. Appl. Phys.30,L845 (1991).
[4] J.Eckert, J.C.Holzer, C.E.Krill III, W.L.Johnson, J.Appl.Phys., 73, 2794 (1993).
[5] G.Gente, M.Oehring, R.Bormann, Phys.Rev.B, 48, 13244 (1993).
[6] A.R.Yavari, P.J.Desré, R.Benameur, Phys.Rev.Lett, 68, 2235 (1992).
[7] A.R.Yavari, P.J.Desré, Mater. Sci. Forum, 88-90, 43 (1992).
[8] F.Cardellini, G.Mazzone, M.Vittori Antisari, Acta Metall.Mater. in press
[9] M.Angiolini, F.Cardellini, M.Krasnowski, G.Mazzone, A.Montone, M.Vittori-Antisari Micros. Microan. Microstr. in press
[10] J.F.Nicholas, Acta Metall. 7, 544 (1959).
[11] C.W. Berghout, Acta Metall., 6, 613 (1958).
[12] U.Gosele and K.N.Tu, J.Appl. Phys. 53, 3252 (1982).
[13] J.Z.Jiang, U.Gonser, G.Gente, R.Bormann, Appl.Phys.Lett. 63,(8),1056 (1993).
[14] A.Hutten, G.Thomas,Ultramicroscopy, 52,581 (1993).
[15] M.Magini, A.Iasonna Materials Transactions, JIM, 36,123 (1995).

Materials Science Forum Vol. 195 (1995) pp. 19-24

Synthesis of Nanocrystalline Iron Aluminides by Mechanical Alloying

F. Cardellini[1], V. Contini[1,2], G. Mazzone[1,2] and A. Montone[1]

[1] Enea, Centro Ricerche Energia Casaccia, Settore Nuovi Materiali C.P. 2400, I-00100 Roma, Italy

[2] Istituto Nazionale per la Fisica della Materia, Unità di Perugia, Perugia, Italy

Keywords: Mechanical Alloying, Ball Milling, Al-Fe Alloys

ABSTRACT

This study describes the effect of ball milling and heat treatment on the phases synthesized in a SPEX mill starting from mixtures of Al and Fe powders containing 75 and 80 at.% Al. Upon milling of the 75% Al mixture we have observed the formation of the Al_5Fe_2 phase. At the 80% Al composition we have observed the formation of an Al-rich ultrafine phase with a size of the coherent domains of the order of 1 nm.The thermal behaviour of the samples depends on the time of milling. Upon heating of samples milled for short times we have observed at both compositions the nucleation of the metastable Al_6Fe compound at about 320°C. At the 75% Al composition the Al_5Fe_2 compound has been observed to nucleate at about 380°C in those samples where it had not formed during milling. Finally, the nucleation of Al_3Fe was observed only at the 80% Al composition in the samples heated above 450°C.

INTRODUCTION

The improvement of physical and mechanical properties like diffusivity and ductility, resulting from grain size reduction, makes nanocrystalline materials potentially attractive for structural applications. For this reason several studies have been made to improve the mechanical properties of intermetallic compounds through control of their microstructure. Mechanical alloying is a powerful technique for the production of a variety of alloy phases in the nanocrystalline state [1]. Particularly interesting for their excellent combination of properties are several light metal-based intermetallic compounds like the iron aluminides. Addition of Fe to Al can increase the use temperature and, because of this, there have been several investigations on the mechanism of phase formation and on the thermal stability of the end product of ball milling of Al and Fe powder mixtures [2-6]. This study extends the investigations reported above with the purpose of providing further information relative to the solid state reactions induced by ball milling of Al-rich Al-Fe powder mixtures.

EXPERIMENT

Al powder of 99.95% purity (particle size 50-100 μm) and Fe powder of 99.999% purity (particle size 5-10 μm) both by Koch Chemicals were mixed at the compositions 75 and 80 at% Al. Milling of these mixtures was performed in steps of 30 min alternated with rest intervals of 5 min, using an air cooled SPEX 8000 mixer/mill equipped with a hardened steel vial and two balls of 13.5 mm diameter and 10 g weight (ball to powder ratio 6:1). To prevent sample oxidation, the powders were sealed in the vial under argon atmosphere and a fresh sample was used for each ball milling run.The structural characterization was performed by X-ray Diffraction (XRD) and Transmission Electron Microscopy (TEM), while the thermal stability of the samples was tested in a Differential Scanning Calorimeter (DSC). XRD patterns were obtained using crystal focused and monochromated Co $K\alpha_1$ radiation in a Debye-Scherrer diffractometer equipped with a curved 120° position sensitive detector.TEM observations were performed with a JEOL 4000 FX microscope equipped with EDS facility. The TEM samples were prepared placing the powder between two Al foils 0.5 mm thick and then pressing the composite to a reduced total thickness of about 0.1 mm. In this way the powder particles were embedded in the Al matrix from which small discs could be punched for further thinning by dimpling and final polishing by Xe ion beam milling on a liquid N_2 cooled stage. Thermal measurements from 100 to 700 °C were performed in a computer assisted Perkin Elmer DSC7 calorimeter operated in the specific heat mode at a constant heating rate of 0.25 K/s.

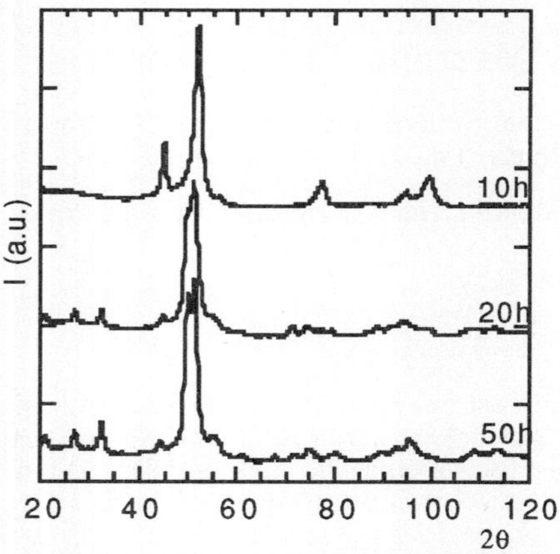

Fig.1a. XRD patterns of Al75Fe25 milled for 10, 20 and 50h

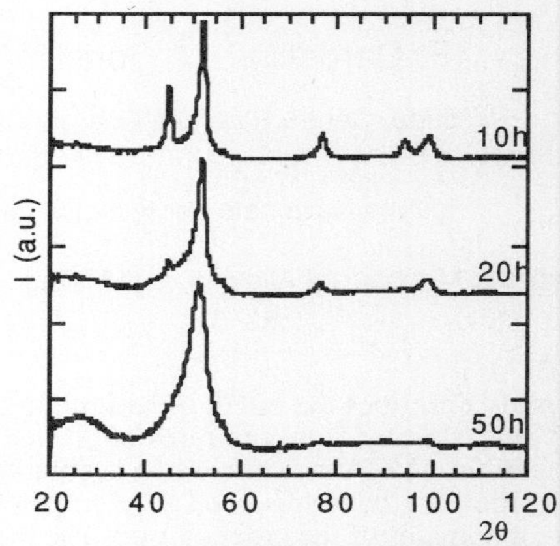

Fig.1b. XRD patterns of Al80Fe20 milled for 10, 20 and 50h

RESULTS

The XRD patterns at both compositions after 10, 20 and 50 h of milling are reported in Fig.1. For shorter milling times the XRD patterns (not reported) show, in addition to broadening of all reflections, a progressive decrease of the Al intensity with the exception of the peak at 2θ~52° which is indexed as Al (200) overlapped with Fe (110). At the 75% Al composition the XRD patterns after 20 and 50 h of milling are very similar showing that apparently these samples have completely transformed to the Al_5Fe_2 phase with a very fine grain size. At the 80% Al composition the XRD pattern of the 20 h sample shows the presence of a small amount of unreacted Al and Fe. In addition, one observes a strong asymmetric reflection peaked at 2θ~52° whose intensity is out of

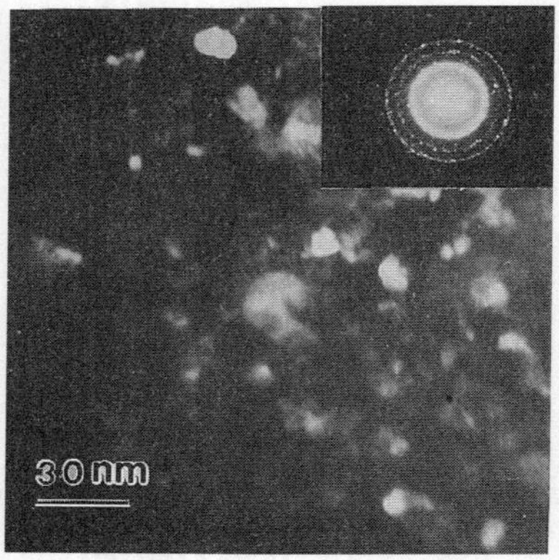

Fig.2a Dark field image of Al75Fe25 milled for 20h. Inset show selected area diffraction pattern.

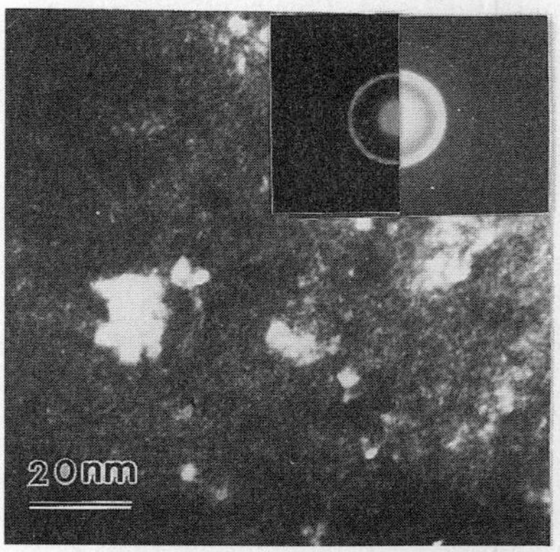

Fig.2b Dark field image of Al80Fe20 milled for 50h. Inset shows micro-diffraction pattern from ultrafine phase.

proportion with that of the other reflections. In the 50 h sample at the same composition, in addition to the strong peak at 52° whose full width at half maximum (FWHM) is of the order of 5° 2θ, one observes several weak bumps.

The microstructure of some of these samples has been examined also with TEM observations. Fig.2a shows a dark field image of the 75% Al sample milled for 20 h together with a selected area diffraction pattern. The sample appears to consist of crystallites with size comprised in the 10 to 50 nm range, while the diffraction pattern is consistent with the XRD observations confirming that the sample is essentially single-phase Al_5Fe_2. Fig.2b shows a dark field image of the 80% Al sample milled for 50 h together with a microdiffraction pattern recorded at two levels of intensity . The dark field image shows clearly the presence of two phases: the first, composed of crystallites of 5-10 nm size, some of which appear fragmented in even smaller pieces; the second, whose volume fraction comprises the vast majority of the sample, composed of ultrafine domains with size of~ 1nm. The microdiffraction pattern obtained with a beam of ~20 nm diameter positioned on the ultrafine phase shows, in agreement with XRD results, a very intense inner ring and three or four weak outer rings.

The effect of heat treatment depends on milling time at both compositions. The DSC thermograms of the 75% Al samples are shown in Fig.3. The sample milled for 10 h shows two exothermal peaks, the first corresponding to the formation of Al_6Fe with residual Fe clearly visible in the XRD pattern, the second corresponding to the formation of Al_5Fe_2 (Fig.4).Further heating to 700°C sharpens the peaks of Al_5Fe_2 without any phase change. The thermogram of the 50 h sample (Fig.3) shows a much smaller heat release, partially masked by a broad endothermal feature, arising essentially from sample homogenization, grain growth and strain release as confirmed by the XRD pattern (not shown) which, after DSC heating to 700°C, is very similar to that of the 10h sample heated to the same temperature. The thermogram of the 80% Al sample milled for 10 h (Fig.5) shows two exothermal peaks, the first corresponding to the formation of Al_6Fe (Fig.6) as observed at the other composition and the second to the formation of Al_5Fe_2 as demonstrated by the Al_5Fe_2 reflections showing, together with those of Al_6Fe, in the XRD pattern recorded after DSC heating to 450°C (Fig.6). Further DSC heating to 700°C transforms the sample to Al_3Fe which is the only phase showing in the XRD pattern. The sequence of transformations induced in the 80% Al samples milled for 20 and 50 h by DSC heating to 700°C is very similar notwithstanding the differences in the DSC thermograms (Fig.5). In particular in the 50 h sample one observes a low temperature endothermal feature which overlaps with the beginning of the exothermal release. The XRD patterns of the two samples, recorded after DSC heating to 400°C are shown in Fig.7. Comparing the XRD pattern of Fig.7 (curve a) with that of Fig.1b, one observes that after DSC heating the reflections corresponding to residual Al and Fe are no longer

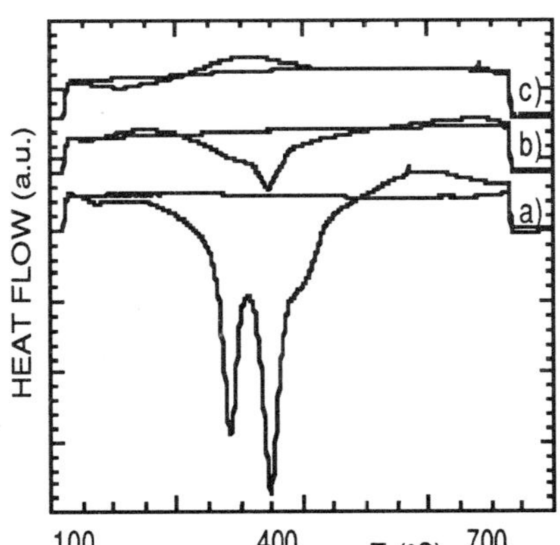

Fig.3. DSC thermograms of Al75Fe25 milled for: a)10h; b)20h; c)50h

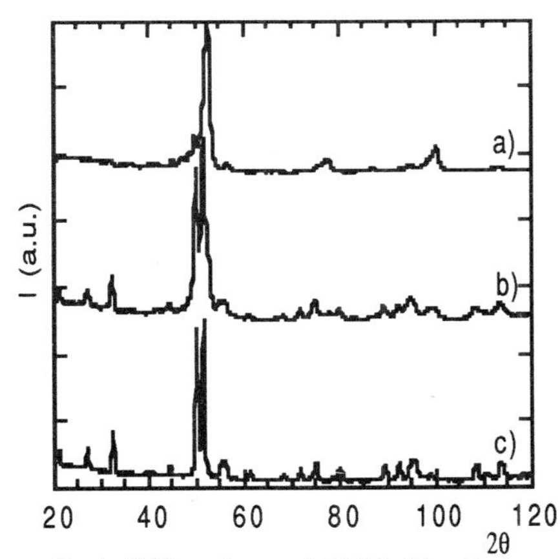

Fig.4. XRD patterns of Al75Fe25 milled for 10h after DSC heating to: a)340°C; b)400°C; c)700°C

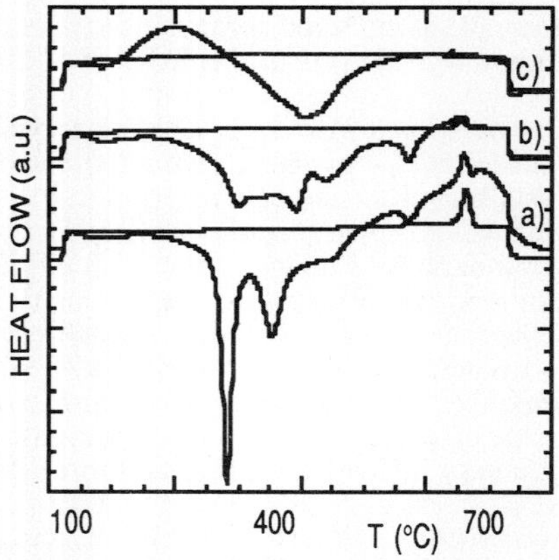

Fig.5. DSC thermograms of Al80Fe20 milled for: a)10h; b)20h; c)50h

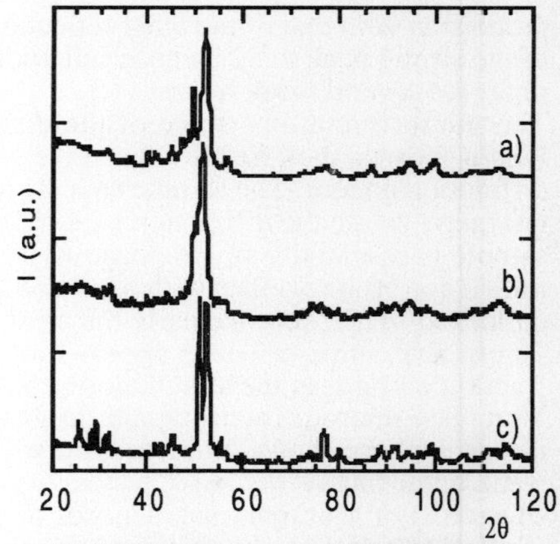

Fig.6. XRD patterns of Al80Fe20 milled for 10h after DSC heating to: a)340°C; b)450°C; c)700°C

visible. One has therefore, that the dissolution of the residual Al-rich and Fe-rich phases present in the 20 h sample causes a thermal release absent in the 50 h sample. In addition to the strong peak at 2θ~52° both XRD patterns show an increased amount of structure which, even if it cannot be resolved in the individual reflections, points to a higher degree of cristallinity in the heat treated samples. The further evolution of the 20 h sample is shown in Fig.8. At 430° C one observes the presence of Al_5Fe_2 while at 530°C the XRD pattern can be indexed as a mixture of Al_5Fe_2 and Al_3Fe. Finally, at 700°C the reflections of Al_3Fe are well resolved and the whole XRD pattern is indistinguishable from that of the sample milled for 10 h and shown in Fig.6. As mentioned above, the same reactions at the same temperatures are shown by the 50 h sample.

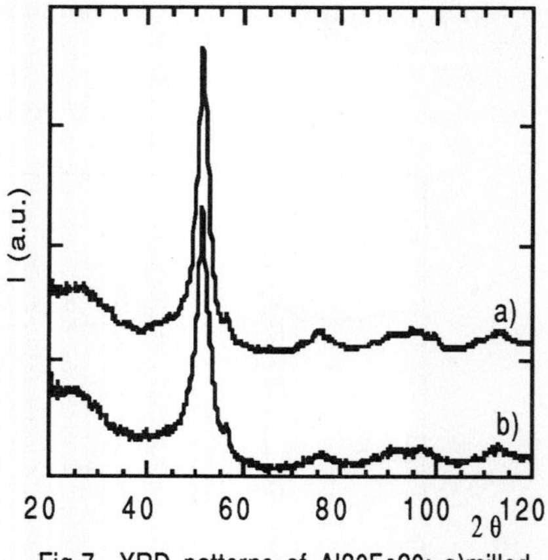

Fig.7 XRD patterns of Al80Fe20: a)milled for 20h and heated to 400°C; b)milled for 50h and heated to 400°C

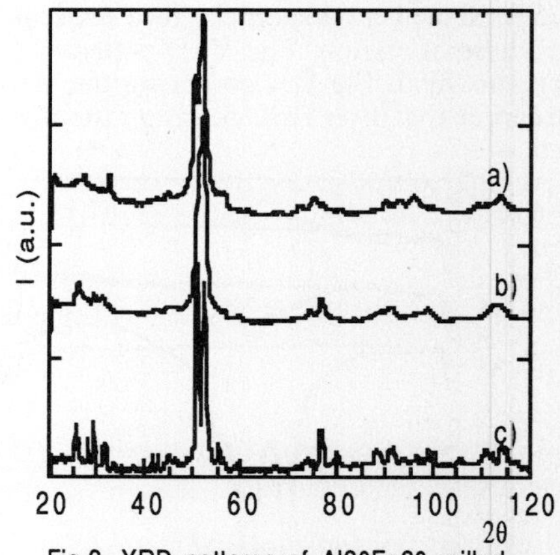

Fig.8. XRD patterns of Al80Fe20 milled for 20h after DSC heating to: a)430°C; b)500°C; c)700°C

DISCUSSION

The results of the present study help to define the sequence of solid state reactions induced by ball milling of Al-Fe powder mixtures. Particularly significant is the comparison of these results with those of previous studies [2-6]. The first conclusion which can be drawn from the comparison is that in this range of composition the end result of the ball milling process depends critically on the type of mill. Actually, at the 75% Al composition, Shingu and coworkers, using a conventional low energy ball milling apparatus [2], on the basis of XRD observations report the formation of an amorphous phase after 454 h of milling. Similar results have been obtained by Dong et al.[3] who, using a planetary mill, on the basis of XRD observations report the formation of an amorphous phase after 180 h of milling in the 50 to 80 at.% Al range. On the contrary Mukhopadhyay et al.[6] using a SPEX mill, have observed at the same composition the formation of Al_5Fe_2 after ~15 h of milling in close agreement with our observations. Finally, Morris and Morris [4] who have used a planetary ball mill, after 40 h of milling at the 75% Al composition report the formation of Al_3Fe. From these results it is possible to conclude that ball milling of the 75% Al mixture at medium-high levels of intensity, corresponding to a milling time of 10-40 h, leads to the formation of an intermetallic compound (Al_5Fe_2 or Al_3Fe) while low intensity ball milling for much longer milling times leads to the formation of an amorphous phase.

At the 80% Al composition, Shingu and coworkers [2] report the formation after milling times of several hundred hours of an amorphous phase as in the case of Al75Fe25. However, the same Authors report that the Mössbauer spectra at the two compositions are totally different, showing a broadened ferromagnetic sextet at 80% Al and a paramagnetic doublet at 75% Al. These observations are in contrast with the Mössbauer measurements of Dong et al. [3] who, at the 80% Al composition, report that the amorphous phase exhibits only the paramagnetic doublet with no trace of residual ferromagnetic sextet. These two experiments, therefore, show quite clearly that unequivocal identification, based on XRD patterns, of a phase formed during ball milling is sometimes very difficult. At the same composition Morris and Morris [4] report the formation of an amorphous phase after 40 h of milling and of Al_3Fe after 65 h, so that it is possible to conclude that at both compositions the formation of an amorphous phase is favoured by low energy ball milling, in agreement with a trend already established in a number of other cases.

The DSC thermograms presented in Figs.3 and 5 show that the efficency of ball milling in promoting atomic level mixing is remarkably different at the two compositions. In fact, while at 75% Al the heat released by the sample milled for 20 h is a small fraction of the heat released by the 10 h sample, at the 80% Al composition the heat released by the 20 and the 10 h samples is practically the same, thus pointing to a considerable decrease in the rate of interdiffusion. The same trend can also be observed for a milling time of 50 h even if in these samples the presence of an endothermal effect prevents a quantitative evaluation of the heat release.

At both compositions the first effect of DSC heating on samples milled for short times (10 h) is to promote the nucleation of the Al-richest compound (metastable Al_6Fe) at about 320° C thus suggesting that, at least in the initial stages of ball milling, interdiffusion proceeds through diffusion of Fe into Al, in agreement with the results reported by Niu et al.[5]. Upon further DSC heating of the samples milled for 10 h, one observes at about 400° C an exothermal peak corresponding to the formation of Al_5Fe_2, with Al_6Fe still present in the 80% Al sample heated to 450 °C. At this composition the reaction between Al_6Fe and Al_5Fe_2 which leads to the formation of Al_3Fe occurs between 450 and 700 °C as shown by the XRD patterns of Fig.6. After heating to 700°C the composition of the final reaction products is very near to the overall composition of the samples so that their formation is clearly a result of thermally induced interdiffusion.

The present study shows that the 80% Al samples have a much higher tendency than those at the lower Al concentration to form upon milling a disordered or amorphous structure even in a high energy ball mill. This tendency is clearly borne out by the XRD patterns and the TEM micrographs shown in Figs.1 and 2. The scale of the TEM contrast and the FWHM of the main XRD peak of the 80% Al sample milled for 50 h are consistent with a size of the coherent domains of the ultrafine phase of ~1nm, which is comparable with or a small multiple of the unit cell dimensions of the intermetallic compounds Al_6Fe, Al_3Fe and Al_5Fe_2. Accordingly, it is difficult to define the nature of this phase which seems to be composed of more or less ordered clusters of about 10^2 atoms. As the temperature is increased, the degree of crystalline order and/or the size of the coherent domains tend to increase, as shown by the XRD patterns of Fig.7. Still, after DSC heating to 400° C it is not possible to define unequivocally the state of the samples. However, considering that after DSC heating to 430° C one observes the formation of Al_5Fe_2 , one may reasonably suppose that either the structure of the clusters formed during ball milling is similar to that of Al_5Fe_2 or that the structure of the clusters resembles none of the two phases and that Al_5Fe_2 nucleates first because its nucleation barrier from the clusters is lower than the nucleation barrier for Al_3Fe. The second conjecture is supported by the fact that also in the 80% Al sample milled for 10 h, Al_3Fe nucleates from Al_5Fe_2. Even if the reason for the intermediate step through Al_5Fe_2 is not clear it is probably linked with the fact that while the unit cell of Al_3Fe contains about 100 atoms, the unit cell of Al_5Fe_2 contains about 14 atoms. Further heating to about 500° C starts the nucleation of Al_3Fe which in the samples at the 80% Al concentration is the only phase present after DSC heating to 700° C.

ACKNOWLEDGEMENTS
We thank M.Federici for operating the ball mill and R.Marazzi for preparing the samples for electron microscopy.

REFERENCES
[1] C.C. Koch in Materials Science and Technology. A Comprehensive Treatment.Vol. 15, pag.193. VCH, Weinheim, Germany (1991)
[2] P.H. Shingu, B. Huang, J. Kuyama, K.N. Ishihara and S. Nasu in New Materials by Mechanical Alloying Techniques, E. Artz and L.Schultz Editors, Deutsche Gesellschaft für Metallkunde, Oberusel, Germany (1989); B. Huang, N. Tokizane, K.N. Ishihara, P.H. Shingu and S. Nasu, J. of Non-Cryst. Solids 117/118, 688 (1990); S. Nasu and P.H. Shingu in Ordering and Disordering in Alloys, A.R. Yavari Editor, Elsevier, London (1992); S. Nasu, S. Imaoka, S. Morimoto and H. Tamimoto, B. Huang, T. Tanaka, J. Kuyama, K.N. Ishihara and P.H. Shingu, Mater. Sci. Forum 88-90, 569 (1992).
[3] Y.D. Dong, W.H. Wang, L. Liu, K.Q. Xiao, S.H. Tong and Y.Z. He, Mater. Sci. Eng. A134, 867 (1991)
[4] M.A. Morris and D.G. Morris, Mater. Sci. Eng. A136, 59 (1991)
[5] X.P. Niu, A. Mulaba-Bafubiandi, L. Froyen, L. Delaey and C. Peytour, Scripta metall. mater. 31, 1157 (1994)
[6] D.K. Mukhopadhyay, C. Suryanarayana and F.H. Froes, Scripta metall. mater. 31, 333 (1994).

Materials Science Forum Vol. 195 (1995) pp. 25-30

Mechanosynthesis of Metal-Silica Nanocomposites

A. Corrias, G. Ennas, G. Paschina, G. Piccaluga and D. Zedda

Dipartimento di Scienze Chimiche, Università di Cagliari,
Via Ospedale 72, I-09124 Cagliari, Italy

Keywords: Nanocomposites, Mechanosynthesis

Abstract

A mechanochemistry route has been experimented to synthesise metal-silica nanocomposites with different volume fractions of nickel or iron through oxygen transfer reactions from the corresponding metal oxides to silicon.

In nickel containing samples, oxygen transfer occurs in a direct way and Si can be completely transformed into silica, while the stoichiometric quantity of NiO is reduced to Ni(0) nanometer crystallites. When NiO is present in a larger content, the excess, present at the end of the milling process, is reduced by a thermal treatment in H_2 flux. This does not produce a significant growth of the nanocrystallites.

Similar process conditions applied to Fe containing samples give rise to a more complex reaction path which prevents the complete conversion of Fe(III) to Fe(0).

Introduction

Materials consisting of metallic granules embedded in an immiscible medium (insulating ceramics or glasses or, less frequently, metals) display a rich variety of interesting chemical and physical properties [1-2], which have recently attracted a large scientific and technological interest. If either metallic granules or medium particles or both have dimensions on the nanometer scale, such materials belong to the family of nanocomposites [3-4] and their properties, besides reaping benefits from individual components, can also be enhanced by controlling particle size and metal volume fraction x [1,3,5-6].

Nanocomposite materials can be prepared by a number of chemical, physical and mechanical techniques as deposition methods, sol-gel process, impregnation and reduction, etc. [1,5,7].

Mechanical methods and in particular high energy ball milling, which has been largely used for mechanical alloying [8,9], have been recently proposed for preparing metal-ceramic nanocomposites by the activation of displacement reactions such as $M_xO_y + zR \rightarrow xM + R_zO_y$ [10,14]. So far, studies have mostly dealt with solid state reactions between aluminium and metal oxides whereas silicon has less frequently been used as oxygen acceptor [12]. On the other hand, M-SiO_2 nanocomposites have attracted much attention because of their magnetic, electric and catalytic properties [5,15-18]. We have applied the ball milling technique with the aim of preparing a series of metal-silica nanocomposites with different nickel or iron content. As material properties can largely vary with the metal volume fraction around the percolation threshold [1], mixtures with a high metal oxide content have been investigated for both iron and nickel systems.

Experimental

Mixtures of silicon (Fluka 99.9%) powder with either iron (III) oxide (Carlo Erba 99%) or nickel (II) oxide (Aldrich 99%) powders were milled in a Fritsch Pulverisette 5 ball mill. 20 g. of each mixture were sealed in a 250 ml stainless steel vial with balls of the same material (diameter 8mm) in argon atmosphere. Full/void volume ratio and ball/powder weight ratio were fixed to 1/8 and 10/1 respectively; rotation speed of the mill was kept equal to 290 rpm. Overheating was prevented by a cold air jet and alternating 5 minute milling and rest periods.

The mixtures will be hereafter referred to as xFeSiO and xNiSiO where x indicates the metal volume fraction percentage.

The effect of milling on powder mixtures was monitored by X-ray diffraction (XRD) and transmission electron microscopy (TEM) performed on small portions of the powders, sampled at different milling times in argon atmosphere. XRD spectra were recorded on a θ-θ Seifert diffractometer in the range $10° < \theta < 60°$ using monochromatic CuKα radiation. TEM micrographs were taken on a JEOL 200CX microscope.

Results and Discussion

Table 1 shows the molar compositions of the starting mixtures and the corresponding metal volume fractions.

SAMPLE	nNiO	nFe_2O_3	nSi	x(%)
35NiSiO	2		1	35
45NiSiO	3		1	45
66NiSiO	7		1	66
25FeSiO		2	3	25
41FeSiO		4	3	41
67FeSiO		12	3	67

Table 1: Compositions of the samples. n = mole number; x = metal volume fraction assuming a complete reduction of metal oxide.

Figs. 1 and 2 show the XRD spectra referring to the most meaningful steps of the milling process carried out on 35NiSiO and 25FeSiO respectively. The compositions correspond to the molar ratios of the exchange reaction stoichiometry between either NiO or Fe_2O_3 and Si. The pattern evolution with milling seems to indicate that the exchange reaction is completed in both systems. In fact, while the spectra of the starting powders show peaks of crystalline Si, and either NiO or Fe_2O_3, the final patterns only display peaks of either fcc Ni or bcc Fe [19]. In both systems no peaks due to any SiO_2 crystalline form appear; SiO_2 is therefore obtained as an amorphous phase which contributes with a background to the XRD spectra. At the end of the milling, both Fe and Ni peaks are quite broad indicating that the metal crystallites are very small. From Warren-Averbach peak profile analysis [20], performed on (110) and (220) reflections of Fe and (111) and (222) reflections of Ni, an average size of about 10 nm for both metal crystallites was obtained. These results have been confirmed by direct TEM analysis. In fact, TEM micrographs of 25FeSiO and 35NiSiO indicate a broad size distribution of medium particles from 20 to 2000 nm, while the metal crystallites have a narrow size distribution ranging between 5 and 10 nm in both cases. This confirms that high energy ball milling techniques are suitable for the synthesis of nanocomposites.

Although the final results seem the same, the kinetic of the reactions and their mechanism are different in the two cases. NiO is reduced to metallic Ni after 43 hours, while Fe_2O_3 is converted after 33 hours. Different rates are a possible consequence of different reaction mechanisms. In fact, in 35NiSiO patterns only Ni peaks appear in addition to those of NiO and Si which progressively disappear with milling time, indicating a direct transfer of oxygen from NiO to Si. On the contrary, in 25FeSiO spectra Fe peaks appear together with those of an intermediate crystalline phase which has been identified as an iron silicate, Fe_2SiO_4 [19], whose peaks disappear with further milling.

Mossbauer measurements, performed on this sample, indicate a substantial amount of Fe(II) at the final stage of the milling process. The Fe(II) amount is not significantly affected by a thermal treatment in H_2 flux. Fe(II) is probably present in a highly disordered form of the Fe_2SiO_4 phase

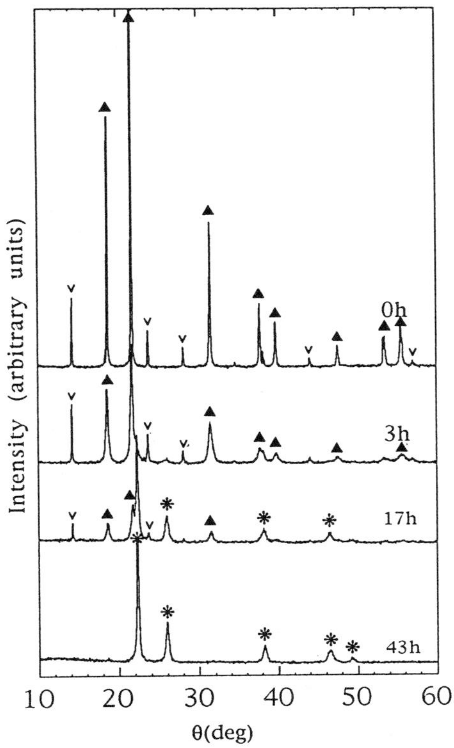

Fig. 1: XRD spectra at selected milling times for 35NiSiO sample. Si (v); NiO (▲); Ni (❋).

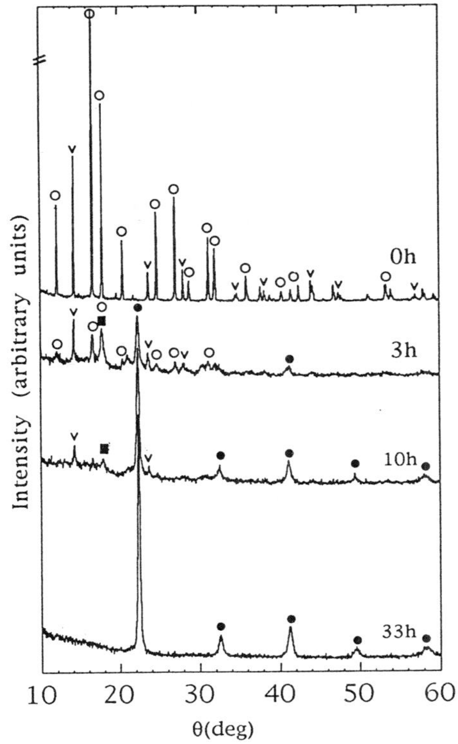

Fig. 2: XRD spectra at selected milling times for 25FeSiO sample. Si (v); Fe_2O_3 (o); Fe_2SiO_4 (■); Fe (•).

which contributes to XRD pattern with a background intensity. In the 35NiSiO sample at the end of the milling significant amounts of Ni(II) can be ruled out. In fact, no intermediate compound forms during milling and the presence of a highly disordered NiO phase can also be excluded because NiO peaks decrease with milling without broadening until they vanish.

The differences in reaction mechanism and final results of milling have been confirmed in the samples containing a higher metal oxide content. In fact, the patterns of 45NiSiO and 66NiSiO at the end of the milling process, reported in Fig. 3b and 3d respectively, show only peaks of Ni(0) together with the excess of NiO. The residual oxide is almost completely converted into Ni(0) after thermal treatment in H_2 flux at 400 °C, a temperature which is low enough to prevent crystallite growth, as can be inferred from the half height width of the pertinent XRD peaks (Fig. 3c and 3e) and from TEM measurements.

In the iron containing samples the formation of intermediate compounds, which are still present at the final stages, has been observed confirming that oxygen transfer to Si occurs in a very complex way. The intermediate and final products depend on the metal oxide concentration. In fact in 41FeSiO pattern, reported in Fig. 4b, peaks due to Fe(0), Fe_2SiO_4 and unreacted Fe_2O_3 are present while in 67FeSiO (Fig. 4d) unreacted Fe_2O_3 peaks are accompanied by those due to Fe_3O_4. Intermediate compounds can form in this case because iron can exist in an Fe(II) oxidation state and they are sufficiently stable to hinder complete reduction of Fe(III) to Fe(0). In fact after thermal treatment in H_2 atmosphere, both Fe(0) and Fe_2SiO_4 peaks increase in intensity in 41FeSiO pattern (Fig. 4c). This indicates that Fe(0) is obtained from the reduction of Fe_2O_3 excess, while Fe_2SiO_4, which is not affected by the reduction treatment, recrystallize from the highly disordered fraction formed during milling. In 67FeSiO, thermal treatment gives rise to the complete reduction of unreacted Fe_2O_3 into Fe(0) while Fe_3O_4 content does not change significantly (Fig. 4e). In both samples peaks sharpen after thermal treatment due to the growth of crystallite size, unlike what observed in Ni samples.

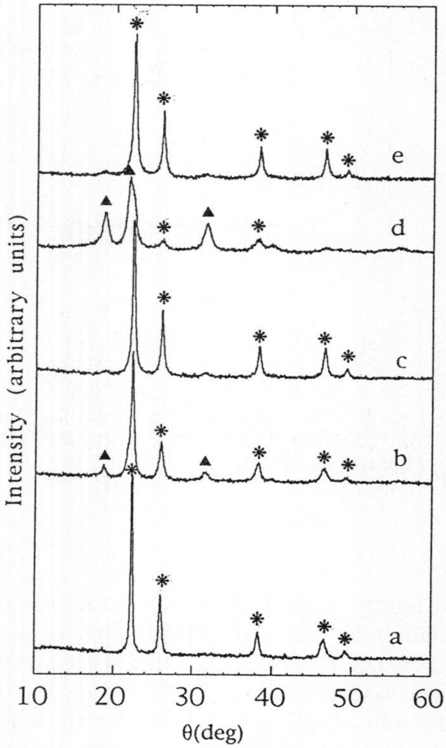

Fig. 3: XRD spectra for 35NiSiO after 43h milling, a); 45NiSiO after 50h milling, b) and after thermal treatment in H_2, c); 66NiSiO after 50h milling, d) and after thermal treatment in H_2, e). NiO (▲); Ni (✱).

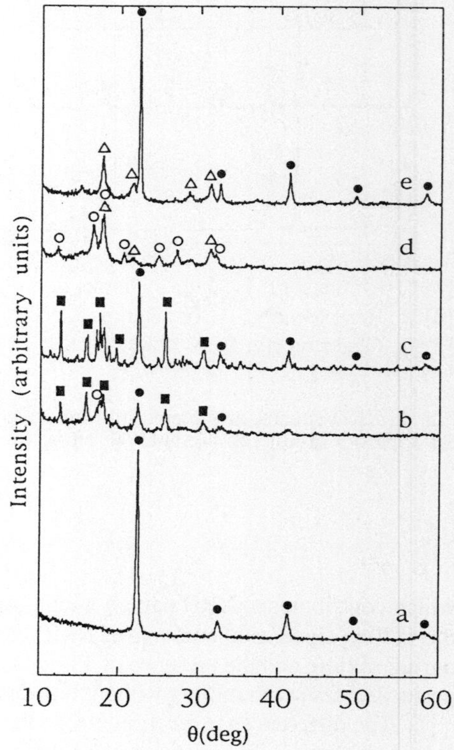

Fig. 4: XRD spectra for 25FeSiO after 33h milling, a); 41FeSiO after 40h milling, b) and after thermal treatment in H_2, c); 67FeSiO after 40h milling, d) and after thermal treatment in H_2, e). Fe_2O_3 (○); Fe_2SiO_4 (■); Fe_3O_4 (△); Fe (●).

Conclusion

Ball milling has been used with the aim of synthesising metal-silica nanocomposites with different metal volume fractions via solid state exchange reactions such as M_xO_y + y/2Si → xM + y/2SiO_2.

In nickel containing samples, oxygen transfers directly from NiO to Si, so that the latter is completely transformed into silica, while the stoichiometric quantity of NiO is reduced to Ni(0). When NiO is present in a ratio larger than the stoichiometric one its excess can be reduced by a thermal treatment in H_2 flux. Nickel crystallites are obtained with nanometer size in the milling process and there is no significant growth during thermal treatment. Therefore Ni-silica nanocomposites can be obtained in a large composition range.

On the other hand, similar process conditions applied to Fe containing mixtures give rise to a more complex reaction path which prevents the complete conversion of Fe(III) to Fe(0).

Acknowledgement

This work has been supported by MURST and CNR.

References

[1] C.L. Chien in Science Technology Nanostructured Magnetic Materials NATO ASI Ser. B 259, 477-96, (1991).
[2] K. M. Unruh, C. L. Chien and P. Sheng in On Cluster and Clustering, ed. P. J. Reynolds, 303, (1991).
[3] S. Komarneni, J. Mater. Chem., 2, 1219 (1992).
[4] R. Roy, Mat. Res. Soc. Symp. Proc. 286, 241 (1993).
[5] D. Chakravorty, Bull. Mater. Sci. 15, 411 (1991).
[6] R. E. Newnham, S.E. Trolier-McKinstry, Ceram. Trans. 8, 235 (1990).
[7] J. W. E. Coenen, Appl. Catal. 54, 65 (1989).
[8] Mechanical Alloying for Structural Applications, Proc. 2nd Int. Conf., Vancouver, Canada, 1993, ed.: J. J. de Barbadillo, F. H. Froes and R. Schwarz (ASM Intern., Ohio, 1994).
[9] A. Corrias, G. Ennas, M. Marongiu, A. Musinu, G. Paschina and D. Zedda, Mater. Sci. Eng. A (1995), in press.
[10] Kaneko, D. G. Kim and M. Sugamata, in Mechanical Alloying for Structural Applications, Proc. 2nd Int. Conf., Vancouver, Canada, 1993, ed.: J. J. de Barbadillo, F. H. Froes and R. Schwarz (ASM Intern., Ohio, 1994) p.261.
[11] G. B. Schaffer and P. G. Mc Cormick, Metall. Trans. A21, 2789 (1990).
[12] P. Matteazzi and G. L. Le Caër, Hyperf. Inter., 68, 177, (1991).
[13] D. Basset, P. Matteazzi and F. Miani, Mater.Sci.Eng., A168, 149 (1993).
[14] C. Laurent, J. J. Demai, A. Rousset, K. R. Kohnan, C.N.R. Rao, J. Mater. Res. 9, 229 (1994).
[15] S. Roy, D. Das, D. Chakravorty and D. C. Agrawal, J. Appl. Phys., 74, 4746 (1993).
[16] S. Roy, D. Chakravorty, J. Mater. Res., 9, 2314 (1993).
[17] J. P. Wang and H. L. Luo, J. Appl. Phys., 75, 7425 (1994).
[18] R. D. Shull, J. J. Ritter, A. J. Shapiro, L. J. Swartzendruber, L. J. Bennett, Mat. Res. Soc. Symp. proc. 132,179 (1989).
[19] PDF file, Int. Center for Diffraction Data, 1601 Park Lane, Swarthmore PA.
[20] B. E. Warren "X-Ray Diffraction" (Addison-Wesley, Reading, Massachussetts,1969) p. 264.

Materials Science Forum Vol. 195 (1995) pp. 31-36

Mössbauer and XRD Study of Thermally Treated Fe-Al Nanocrystalline Alloys

S. Enzo[1,2], R. Frattini[2], R. Gupta[3,5], P.P. Macrì[2], G. Principi[3], L. Schiffini[1] and G. Scipione[4]

[1] Dipartimento di Chimica, via Vienna 2, I-07100 Sassari, Italy

[2] Istituto Nazionale per la Fisica della Materia, Dipartimento di Chimica Fisica, D.D. 2137, I-30123 Venezia, Italy

[3] Istituto Nazionale per la Fisica della Materia, Dipartimento di Ingegneria Meccanica, via Marzolo 9, I-35151 Padova, Italy

[4] Istituto Nazionale per la Fisica della Materia, Dipartimento di Fisica, via Berti Pichat 6/2, I-40126 Bologna, Italy

[5] On leave from School of Instrumentation, D.A. University, Khandwa Road, Indore-452001, India

Keywords: Mechanical Alloying, Iron Aluminides, Mössbauer Spectroscopy, Structure, Refinement

ABSTRACT

Equiatomic Fe–Al powders were synthesized by mechanical alloying of pure elements in a high energy ball mill. X–ray diffraction and Mössbauer spectroscopy are used to study the long–range order and the local atomic arrangement of the products precipitated after heat–treating in a differential scanning calorimetry furnace. Although the calorimetric traces of specimens mechanically alloyed for 4 and 16 h show some similarities, with at least two exothermic peaks, the nature of the phases, precipitated after intermediate stages of heat treatment, is quite different. This is the result of different degrees of crystallites fragmentation, disorder, chemical homogeneity and magnetic properties induced by the mechanical treatment. The structural analysis performed on the alloys treated at intermediate stages of annealing (just after completion of a first exothermic peak in the calorimetric trace), shows remarkable differences both on local and long–range order scale. The specimen MA 4 h, annealed at 718 K, shows the early precipitation of an Al–rich phase and the presence of two constituents without substantial change in their lattice parameters. On the other hand, the sample MA 16 h, annealed at 663 K, shows the disappearance of pure aluminium peaks and a sequence of bcc peaks with a strong asymmetry that can be interpreted, in agreement with Mössbauer data, as the contribution of two phases. By further treating the specimens at 873 K, the sample MA 4 h shows urreacted α–iron and aluminium phases and a minority of Al_5Fe_2 non–magnetic intermetallic compound. However, the sample MA 16 h, subjected to the same annealing conditions, displayed FeAl ordered nanocrystalline compound and a small fraction of Fe_3AlC carbide.

1. INTRODUCTION

Since the recent use of the mechanical alloying (MA) techniques to produce metastable pnases and to improve the mechanical properties of metallic materials, several studies have been undertaken on the Fe–Al system, aiming at defining the processing parameters which mainly affect the tech ılcal properties [1–5]. To our knowledge the first report on MA of the Fe–Al system dates back to the work of Taylor and Jones [6]. During their revisitation of the Fe–Al equilibrium phase diagram, it was noticed that the cold working of pure metals by ball milling affected magnetic properties. Our recent efforts to characterize of the structural and microstructural changes brought by the MA of Fe–Al powders have considered the equiatomic composition as a function of the time of treatment. The structure evolution seems to be related in a complex way to the amount of energy transferred to the metal powders and alloys [7]. The previous work pointed out the various morphological aspects of the alloys subjected to different mechanical treatments [8]. This was also evident from the calorimetric traces recorded to assess the thermal stability of the MA products [9]. In this work we want to detail, in selected $Fe_{50}Al_{50}$ specimens obtained by MA, the mechanism of decomposition of the Fe–Al alloys in correspondence to critical points of the thermograms. For this, we have studied the long–range ordering in the samples with the X–ray diffraction (XRD) technique and the local order around iron atoms with the Mössbauer spectroscopy.

2. EXPERIMENTAL

Pure Al (ALFA products, 325 mesh, 99.9% purity) and pure Fe (325 and 200 mesh, 99% purity) powders were milled in the composition $Fe_{50}Al_{50}$ in a hardened tool steel vial of a SPEX mixer/mill model 8000 for 4 and 16 hours at room temperature. The weight of the mixed powders was about 10 g each time and the ball–to–powder weight ratio was 3:1. To minimize oxygen contamination the milling treatments were performed under argon atmosphere (oxygen < 5 ppm). To avoid the sticking of aluminium to the milling media during the process we added ethanol as a lubricant agent, in the concentration of 0.05 ml/g of powder.

X–ray Diffraction measurements were performed with a Bragg–Brentano goniometer, using CuKα radiation (λ = 1.5418 Å) and a monochromator in the diffracted beam. The data were analysed with the Rietveld code FULLPROF, available from J. Rodriguez–Carvajal at the ILL Laboratory of Grenoble (France) [10]. The Mössbauer spectra were measured at room temperature using a conventional spectrometer in transmission geometry. A ^{57}Co:Rh source of about 5 mCi was used and the spectra were analysed by means of the NORMOS code developed by Brand [11]. The isomer shifts are referenced to metallic iron. DSC analyses were carried out with a DSC7 Perkin Elmer calorimeter. The powders were crimped in aluminium pans and heated at a constant rate of 20 K/min under flow of purified argon.

3. RESULTS AND DISCUSSION

A full account of the structural evolution induced by the mechanical alloying of a equiatomic Fe–Al mixture as a function of the mechanical treatment is reported in the work previously mentioned [8].

The DSC traces of samples MA 4 and 16 h, reported in figure 1, present complex exothermic features, consisting of at least two partially overlapped peaks. It can be noted that the heat released is larger for the specimen MA for 4 h, but the center of gravity of the exothermic feature is located at a lower temperature in the specimen MA 16 h. In an attempt to study the transformations induced by the thermal treatment, we have annealed the specimens at intermediate temperatures (718 and 663 K for 4 and 16 h respectively), after the completion of the first exothermic component (full circles of figure 1).

Figure 2 shows the Mössbauer patterns of the 4 h MA specimens before and after annealing. The spectrum of the as–milled sample shows a sextet, typical of α–iron and a small broad doublet, attributed to a Fe–Al

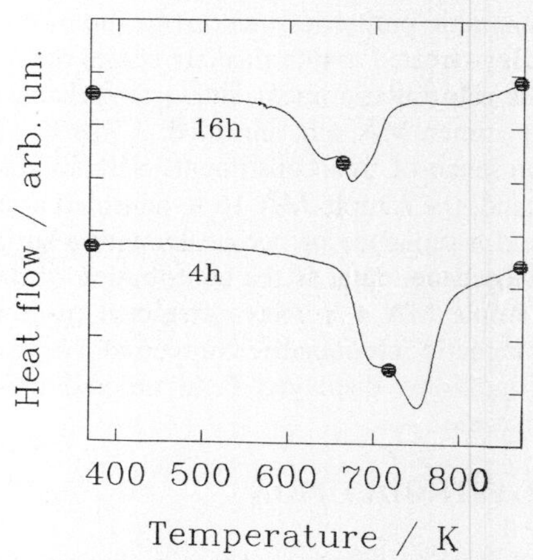

Figure 1. DSC curves of the $Fe_{50}Al_{50}$ MA for 4 (bottom) and 16 h (top). The full circles represent the temperatures to which the specimens were annealed.

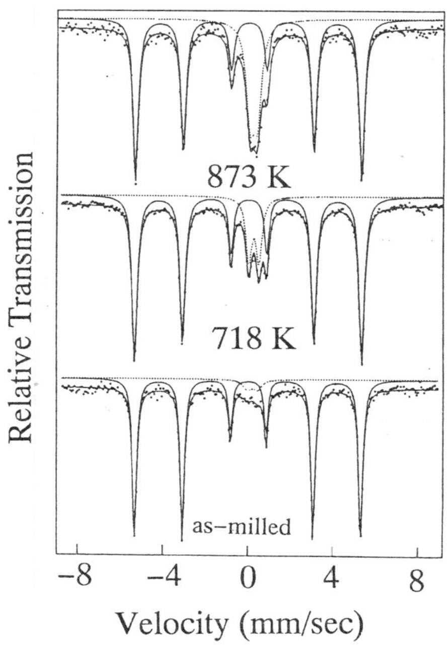

Figure 2. Room temperature Mössbauer spectra of the $Fe_{50}Al_{50}$ powders MA for 4 h as-milled (bottom), after thermal treatment at 718 K, corresponding at the end of the first exothermic peak in the calorimetric trace, and at 873 K (top). The full lines are the best-fitted individual and total components which are slightly shifted to the top with respect to the experimental data points.

solid solution. The relative intensity of lines 2 and 5 of the sextet suggests a textured condition. In the pattern of the specimen thermally treated at 718 K the sextet of α–iron is restored to its typical relative line intensity, while a doublet of the FeAl phase appears evident. The doublet is more intense in the specimen annealed at 873 K and presents a lower quadrupole splitting, indicating also the presence of phases richer in aluminium.

Figure 3 shows the X-ray diffraction spectra of the same specimens as figure 2. The pattern of the as-milled specimen (lower curve) shows only the peaks of Al with hkl odd (111, 311) indices free from overlapping, as those with even indices are superimposed by the α–iron peaks. As a function of the time of mechanical treatment the previous analysis indicated that aluminium was partially dissolved in bcc iron at this stage. Moreover, from the peak broadening, the average crystallite size resulted substantially reduced, while the atomic strain level was increased.

A whole powder-pattern analysis according to the Rietveld method [12] (full line) is able to refine numerically structural and microstructural quantities such as the lattice parameters, the disorder of Debye–Waller kind, the site occupancy (i.e., the chemical disorder), the presence of preferred orientation, the average crystallite size (including dislocations), and the average microstrain. The numerical values of this analysis are reported in ref. [8] for the as-milled specimens and only the relative changes induced by the annealing will be discussed here. The pattern of the specimen thermally treated at 718 K (central curve) shows the presence of α–iron with preferred orientation along the 110 direction similar to the as–milled sample. There is also a decrease and broadening of Al peaks and the appearance of weak extra peaks attributable essentially to the Al_5Fe_2 phase, whose sequence of peaks is reported at the top of the figure. This phase can nucleate from a small amount of fcc Al(Fe) solid solution. The ability of the Rietveld method to fully account for the details of the experimental pattern gives support to the structure assignment in terms of three phases. In the pattern of specimen thermally treated at 873 K, the Al peaks have almost disappeared and the Al_5Fe_2

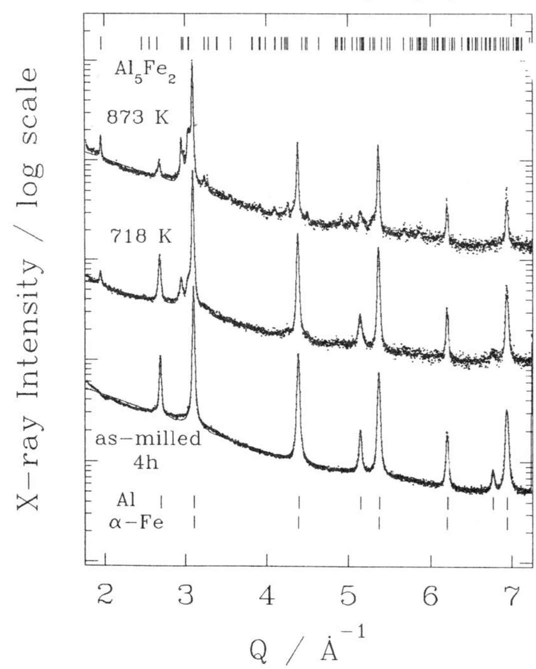

Figure 3. XRD intensity (logarithmic scale) vs Q of the $Fe_{50}Al_{50}$ powders MA for 4h. Dots are experimental data points, full lines are the fit according to the Rietveld method and the bars mark the best-fitted peak positions for the individual phases.

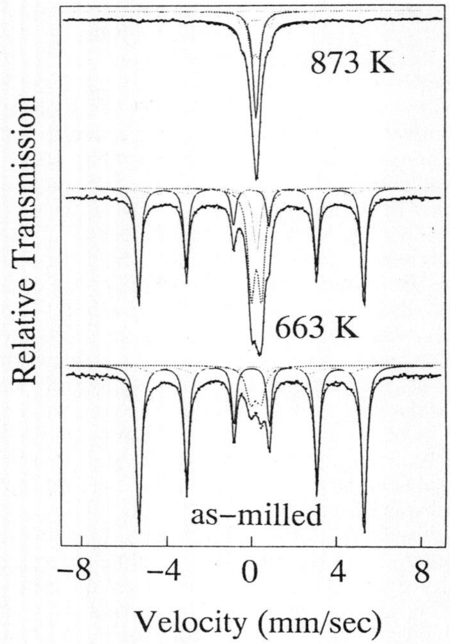

Figure 4. Room temperature Mössbauer spectra of the $Fe_{50}Al_{50}$ powders MA for 16 h as–milled (bottom), after thermal treatment at 663 K (central curve) and at 873 K (top).

phase is more evident. There is 'he additional possibility of indexing some extremely weak extra peaks with the structure factor of Al_3Fe intermetallic compound. This can be related to the Mössbauer component with a low quadrupole splitting. However, it must be said that at this stage of the numerical analysis, there is no substantial benefit of improving the goodness of the fit at the expense of using more parameters.
The Mössbauer spectra showing the evolution of the specimen MA for 16 h are reported in figure 4. In the pattern of as–milled alloy, a broad and weak magnetic component, attributed to a disordered bcc local environment coexists with the magnetic sextet of α–iron and with the broad doublet of the Fe–Al solid solution. The spectrum of the sample treated at 663 K shows the growth of a central contribution which can be assigned to an heterogeneous extended solid solution. The latter local environment can be further decomposed in terms of a singlet (FeAl intermetallic type contribution) and a doublet (solid solution). The pattern of the specimen annealed at 873 K presents the singlet of the intermetallic FeAl, a small doublet due to the disordered solid solution and a sextet with a low field (about 4.2 T).
The structural analysis by X–ray diffraction of samples 16 h MA is reported in figure 5. In the pattern of the as–milled specimen, the fcc Al peaks are low and all the peaks appear broader than the corresponding pattern of sample MA 4 h, due to the fragmentation of crystallites, stacking faults and to microstrain [13]. The pattern of the specimen thermally treated at 663 K shows just the broad peaks of a bcc solid solution. Therefore it is substantially different from the specimen MA 4 h, annealed at 718 K. A careful inspection of the peak shape shows that it is skewed to the low–Q side. There are various mathematical ways to account for this skewedness, corresponding to various interpretations of the microstructure of this sample. Asymmetry of the peak shape is usually explained in terms of stacking faults, but this physical effect would change the sign of skewness across the various hkl in a bcc system, whilst this is not true in our case. The fit reported in the figure for the sample

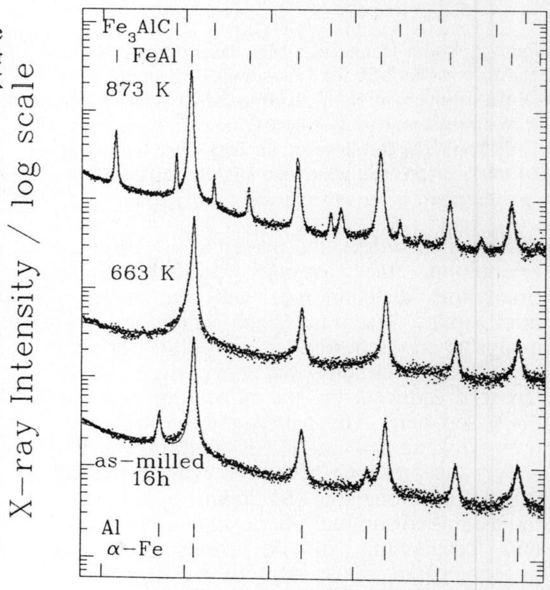

Figure 5. XRD intensity (logarithmic scale) vs Q of the $Fe_{50}Al_{50}$ powders MA for 16 h and subjected to the thermal treatment as reported in fig. 4. The sequence of bars referring to the structure of the precipitated alloys is reported at the top of the figure.

annealed at 663 K was obtained using two bcc phases, the first having the lattice parameter of pure α–iron (2.866 Å), and the second having a larger value of 2.920 Å. This may reflect the insertion of aluminium atoms in the bcc cell of iron. The lattice parameters reported for equilibrium Fe–Al solid solutions are around 2.900 Å [14], so we can surmise that the relative component used in the fit is rich in Al, as it comes out after considering that the iron contamination from the vial is reduced and the overall composition must be equiatomic. The pattern of the specimen annealed at 873 K shows the FeAl superlattice phase and the $AlFe_3C$ cubic phase. The latter can be related to the small magnetic component at low field observed in the correspondent Mössbauer spectrum. The occurrence of two phases after the thermal treatment at 873 K is a further confirmation that the 16 h MA system is driven to a different condition with respect to the specimen MA 4 h and that the apparent similarity of the DSC traces hides a large difference in terms of microstructure and free energy content. The presence of a carbide phase anti–isomorphous with perovskite was also reported for quenched specimens by Taylor and Jones [6], who noted that this structure can be stabilized epitaxially with the FeAl structure of CsCl type. The occurrence of the carbide phase is certainly ascribable to the use of ethanol as a lubricant agent, as it was reported for other MA alloys synthesized using alcohols and hydrocarbons [15]. In any case, the fact that no carbide phase was observed in the annealed products of the specimen MA 4 h leads to the inescapable conclusion that the structural and morphological condition of the specimen MA 16 h is rather different. Another interesting point is the precipitation of an almost ordered FeAl phase (long range order parameter 80% determined with the Rietveld method) from the completely disordered Fe(Al) solid solution observed at 663 K. In addition, the peak broadening of the FeAl intermetallic remains quite large in the FeAl sequence even after annealing at 873 K, suggesting that this method can be used successfully to produce nanocrystalline ordered intermetallics. Of course, the inclusion of carbon over a certain level can be detrimental to technical properties, as reported previously for ball milling of Al_3Fe composition [2]. We are currently investigating the lowest concentration of ethanol that can be tolerated to avoid the carbide precipitation.

4. CONCLUSIONS

We can make the obvious conclusion that a prolonged mechanical treatment by MA of the equiatomic Fe–Al system induces a variety of local situations which are responsible for the different thermal behaviour observed in the furnace of a differential scanning calorimeter. It is then possible to drive the system into selected metastable or stable phases by a proper set–up of the reactor. In the present case we have used a Spex mixer–mill, with fixed shaking frequency and vial geometric path, inside which the solid state reaction is conducted. Thus, the energy tranferred to the metal powders can be changed only by varying the time of cold working and the milling media, typically the size, density and number of balls with respect to the powder volume. Our structural analysis on the local and long–range scale orders has pointed out that the MA of $Fe_{50}Al_{50}$ powders introduces various kinds of structural, static and dynamic disorder, changes in the magnetism of iron atoms and fragmentation of the coherent domains of diffraction. However, no crystal–to–amorphous transformations were observed, nor quasicrystalline phases of pentagonal symmetry [16]. MA alloys, when subjected to annealing, give a complex exothermic trace below 873 K, that hides the quite different nature of the transformation in progress. At temperatures below 718 K, the specimen MA 4 h quickly precipitates a small quantity of the Al_5Fe_2 phase, which subsequently increases, even though remaining in a lower quantity with respect to the unreacted elemental blend. Conversely, in the specimen MA 16 h, we observe a diffusion of unreacted aluminium in the α–iron lattice for temperatures below 663 K. Further annealing at 873 K precipitates the nearly ordered FeAl intermetallic compound, which seems to increase the epitaxialstabilization of a carbide phase due to the ethanol used as a lubricant agent. We intend to develop our studies on the Fe–Al system toward the Al–rich side of the phase diagram to search for the composition at which the reverse solid state diffusion of iron in fcc aluminium becomes observable.

ACKNOWLEDGMENTS

We thank Prof. G. Cocco for providing the facilities of the MA synthesis and for useful discussions. RG acknowledges a fellowship grant by the ICTP, Trieste (Italy). We acknowledge the financial support from the National Institute for Physics of Matter (INFM) for a post–doc fellowship to PPM and the National Consortium of Materials Chemistry (INCM). Thanks are due also to the MURST (Ministry of the University, Science, Research and Technology of Italy). This research is part of a long–lasting collaboration with prof. E. Bonetti on iron aluminides.

REFERENCES

1) P. H. Shingu, B. Huang, J. Kuyama, K. N. Ishihara and N. Nasu, in Proc. Conf. "*New materials by mechanical alloying techniques*" Calw–Hirsau, Germany, 1988, E. Artz and L. Schultz eds. p. 319
2) M. A. Morris and D. G. Morris, Mat. Sci. Engin. **A136**, 59 (1991)
3) M. A. Morris and D. G. Morris, Mat. Sci. Forum **88–90**, 529 (1992)
4) E. Bonetti, G. Scipione, G. Valdrè, G. Cocco, R. Frattini and P. P. Macrí, J. Appl. Phys. **74**, 2053 (1993)
5) *Processing, Properties, and Application of Iron Aluminides*, edited by J. H. Schneibel and M. A. Crimp, (TMS Pub., 1994)
6) A. Taylor and R. M. Jones, J. Phys. Chem. Solids **6**, 16 (1958)
7) G. Cocco, G. Mulas and L. Schiffini, Mat. Trans. JIM **36**, 312 (1995)
8) R. Gupta, P. P. Macrì, R. Frattini, G. Principi, G. Scipione, L. Schiffini and S. Enzo, submitted to Acta Metall. Mater.;
E. Bonetti, G. Scipione, R. Frattini and L. Schiffini, submitted to J. Appl. Phys.
9) R. Frattini, L. Schiffini, G. Scipione, E. Bonetti and S. Enzo, in ref. 1, p. 79
10) J. Rodriguez–Carvajal, Physica B **192**, 55 (1993)
11) R. A. Brand, Nucl. Instrum. & Meth. **B28**, 398 (1987)
12) *The Rietveld Method*, R. A. Young ed., International Union of Crystallography, Oxford University Press, 1993
13) B. E. Warren, *X–Ray Diffraction*, Addison–Wesley, Reading, 1969
14) W. Hume–Rothery, R.E. Smallman and C.W. Haworth, *The structure of metals and alloys*, (The Institute of Metals, London, 1969)
15) S. Nasu, P. H. Shingu, K. N. Ishihara and F. E. Fujita, Hyp. Inter. **55**, 1043 (1990)
16) V. I. Fadeeva, A. V. Leonov and L. N. Khodina, Mat. Sci. Forum **179–181**, 397 (1995)

Materials Science Forum Vol. 195 (1995) pp. 37-42

Wide Angle X-Ray Diffraction (WAXS) Characterisation of Nanoscale Silicon Particles Produced by Laser Synthesis from Silane Gas

E. Borsella[1], S. Botti[1], M.C. Cesile[1], P.E. Di Nunzio[2], S. Martelli[1] and A. Nesterenko[1]

[1] ENEA C.R.E. Frascati, Div. INN-FIS-SPET, Via E. Fermi 37, PO Box 65, I-00044 Frascati (Roma), Italy

[2] CENTRO SVILUPPO MATERIALI, Via di Castel Romano 100-102, PO Box 10747, I-00129 Roma-EUR, Italy

Keywords: Nanomaterials, X-Ray Diffraction, Laser Synthesis

ABSTRACT

Silicon nanoparticles have been produced by means of laser synthesis from silane gas. The average size and size distribution of the particle can be controlled by the synthesis parameters. To optimise the process, in order to reduce the particle size, a wide angle x-ray diffraction (WAXS) investigation method has been developed to rapidly and easily characterise the produced powder batches. The method is based on a Montecarlo interference function (MIF) fitting of the experimental peaks. The results of the application of the MIF procedure on powder batches with different particle sizes are presented and compared with Brunauer-Emmett-Teller specific surface measurements and transmission electron microscopy.

1. INTRODUCTION

Nanophase materials have received special attention in the last years, since they are expected to realise surprising performances[1]. For instance, ceramic nanocomposite materials have proved superior hardness and toughness properties up to the attainment of superplastic behaviour. The process of nanosized powder production by laser induced synthesis in the gas phase[2] is one of the most promising for massive applications. As a matter of fact laser synthesis permits the attainment of powder particles, which meet most of the criteria for an ideal nanopowder, i. e. nanoscale dimension, spherical shape, narrow size distribution, chemical purity. In particular, due to the good absorption of the SiH_4- silane gas at the wavelength of powerful CO_2 continuos lasers (10.6 μm), silicon based ceramic powder, like SiC[3], Si_3N_4 and Si/C/N compounds[4], have been already processed in a pre-industrial scale. Very recently it has been shown that nanosized pure Si particles are able to originate visible[5]photoluminescence, in analogy with porous silicon, when their dimension is reduced beneath a critical value estimated in 4-6 nm[5,6]. The undertaking of producing nanosized Si particles by the laser synthesis method is, therefore, of relevant interest, both from a basic research point of view and for technological application in the opto-electronic industry. The size of the formed particles is strongly dependent on the process parameters and, up to now, a minimum size of about 8 nm have been reached. The present effort is directed to further diminish the silicon particles to get to the appropriate dimension for photoluminescence. In order to tune up the process variables, characterisation methods should be selected, which allow easy and rapid measurements of the chief powder features, i.e. structure, crystallinity, mean size and size distribution. Among the several techniques, used for samples characterisation, conventional wide angle x-ray diffraction (WAXS) has been preferentially applied, because of the non destructive character and the real simplicity of the measurements. With the aim to obtain information, not only on the crystalline structure of the particles, but also on the average size and size distribution of the powder grains, a Montecarlo interference function (MIF) procedure[7] has been applied. This procedure, originally developed for plastically deformed materials, has been tailored for the present purposes. The results obtained by this method have been compared with the ones from the other used techniques: electron microscopy

(TEM) and specific surface (Brunauer-Emmett-Teller [B.E.T.] method) measurements. The cross comparison of the obtained results allowed to determine the limits of validity of the MIF procedure as well as to gain a deeper knowledge on the particle growth mechanisms.

2. EXPERIMENTAL

The powder synthesis process can be outlined as a stream of reactant gas passing through a CO_2 c.w. laser beam inside a reaction chamber. The reactant gas, silane, intersecting the laser beam at the centre of the reaction cell, is heated rapidly up to the dissociation temperature and reacts to form silicon particles. The silane gas is converted into silicon by the simply overall reaction:

$$SiH_4(g) \rightarrow 2H_2(g) + Si(s) \quad (1)$$

A detailed description of the experimental set-up has been given elsewhere[3,4]. The dimension of the formed particles is essentially governed by the time, during which Si nuclei remain in the molten state and crystal growth takes place, due to collision and coalescence mechanisms[8] In the present case to produce the large and small particle powder batches, hereafter referred as *lpb* and *spb*, the silane gas was injected pure (*lpb*) or previously mixed with helium gas (*spb*) to enhance the heat exchange with the environment and thus reducing the allowable coalescence time. In both cases a coaxial argon flow (1.5÷2 l/m) was used to keep the formed particles entrained in the gas stream to the cell exit and to the collecting unit. The reaction was accompanied by a visible flame and the reaction temperature could be estimated by an optical pyrometer pointing to the flame through a quartz window.
The chosen experimental conditions of laser synthesis are summarised in table 1.

TABLE 1

Sample	SiH_4 (sccm)	He (slm)	P_{laser} (W)	T_{flame} (°C)	d_{XRD} (nm)	d_{BET} (nm)
lpb	500	—	500	1130	20	60
spb	660	1.5	400	~800	8	10

TABLE 1. Synthesis parameters of Si nanopowders. In the last two columns the average sizes measured from Eq. 2 and as B.E.T. equivalent spherical diameters are reported.

Wide angle x-ray diffraction (WAXS) was performed with a conventional θ-2θ instrument, equipped with a graphite monochromator on the diffracted beam to lower the background arising from parasitic scattering. The recorded intensity were corrected for angular dependent terms, like Lorenz and polarisation factors and rescale in transferred momentum unit $b = 2\sin\theta/\lambda$ (nm^{-1}).
Transmission electron microscopy was performed with a Jeol 200 CX microscope operating at 200 kV. The powders were previously dispersed in a isopropyl alcohol solution. A solution drop was deposited onto a standard copper grid cover by an evaporated carbon layer.
Specific surface measurements were carried out by nitrogen absorption desorption according to the Brunauer-Emmett-Teller method.

3. RESULTS AND DISCUSSION

3.1 WAXS analysis

In fig. 1 the x-ray diffraction patterns of the *lpb* and *spb* sample are shown. The inverse dependence of the diffraction peak broadening on the particle dimension is clearly visible, confirming the strong influence of the synthesis parameters. According to the Williamson Hall analysis[9] the full width at half maximum (FWHM) is related to particle dimension and to lattice defects by the following equation:

$$\delta b_{hkl}=1/L_{hkl} + \alpha_{hkl}b + \beta_{hkl}b^2 \tag{2}$$

where L_{hkl} is the coherence length or x-ray particle size along the <hkl> crystallographic direction; the α_{hkl}, β_{hkl} coefficients are related to the presence of lattice defects and $b = 2\sin\theta/\lambda$. The term particle size causes often misleading interpretations. As a matter of fact x-ray diffraction measures the spatial extension of coherent (perfect) regions inside the sample. The obtained value is therefore always less or equal than the physical dimension of the particles. A coincidence between WAXS measurements and other morphological investigation methods is expected only in the case of single crystal particles. The intercept of the extrapolated curve towards $b\rightarrow 0$ gives the coherence length while the slope is proportional to the lattice defects density. The almost constant FWHM values for all the recorded reflections indicate, not only, that the particles are defect free: $\alpha = \beta \approx 0$, but, moreover, that the dimension is the same along the different crystallographic directions <hkl> (8 ÷ 10 nm), as awaited for spherical particles.

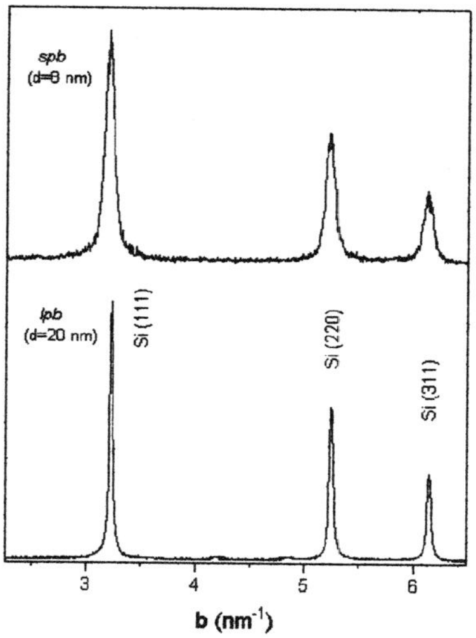

Fig 1 X- ray diffraction pattern of the large particle sample *spb* (lower) and of the large particle one *lpb* (upper).

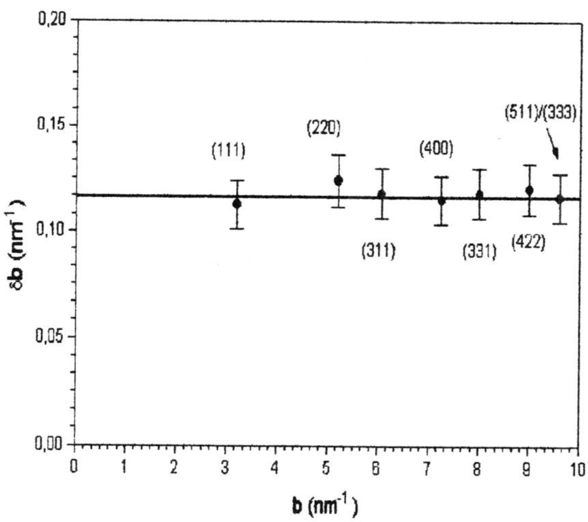

Fig 2. Small particles sample: peak integral widths as a function of b. The straight line indicates the Williamson Hall analysis of Eq. 2: $\delta b_{hkl}=1/L_{hkl} + \alpha_{hkl}b + \beta_{hkl}b^2$

Assuming a defect free and spherical powder the more general MIF[7] procedure can be quite simplified. The powder particles are considered as constituted by columns, expressing the coherence length, perpendicular to the diffracted planes. The highs of the column are supposed to be bounded by the physical dimensions of the particles themselves. Since in the present case the peak broadening due to lattice defects can be reasonably neglected, the experimental x-ray line profile can be described in the "column like" approximation as the sum of single scattering contributions:

$$I(b-b_o) = \sum_i \Phi(b-b_o, L_i)W(L_i) \tag{3}$$

where $b = 2\sin\theta/.\lambda$, $b_0 = 1/d_0$ is the centre of gravity of the reflection, d_0 is the interplanar spacing of the diffracting planes (hkl); L_i is the high of the i column and $\Phi(b - b_0, L_i)$ is the correspondent column scattering function:

$$\Phi(b - b_o, L_i) = L_i^2\left[\frac{\sin^2(\pi L_i b)}{(\pi L_i b)^2}\right] \tag{4}$$

The term $W(L_i)$ is the frequency distribution function of the column, i. e. number of column of high L_i, which is computed by the MIF procedure by least square fitting of (111)-Si reflection..
Once determined the $W(L_i)$ distribution, the volume averaged coherence length distribution is

$$<W(L_i)>_{vol} = W(L_i) \cdot L_i \tag{5}$$

since the column have been defined with a unit base.

The calculated distributions are shown in Fig.3 a and b. It can be observed that , while for the *spb*-sample an uniform and rather narrow distribution, centred at 6 - 8 nm , is obtained. Otherwise the *lps* sample shows a quite different behaviour. In this latter case a "bimodal" distribution has been computed, which indicates the presence of two kinds of particles, the first one having an average x-ray size of 12 ÷ 15 nm the second of 40 ÷ 45 nm. Bimodal distributions are rather unlikely in continuos particle growth processes and suggest the occurrence of two distinct mechanisms of formation. Comparing the x-ray particle size with the one evaluated from the specific surface in spherical approximation, the good agreement, for the *spb*, powder indicates that the powder grains are made of Si single nanocrystals (see Table 1). For the *lpb* sample, the surface derived equivalent diameter (≈60 nm) greatly exceeds the x-ray value (≈20 nm) and it suggests that the powder is composed by polycrystalline grains.

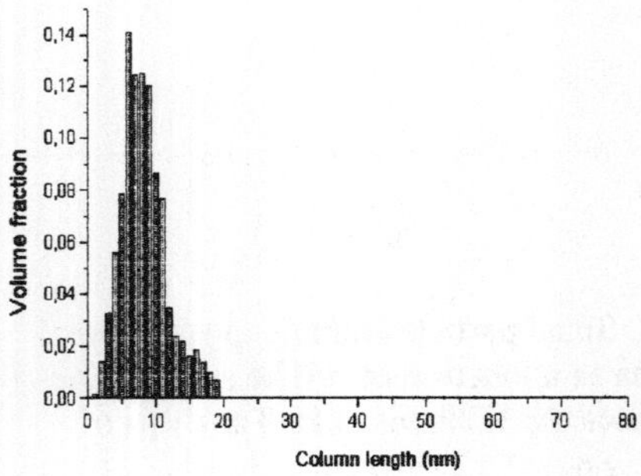

Fig 3a. Volume averaged column length distribution (x-ray particle size) of the small particle sample

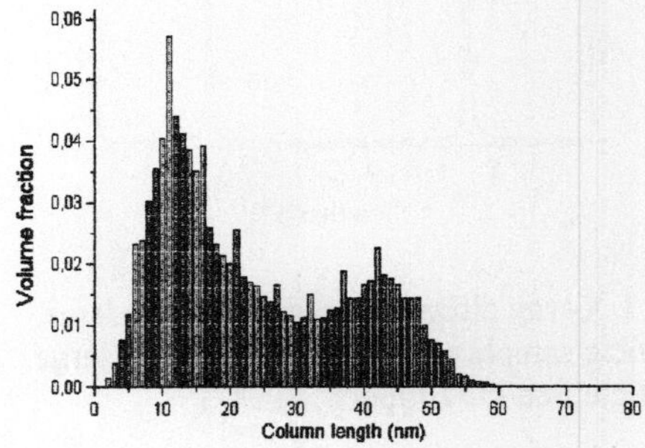

Fig. 3b. Large particle sample

3.2 TEM analysis

Fig. 4a and 4b show the bright field micrographs recorded on both the investigated samples. Nice spherical particle can be observed in the *spb* sample. The mean diameter ranges between 8-12 nm in perfect agreement with WAXS analysis. A quite different picture has been recorded on the *lpb* powder. The simoultaneous presence of a large variety of small and big grains can be easily observed. Moreover various particles show a structure inside their bodies, which confirms their polycrystalline character, as suggest by the different sizes measured by the used techniques (Tab.1). Combining together WAXS and TEM results it may be concluded that the lower diameter of the distribution corresponds to the averaged dimension of the particle subgrains. The appearance of a bimodal distribution may be related to an imperfect particle coalescence. Some larger particles are really single crystals, whereas others are originated by the sintering of small units having thus a polycrystalline structure. TEM investigation points also to the limits of the applied MIF procedure. Particles with a dimension larger than 50 nm give rise to very narrow diffraction peaks, which can not be detected due to the limited resolution power of the instrument. Therefore by computing the size distribution function the bigger grains (≥ 50 nm), which can be clearly observed in fig. 4b , become invisible causing an artificial drop of the distribution in the large size part.

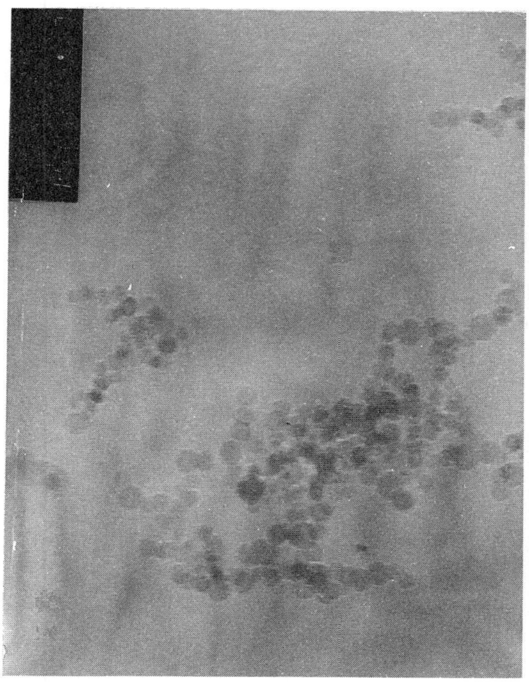

100 nm

Fig. 4a Brigth field image (225000x) of the small particles *spb* sample

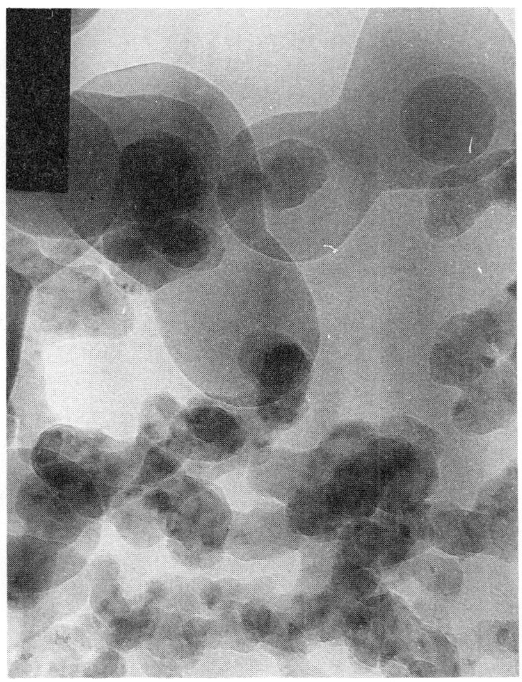

100 nm

Fig. 4b Brigth field image (225000x) of the large particles *lpb* sample

4. CONCLUDING REMARKS

From the present results the following conclusion can be drawn:
Wide angle x-ray diffraction can be applied as an easy and rapid method to characterise the Si nanopowders produced by laser synthesis.
The size distribution function, computed according to the MIF procedure gives, in case of single nanocrystals powder, a correct estimate of the particle size.
The appearance of bimodal distributions signalises that the particle growth follows different mechanism as confirmed by transmission electron microscopy.
The MIF procedure looses its validity for large particle sizes, being well suited for the nanocrystalline powder with an average dimension $\leq$ 50 nm.

REFERENCES

[1] R. Dagani, Chem. Eng. News **23**, 16 (1992)
[2] J. S. Haggerty, and R. W. Cannon, "Laser Induced Chemical Processes", Ed. J. I. Steinfeld, Plenum Press, New York (1981).
[3] R. Fantoni, E. Borsella, S. Piccirillo, R. Ceccato, and S. Enzo, J. Mat. Res. **5**, 143 (1990)
[4] E. Borsella, S. Botti, R. Fantoni, R. Alexandrescu, R. I. Morjan, C. Popescu, T. Dikonimos-Makris, R. Giorgi, and S. Enzo, J. Mat. Res. 7, 2257 (1992).
[5] H. Takagy, H. ogawa, Y. Yamazaky, A. Ishizaki, and T. Nakagiri Appl. Phys. Lett. **56**, 2379 (1990)
[6] Y. Kanemitsu, Phys. Rew.**B49**, 16845 (1994)
[7] P. E. Di Nunzio, S. Martelli, R. R. Bitti, J. Appl. Cryst. **28**, 146 (1995).
[8] J-M Lihrmann, and M. Cauchetier, J. Eur. Cer. Soc. **13**, 41 (1994)
[9] G. K. Williamson, and W. H. Hall, Acta Metall. **1**, 22 (1953)
[10] A. Guinier "X-ray Diffraction in Crystals, Imperfect Crystals and Amorphous Bodies", San Francisco: W. H. Freeman, 151 (1963)

Materials Science Forum Vol. 195 (1995) pp. 43-48

Pulsed Vaporization Sources for High Efficiency Production of Metal Clusters

P. Milani[1], P. Piseri[1], E. Barborini[1] and S. Iannotta[2]

[1] INFM-Dipartimento di Fisica, Università di Milano, Via Celoria 16, I-20133 Milano, Italy

[2] CEFSA-Centro CNR-ITC, I-38050 Povo (TN), Italy

Keywords: Metal Clusters, Molecular Beams, Pulsed Source, Laser Vaporization, Arc Discharge

Abstract: We describe a pulsed Laser Vaporization Source (LVS) and a Pulsed Arc Ion Cluster Source (PACIS) for the production of intense beams of neutral and ionized clusters. The PACIS source has been tested in a molecular beam apparatus equipped with a time of flight spectrometer in order to characterize the cluster mass distribution. We show that it is possible to control the mass distribution of the aggregates, their kinetic energy and their internal temperature. These parameters are important for the deposition of metallic and composite thin films with controlled properties.

The synthesis of nanoscale particles is of relevant importance for the realization of nanostructured materials with tailored properties. The characterization of the properties of the nanoparticles can ultimately lead to identifying many potential uses of fundamental and technological interest [1, 2].

The cluster size turns out to be one of the key parameters to control the properties of nanoscale granular materials. Conspicuous efforts have been devoted to the development of cluster growth techniques allowing a control of the size distribution and of the nucleation and deposition rates. Various physical and chemical methods have been applied to this purpose, however many key issues such as size distribution, elimination of contaminants, control on stoichiometry are still not completely managed [3].

The production of supersonic beams of clusters by laser vaporization sources (LVS) has received an increasing attention because of its versatility and simplicity and because clusters of even the most refractory materials can be obtained [4]. High-intensity cluster beams free from contamination and in a wide mass range can be produced, allowing the synthesis of new materials using neutral and ion cluster beam deposition [5].

The principle of operation is very simple: an intense pulse of laser radiation is focused on a target, vaporising a small amount of material in a flow of an inert carrier gas produced by a pulsed valve. The inert gas quenches the vapour and then the mixture is expanded into a vacuum. It is thus possible to obtain pulsed beams of neutral and ionized clusters cooled to low temperatures by the thermalization with the carrier gas and the supersonic expansion. A further improved of this kind of sources consists in the introduction of a thermalization cavity before the expansion channel. This favours the aggregation of clusters and improves the stability and the intensity of the source [6].

The LVS can be coupled with any pulsed spectroscopic technique. In particular Time of Flight Mass Spectrometry can be used to characterize the mass distribution in the molecular beam before the cluster deposition. This allows a careful control on the mass distribution and the possibility of correlating the final properties of the material with the precursors [7].

Another pulsed cluster source which can be used for cluster production of materials in as wide a range as the LVS is the so-called Pulsed Arc Cluster Ion Source (PACIS) [8]. In this source the vaporization is obtained by pulsing an electric arc between two electrodes instead of a laser pulse. The advantages of a PACIS compared to the LVS are a higher intensity and flexibility, moreover the cost is only a fraction of the LVS. However, despite its merits, improvements are possible and necessary if better stability and control over the cluster physical parameters are required in order to use this source for the synthesis of nanostructured materials. Here we present an improved PACIS source which can produce intense and stable cluster beams and it can offer the possibility to control the mass distribution, the kinetic energies and the temperatures of the produced clusters.

Our source consists of a main ceramic body with two channels perpendicularly intersecting each other in the center. Electrodes can be inserted from opposite sides in the larger channel and blocked to form a gap of around 1 mm. Just after the gap the second channel ends into a cylindrical cavity with a volume on the order of several cubic centimeters. The cavity is closed at the opposite end with a stainless steel cap in which a coaxial expansion channel and a nozzle is drilled. A timing unit allows to control the delay between the arc ignition and the opening of the gas valve.
A mass spectrum of ionized aluminum clusters is shown in figure 1.

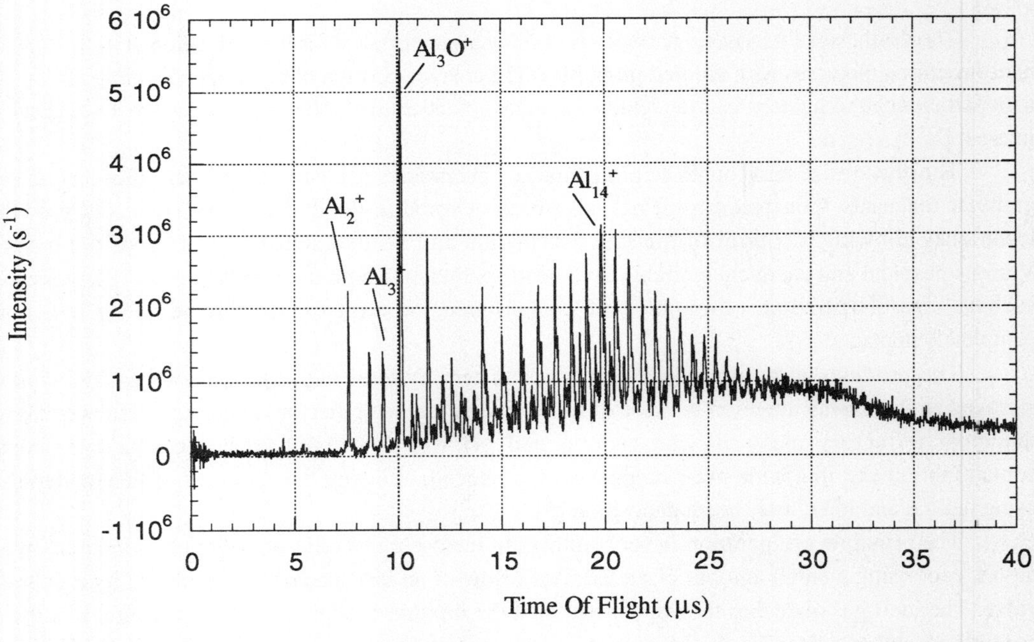

Fig. 1 Plot shows a TOF mass spectrum of Aluminium Clusters produced with PACIS. Thermalization volume in the source is 0.65 cm^3 and expansion takes place through a 1 mm Ø nozzle. The delay between pulsed valve and the arc is 450 μs.

Single cluster peaks are visible up to mass n=25, larger clusters are present but not resolved due to the low resolution of the mass spectrometer. This spectrum is produced by creating an arc discharge between two aluminum electrodes. The intense current between the anode and the cathode erodes the latter by ion sputtering producing a metal vapour. During the discharge a pulse of helium is flushed in the electrodes gap, it mixes with the metal vapour and then expands into a vacuum through the expansion channel. In order to improve the stability of the source and to promote the aggregation we introduced a thermalization chamber where the discharge takes place. In this way turbulence affecting the expansion are eliminated; moreover we prevent the formation of deposits of sputtered material on the walls of the channel thus avoiding clogging and intensity fluctuations.

In the PACIS a competition between the electric field force and the pressure of the carrier gas stream takes place. Only if a sufficiently high gas pressure and flow are obtained within the gap, a large amount of material is transported into the thermalization chamber. The evolution of the pressure in the thermalization chamber is also an important factor for cluster formation: a high pressure favours the collision which are necessary for the cooling and the neutralisation of the plasma. Moreover the expansion conditions i.e. the cooling of the aggregates are controlled by the gas pressure in the source. It is thus very important to control the energy delivered in the arc, the pressure in the gap during the discharge, the pressure in the thermalization region. These parameters can be controlled by varying the voltage applied on the electrodes, the delay between the discharge and the opening of the pulsed valve, the volume of the thermalization zone, the diameter of the nozzle. In figure 2 we show two mass spectra taken for clusters produced with different delays between the discharge and the opening of the pulsed valve. As can be seen the effect of the mass distribution is dramatic. In particular a longer delay causes the formation of bigger clusters due to the intense flow realised between the electrodes during the spark.

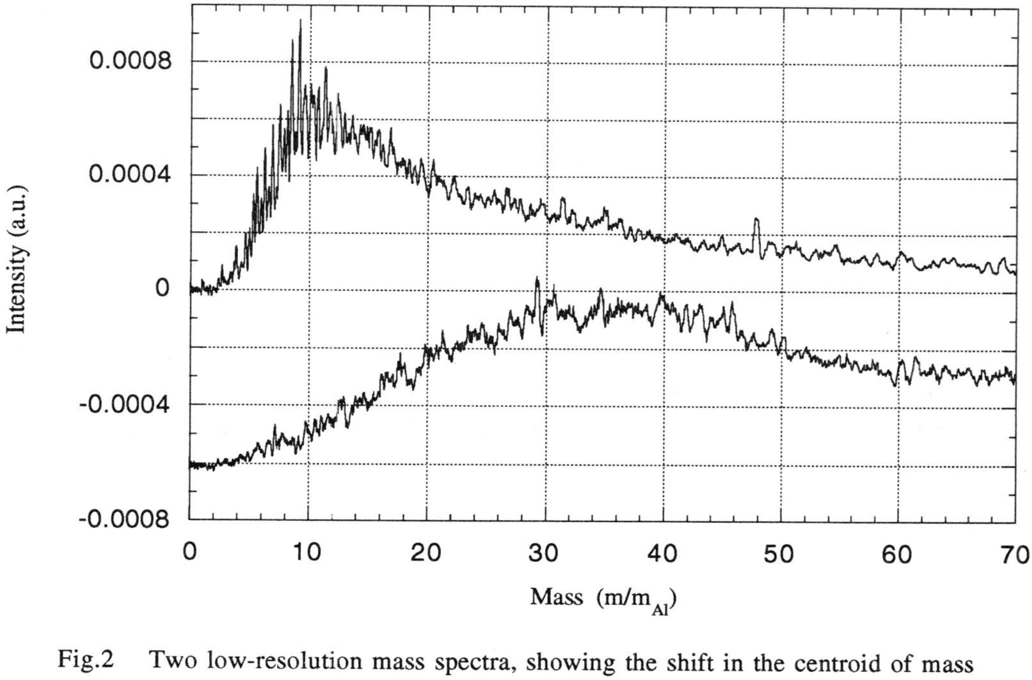

Fig.2 Two low-resolution mass spectra, showing the shift in the centroid of mass distribution obtained using different valve-spark delays. Upper curve is taken with a 300 μs delay and lower with 400 μs. Time of flight over the source-MS distance is 1300 μs for both.

The gas pressure in the thermalization region is also critical in determining the dimension of the clusters, their internal temperature and also their kinetic energy [9]. On one hand a high number of collisions favours the growth and the thermal equilibration of large aggregates, on the other hand the clusters are further cooled and accelerated if the high pressure of the source can produce a supersonic expansion. It is thus very important a characterization and control of the gas dynamics in the source in order to master the final cluster production. Many parameters are involved in the gas dynamics: the geometry of the source, the supplying pressure of the carrier gas, the gas pulse shape and duration, the delay between the arc discharge and the gas pulse. Concerning the source geometry the presence of a cavity before the expansion channel must be considered very carefully: this element is characterized by a filling time T_f in which the gas pressure rises and an evacuating time T_e in which the pressure lowers down to the initial values. These two characteristic times are very different and in particular $T_f << T_e$ since the former depends mainly from the gas pulse width (typically on the order of 100 microsec.) and the latter from the volume of the chamber and from the diameter of the nozzle (the evacuating time is on the order of several milliseconds.) [10]. In figure 3 we show the evolution of the cluster mass spectra as a function of these characteristic times. We report the intensity of the cluster signal integrated over the whole mass spectrum versus the delay between the gas pulse and the arc discharge. Several curves are plotted for different time of flights between the source and the mass spectrometers. Variations of the time of flight allow the selection of cluster populations with different velocities or, in other words, different expansion conditions. For fast clusters (TOF=600 microsec.) we see that an efficient cluster production is obtained for a delay of 400 microsec. which correspond to a very intense gas flow between the electrodes. For slower clusters this peak splits in two components: one narrow at earlier time and another more wide and unstructured at longer times. This is an indication of the existence of two different cluster population: the first is formed by clusters produced when the discharge coincides with the arrival of the gas pulse, the second formed by clusters which have spent a long time in the thermalization chamber. Varying the source parameters one can select cluster beams with a different aggregation and expansion story i.e. with different mass distribution, temperature and velocity.

In this preliminary report we have shown that pulse vaporization sources can be used to produce intense beams of neutral and ionized clusters of refractory materials. We discussed the principle of operation and the performance of LVS and PACIS source. In particular the PACIS can be considered a valid alternative to the laser vaporization sources due to its low cost, flexibility and intensity. A modified version of the PACIS has been tested with Al_n clusters. Mass and velocity distributions of the aggregates can be controlled by varying the source parameters. Our results suggest that a PACIS source followed by a mass analysing system can be successfully used in the field of cluster assembled materials.

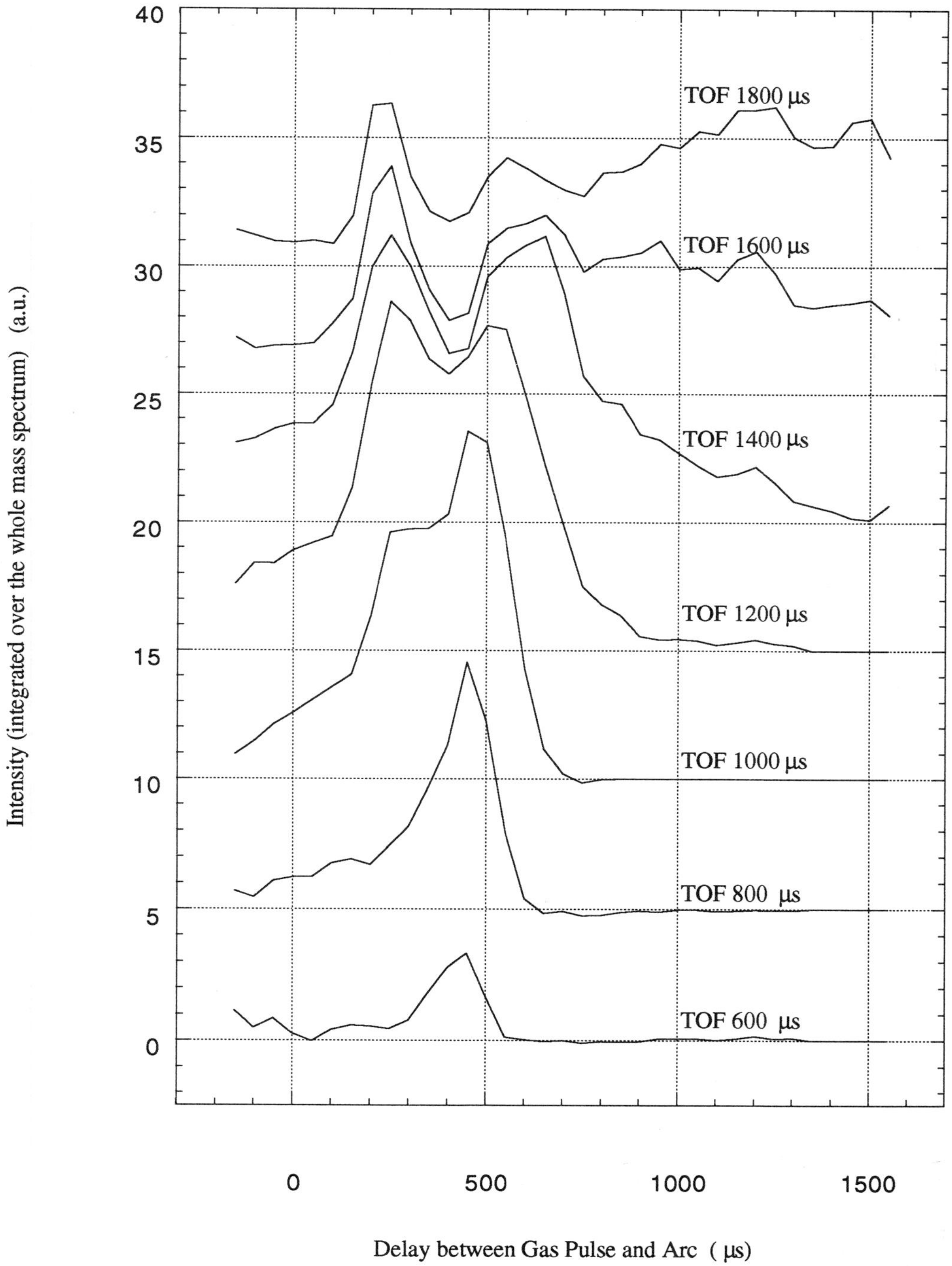

Fig.3 Total area of the mass spectrum is plotted versus delay between gas pulse and arc. Different curves are taken (from the bottom) for increasing Time Of Flight over the distance between source and Mass Spectrometer.

References.

[1] R.W. Siegel, Annu. Rev. Mater. Sci. **21**, 559 (1991)
[2] M. S. El-Shall, et al., J. Phys. Chem. **98**, 3067 (1994)
[3] H. Gleiter, Prog. Mater. Sci. **33**, 223 (1989)
[4] R.E. Smalley, Laser Chem. **2**, 167 (1983)
[5] H. Haberland, et al., J. Vac. Sci. Technol. A **10**, 3266 (1992)
[6] P. Milani, and W.A. deHeer, Rev. Sci. Instrum. **61**, 1835 (1990)
[7] W.A. deHeer, and P. Milani, Rev. Sci. Instrum. **62**, 670 (1991)
[8] C.Y. Cha, G. Gantefoer. and W. Eberhardt, Rev. Sci. Instrum. **63**, 5661 (1992)
[9] P. Milani , and W.A. deHeer, Phys. Rev. B **44**, 8346 (1991)
[10] J. Woenckhaus, and J.A. Becker, Rev. Sci. Instrum. **65**, 2019 (1994)

Materials Science Forum Vol. 195 (1995) pp. 49-54

Chemical Interactions in Titanium-Implanted Fused Silica

R. Bertoncello[1], G. Battaglin[2], E. Cattaruzza[3], F. Gonella[3], G. Mattei[3], P. Mazzoldi[3] and F. Trivillin[1]

[1] Dipartimento CIMA e Unità INCM, Via Loredan 4, I-35131 Padova, Italy

[2] Dipartimento Ch. Fis. e Unità INFM, Calle Larga S. Marta 2137, I-30123 Venezia, Italy

[3] Dipartimento Fisica e Unità INFM, Via Marzolo 8, I-35131 Padova, Italy

Keywords: Ion-Implantation, Ion Beam Deposition, Nanoclusters, XPS, TEM

Abstract

Amorphous nanoclusters of titanium silicides have been synthesized by implanting 30 keV-titanium ions, at a fluence of $1x10^{17}$ cm^{-2} *in amorphous silica. The cluster stoichiometry was [Ti]/[Si]=1.1±0.3, as obtained by energy dispersive spectroscopic (EDS) x-ray micro-analysis and confirmed by x-ray photoelectron spectroscopy (XPS) analysis. Titanium implanted ions react with the host silica matrix leading to the formation of chemical bonds with silicon.*

Implantation of metal ions into glass is of great interest since, depending on their reactivity with the substrate [1], they cause the formation of metallic nanoclusters and/or the synthesis of new compounds. In our recent works [2,3] we observed the formation of oxide and silicide species in titanium-implanted silica, while tungsten implantation in silica causes the formation of tungsten oxides and metallic tungsten precipitates. These different results have been explained by employing a two-step model where the ballistic effects have been separated from the chemical (thermodynamic) driving forces. Other authors [4] observed that titanium implantation into quartz, at higher fluences, produces a highly conductive layer, ascribed to the formation of metallic titanium or of a compound having a metallic character, embedded in the insulating material. This high conductivity could also be connected to percolation mechanisms between nanoclusters formed by metal ion implantation. Indeed at high implantation fluences we expect that the chemical interaction between silicon and reactive metal ions plays an important role in favouring the formation of small precipitates as isolated compounds in the matrix. In the present work we studied a system where the implantation fluence of titanium is higher than those used in our previous investigations [2].

Type II fused silica samples (supplied by Heraeus) were implanted at room temperature with titanium ions at the energy of 30 keV, at fluence of $1x10^{17}$ cm^{-2}. Current densities were maintained lower than 2 μA cm^{-2}. The value for the titanium ion projected range, R_p, calculated using the TRIM code [5], is R_p=32 nm. Samples were mainly characterized by x-ray photoelectron spectroscopy (XPS), x-ray-excited Auger electron spectroscopy (XE-AES) and transmission electron microscopy (TEM). Secondary ion mass spectrometry (SIMS) and Rutherford backscattering spectrometry (RBS) gave complementary informations on the depth profiles of the different elements. XPS and XE-AES measurements were achieved with a VG Fisons Instruments Escalab MK II using non-monochromatized MgK_α radiation (1253.6 eV). The spectrometer was calibrated by assuming the binding energy (B.E.) of the $Ag3d_{5/2}$ line at 368.2 eV with respect to the Fermi level. Owing to surface charging, samples present a shift of the signals towards higher binding energies; the charging effect was corrected using an internal reference. XPS depth profiles of the different elements were carried out by cycles of Ar^+ sputtering. The sputtering rate was about 1.5 Å min^{-1}. Experimental details are reported elsewhere [2,6]. A sample of titanium silicide, obtained after annealing of an evaporated titanium layer (a few tens of nm thick) on a silicon substrate, was used as a reference standard for XPS peaks. The stoichiometry of this silicide was $TiSi_{1.2}$. Pure silicon and titanium standards were also used. A Cameca IMS-4f spectrometer was employed for SIMS measurements. RBS analysis was performed by using a 2.2 MeV $^4He^+$ beam. TEM measurements were performed on a Philips CM30 microscope operating at 300 kV equipped with a GATAN energy dispersive spectroscopic (EDS) x-ray micro-analysis. The specimens were prepared in cross-sectional samples by mechanical pre-thinning, followed by a low-angle milling using argon ions with energy of 2-5 keV. To minimize ion damage, samples were cryogenically cooled during ion milling. The XPS depth profile of the different elements is reported in Fig. 1. This profile coincides with those obtained from SIMS and RBS analyses. Ti2p signals at different depths are reported in Fig 2. XPS and XE-AES values for $Si2p_{(3/2+1/2)}$ and SiKLL signals are reported in Table I. We want to stress here that the synthesis of clean titanium-implanted silica samples presented some difficulties caused by in depth contamination due to radiation-enhanced-diffusion from the surface of carbon atoms during implantation; this effect is typical for implantation of elements that are chemically very reactive with carbon. Carbon atoms are present at every depth of the implanted regions: their concentration does not exceed 3% atomic fraction, and their depth profile (not reported in Fig. 1) follows that of implanted elements.

All over the depth profile the B.E. value of C1s peak is characteristic of carbide species formation [7-10]. Since the $Si2p_{(3/2+1/2)}$ peak as well as the SiKLL peak do not show the characteristic components indicating the formation of silicon carbide [7-11], we conclude that only small amounts of titanium carbide are present into the matrix. The analysis of the titanium implanted samples shows that the maximum titanium atomic fraction reached in the implanted sample is 30% (see Fig. 1). In this sample the unresolved $Si2p_{(3/2+1/2)}$ doublet and the SiKLL Auger peak exhibit two well distinct components (Table I); the Ti2p doublet (Fig. 2) exhibits two distinct pair of signals, characterized by different B.E. values of the $Ti2p_{3/2}$ peak and different energy splitting between the doublet components.

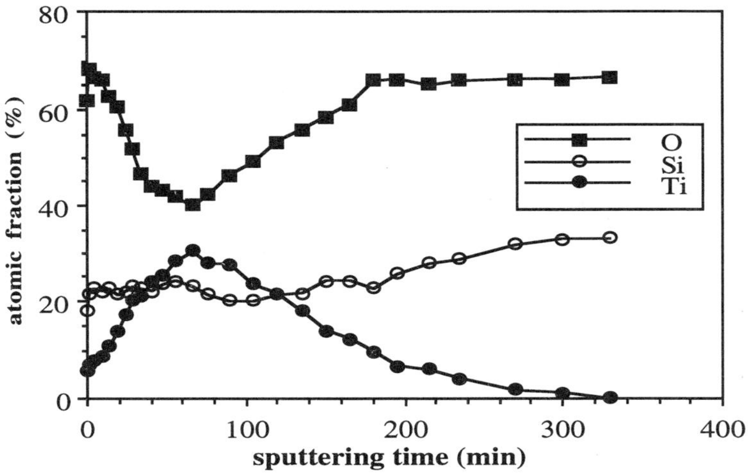

Fig.1. Oxygen, silicon and titanium atomic percent XPS depth profile in titanium-implanted silica.

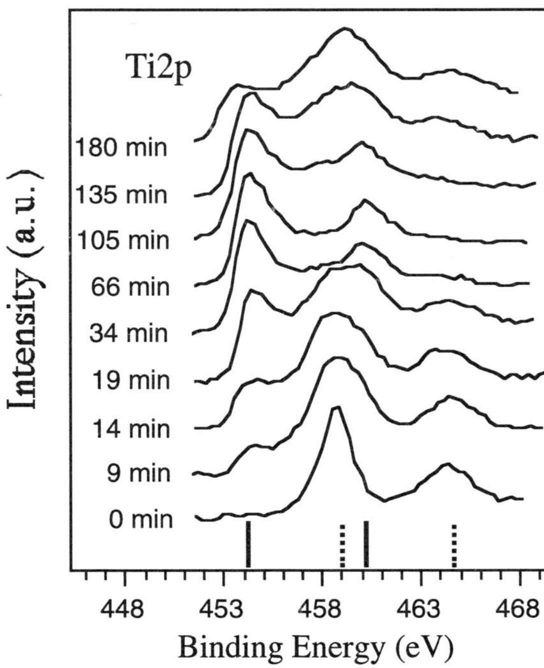

Fig.2. XPS Ti2p doublet, at different depths, in the titanium-implanted silica sample. Plain and dashed lines indicate the characteristic B.E. values of silicides and dioxide species, respectively.

Sample	Si2p (B.E.)	SiKLL (K.E.)	α parameter
Ti:SiO_2 (low B.E.)	99.3 ±0.1 eV	1617.0 ±0.1 eV	1716.3 ±0.2 eV
Ti:SiO_2 (high B.E.)	103.4 ±0.1 eV	1608.9 ±0.1 eV	1712.3 ±0.2 eV

TABLE I. Experimental data of binding energy (B.E.) and kinetic energy (K.E.) of silicon XPS and XE-AES peaks, and α parameter values for different silicon chemical states (α = B.E. of XPS $Si2p_{(3/2+1/2)}$ peak + K.E. of XE-AES SiKLL peak).

Titanium oxide and silicon oxide are the chemical species which originate the high B.E. components of $Si2p_{(3/2+1/2)}$ and Ti2p peaks. At the surface and at depths where the titanium percentage atomic fraction is below 5%, the B.E. of silicon and titanium lines agree with the presence of SiO_2 and TiO_2 species. However the percentages of total silicon and titanium atoms that bond to oxygen atoms are less than 50%. Along the depth profile the intensities of the low B.E. components of $Si2p_{(3/2+1/2)}$ and Ti2p peaks vary in the same way: they both rise up to the maximum titanium concentration depth and decrease at larger depths. The chemical species that originates these components is identified as titanium silicide. In fact, in the titanium-implanted silica sample the value of the silicon α parameter (as defined in the table I captions) is α=1716.3±0.2 eV, well in accordance with the measured α=1716.4±0.2 eV for the titanium silicide standard. Moreover, the energy separation between the $Ti2p_{3/2}$ and $Si2p_{(3/2+1/2)}$ peaks attains to a value of 355.5±0.2 eV for the metallic species and of 354.9±0.2 eV in the $TiSi_{1.2}$ standard. The value we measured in the titanium-implanted silica is 355.0±0.2 eV. These evidences substantiate that there is no metallic titanium and/or silicon formation. The presence of titanium silicide species was inferred, but not detected, in previous works by P. Martin *et al.* [4,12].

The TEM analysis (Fig. 3) reveals the presence of small clusters very close to the surface, over which an amorphous layer of the epoxy resin is visible. The largest clusters have diameters of about 20 nm and they appear to have an amorphous structure from the electron micro-diffraction. The relative concentration obtained from EDS on the clusters is [Ti]/[Si]=1.1±0.3. This is in accordance with the stoichiometry of the titanium silicide compounds as obtained from XPS analysis, i.e. TiSi.

In conclusion, the XPS, XE-AES and TEM analyses revealed the presence of amorphous silicide nanoprecipitates as a result of titanium ion irradiation of silica. In the titanium-implanted sample we detected a large amount of silicide species. Chemical reactions between the irradiated ion and silicon atoms begin to be detectable at very low titanium concentration. These evidences indicate that titanium implanted atoms are very reactive as regard the formation of chemical bonds with silicon of the host silica matrix. In view of future applications of the obtained structures in the field of non-linear optics, the optical characterization of these samples is in progress.

The authors are grateful to F. Caccavale (Dipartimento di Fisica, Padova) for performing the SIMS analyses This work has been partially supported by Italian Ministero dell'Università e della Ricerca Scientifica e Tecnologica (fondi 40%).

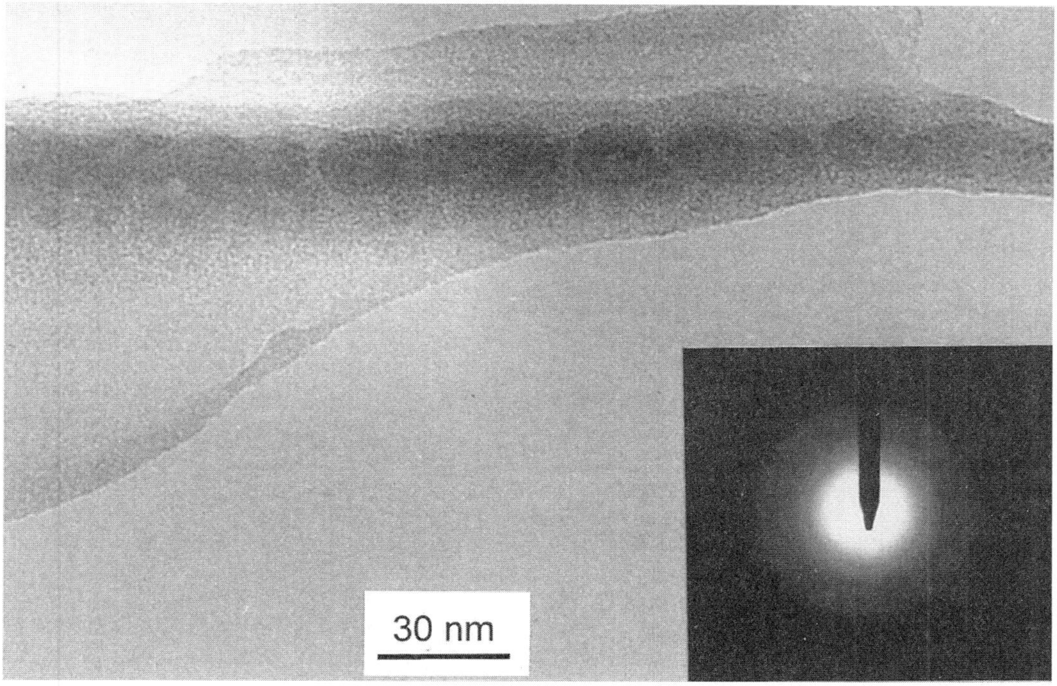

Fig. 3. TEM micrograph of the titanium-implanted silica sample.

References

[1] P. Mazzoldi, G. W. Arnold, G. Battaglin, R. Bertoncello, and F. Gonella, Nucl. Instr. and Meth. **B91**, 478 (1994).

[2] R. Bertoncello, A. Glisenti, G. Granozzi, G. Battaglin, F. Caccavale, E. Cattaruzza, and P. Mazzoldi, J. Non-Cryst. Solids **162**, 205 (1993).

[3] E. Cattaruzza, R. Bertoncello, F. Trivillin, P. Mazzoldi, G. Battaglin, L. Mirenghi, and P. Rotolo, submitted to J. Mat. Res. for publication.

[4] P. Martin, M. Dufour, A. Ermolieff, S. Marthon, F. Pierre, and M. Dupuy, J. Appl. Phys. **72** (7), 2907 (1992).

[5] J. P. Biersack and L. G. Haggmark, Nucl. Instr. and Meth. **174**, 257 (1980).

[6] R. Bertoncello, F. Trivillin, E. Cattaruzza, P. Mazzoldi, G. W. Arnold, G. Battaglin, and M. Catalano, J. Appl. Phys. **77**, 1294 (1995).

[7] D. Briggs and M. P. Seah, *Practical Surface Analysis*, J. Wiley & Sons, Chichester (1983) UK.

[8] *X-ray Photoelectron Spectroscopy Database*, version 1.0, National Institute of Standards and Technology, Gaithersburg, MD 20899, USA (1989).

[9] J. F. Moulder, W. F. Stickle, P. E. Sobol, and K. D. Bomben, *Handbook of X-ray Photoelectron Spectroscopy*, edited by J. Chastain, Perkin-Elmer Corp. (Eden Prairie, USA, 1992).

[10] F. Maury and F. Ossola, Thin Solid Films **207**, 82 (1992).

[11] G. D. Sorarù, A. Glisenti, G. Granozzi, F. Babonneau, and J. D. Mackenzie, J. Mat. Sci. **5**, 1958 (1990).

[12] A. Ermolieff, S. Marthon, P. Martin, F. Pierre, and M. Dufour, Solid State Comm. **82**, 517 (1992).

Materials Science Forum Vol. 195 (1995) pp. 55-60

Structural and Spectroscopic Characterization of Sol-Gel Synthesized Iron (III) Oxide Thin Films Doped with Na^{+} and Mg^{2+}

L. Armelao, G. Depaoli, A. Sandon, S. Daolio and E. Tondello

Dipartimento di Chimica Inorganica - University of Padova - Italy

Istituto di Polarografia ed Elettrochimica Preparativa - CNR Padova - Italy

Centro di Studio Reattività e Stabilità dei Composti di coordinatione - CNR Padova - Italy

Keywords: Iron (III) Oxide, Sol-Gel, Thin Films

Abstract

Sodium and magnesium doped Fe_2O_3 thin films have been synthesized by the sol-gel method and investigated by uv-vis, x-ray photoelectron spectroscopy, secondary ions mass spectroscopy and x-ray diffraction. The results show that the coatings are nanostructured like the undoped film and that sodium segregates at the outermost layers while magnesium in homogeneously distributed. With a suitable surface morphology these films are expected to find applications in non linear optical or sensor devices.

Introduction

Recently [1] we prepared supported nanostructured iron (III) oxide α-Fe_2O_3 thin films in view of their great interest in various technological applications. The coatings were properly tailored *via* the sol-gel process [2] which is particularly suitable for yielding thin oxide films. The chemistry involved in the sol-gel technique is based on hydrolysis and condesation of metal alkoxides, leading to the formation of metal-oxo based macromolecular networks [3]. In particular, the mild conditions of synthesis ("soft chemistry") with respect to other techniques, can allow in principle the preparation of coatings with controlled compositional and nanostructural features.

The obtained nanostructured films have highly optical transparence, high purity and are completely homogeneous in composition and morphology so to be proposed as potential candidates for non linear optical devices [4]. Therefore, some of us [5] investigated by means of quantum mechanical calculations and spectroscopic measurements the electronic absorption spectrum in the uv-vis range. On the basis of theoretical and spectroscopic considerations we concluded that the very strong absorption starting at about 2.5 eV, which was routinely interpreted as originated from an interband transition, mainly consists of charge transfer transitions from the occupied O 2p based levels to the partially empty Fe 3d atom-like orbitals.

In this paper we report the synthesis and the characterization of nanocrystalline Fe_2O_3 thin films doped with Na^+ and Mg^{2+} with the aim to study the relationship between structure and features directly connected with the potential use of such films in catalysis, as sensors or as magnetic or photoelectrochemical devices [6-8]. As it is known the lighter alkaline metal ions, i.e. sodium, generally modify the electrical properties of the high resistive oxides so allowing the technological application of conducting film of iron (III) oxide. On the other hand, the II group metal ions, i.e. magnesium, normally modify only the crystal structure and someone could expect to obtain iron (III) oxide coatings partially crystallized in the γ form, attractive for its magnetic properties.

The coatings were prepared by dipping silica substrates in ethanolic solutions of metal-alkoxides. After thermal treatment the films have been characterized by X-ray Photoelectron Spectroscopy (XPS), UV-VIS spectroscopy and Secondary Ion Mass Spectrometry (SIMS). Further informations were obtained by X-ray Diffraction (XRD).

Experimental

The metal-alkoxides were prepared according to the methods previously described [8]. The precursors for Fe^{3+}, Na^+ anf Mg^{2+} were respectively $Fe(OC_2H_5)_3$, $NaOC_2H_5$ and n-$C_4H_9Mg(OC_3H_7^i)$. Since the products are highly sensitive to oxygen and water, all the handling operations were carried out in dry-box (O_2 < 10 ppm, H_2O < 2 ppm). Furthermore, the solvents used for the reactions were dried with $LiAlH_4$ and K-benzophenone successively. The solutions used for the film deposition were obtained by dissolving the precursors in dehydrated ethanol corresponding to a total concentration of oxides of about 15g x l^{-1}. The sodium etoxide was added to the alcoholic solution as a 1:1 complex with dibenzo-18-crown-6 to rise its solubility in ethanol. The atomic ratio Na^+/Fe^{3+} in the sodium containing solutions was 5, 10 and 15, while for the magnesium containing solutions was 50. The dip-coating procedure was carried out at room temperature with a withdrawal controlled speed of *ca.* 12 cm $\times$ min^{-1}. The samples were prepared by means of eight successive dippings without any treatment between them and, after the heating, coatings 1000 ± 100 Å were obtained, as measured by a Tencor Alpha Step profilometer.

The thermal treatments were performed in a programmable furnace. All the samples were heated for 1h at 400 , 500 and 600°C. The choice of 400°C as lowest treatment temperature is due to the fact that only in these conditions coatings free from hydrocarbons are obtained [9]. After the treatment, the coatings were red-brown and transparent.

The composition of the films at the surface and in the bulk was obtained by XPS. The spectra were run on a Perkin-Elmer Φ 5600ci spectrometer using unmonochromated Al Kα radiation (1486.6 eV). The working pressure was less than 5×10^{-8} Pa. The spectrometer was calibrated by assuming the binding energy (E_b) of the Au $4f_{7/2}$ line at 83.9 eV with respect to the Fermi level. The standard deviation for the E_b values was 0.15 eV. The reported E_bs were corrected for the charging effects by using a flood-gun neutralizer, assigning to the C1s line of adventitious hydrocarbons a value of 284.8 eV. Survey scans were obtained in the 0 - 1200 eV range. Detailed scans were recorded for the Fe 2p, O 1s, Na 1s, Mg 2s, Si 2p and C 1s regions. The atomic composition were evaluated using theoretical cross-sections [10]. Depth profiles were carried out by Ar^+ sputtering at 2.5 - 3.5 keV with an argon partial pressure of 5×10^{-6} Pa. In these conditions the sputtering rate is around 70 - 80 Å $\times$ min^{-1}.

The SIMS profiles were obtained on an instrument custom-built and constituted by : a mass-filtered duoplasmatron ion gun (DP50B, VG Fisons) generating a monochromatic and collimated O_2^+ ion beam, used as bombarding source; a secondary-ion optic based on a three-lens design with central stop and a Balzers model QMA 400 quadrupole mass analyzer (mass range extended up to 2048 u). A secondary electron multiplier (90° off-axis SEM) is used for negative and positive ion detection in counting mode. The samples were exposed to ion beam with different energies and current densities. The sputtering rate was estimated to be from 800 to 2000 Å/min with a primary ion current ranging from 400 to 800 nA.

The uv-vis spectra were recordered in the 200-900 nm range using a double beam Cary5E spectrophotometer with a spectral bandwidth of 1 nm. The contribution due to the silica substrate was subtracted. The X-ray diffraction data were collected using a diffractometer (Cu-Kα radiation, 35 kV, 40 mA) equipped with a thin film attachment.

Results and discussion

To understand and rationalize the role of dopants on the functional properties of oxide thin films, it is first of all necessary to investigate their behaviour in the host matrix. As a matter of fact, the distribution and the chemical interactions with the host matrix can strongly depend on the synthetic procedure. Consenquently, the profile of Na^+ and Mg^{2+} along the film thickness and their effect on the behaviour of iron (III) oxide coatings have been determined.

Let consider first the sodium doped coatings. As far as the ion distribution is concerned it is interesting to consider the XPS depth profile reported in figure 1. The sodium ion is present mainly in the region near the surface, that is to say in the first 100 Å. It rapidly decreases going inside the bulk and the concentration gradient is strongly dependent on temperature, as can be observed by comparing the profiles of the samples heated at 600° C and 400° C. This suggests that the conducting zone of the thin films is mainly formed by the top layers and that it becomes even thinner by raising the temperature. From a structural point of view, there is no difference between the pure and doped Fe_2O_3 thin films since in both cases they crystallize in the haematite form (α-Fe_2O_3). The structural analogy is well evident both from uv-vis (figure 2) and X-ray diffraction investigations [11]. On the other hand, the presence of sodium ions influences the crystallization process of the films. As a matter of fact, whereas the presence of nanometric crystallites is observed starting from 400° C for the pure coatings [1], at least 600°C are necessary for obtaining the same nanosized α-Fe_2O_3 crystals in the doped thin films. Therefore, the role of Na^+ ions is that of influencing the functional features of the host matrix without giving any chemical interaction with it. It is probably for this reason that the dopant is not retained into the bulk of the film but, on the contrary, segregated in the surface region.

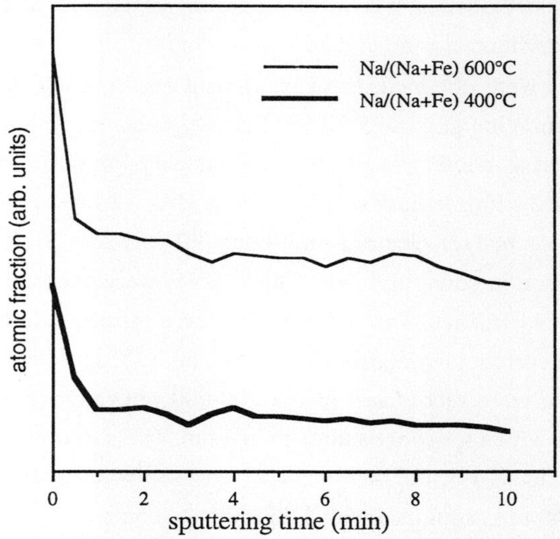

Figure 1 - Na/(Na+Fe) profiles of Na 10% at. samples treated at 400 and 600°C.

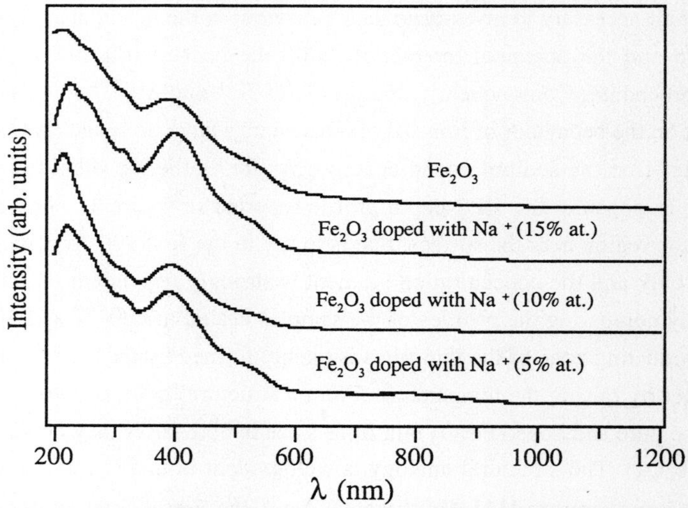

Figure 2 - UV-VIS absorbance spectra of pure Fe_2O_3 films and Fe_2O_3 films doped with Na^+ at various concentrations after thermal treatment at 600°C

For the Mg^{2+} doped thin films the situation is quite different. The XPS depth profile of Mg^{2+}, reported in figure 3, shows a homogeneous distribution of the ion along the film thickness even if the effective Mg^{2+} atomic percentage is much lower (≈ 20 %) than that of the starting solution (≈ 50 %).

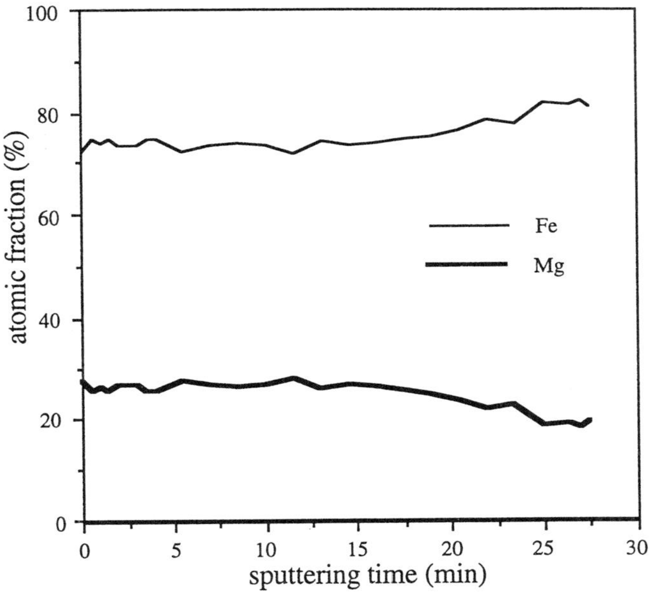

Figure 3 - XPS depth profile of Mg 50% at. samples thermally treated at 400°C.

SIMS investigations evidence that starting from 400° C Fe^{3+} and Mg^{2+} species are neighboring as suggested by the presence in the spectra of clusters (figure 4) with stoichiometry $MgFeO_x$ ($0 \leq x \leq 4$). On the other hand, the local arrangement of the metallic atoms can not be identified since this sample does not show any XRD pattern. Only after a thermal treatment at 800° C nanometric crystallites (≈ 20 nm) are observed and correspond to the spinel phase $MgFe_2O_4$. The absence of any diffraction peak due to the haematite or to the maghemite phase suggests that the remaining iron oxide in the coatings is still amorphous or the crystallites are so small that it is impossible to detect them. Thus the dopant into the host matrix can either create a state of disorder so to delay the crystallization process or it can act as a nucleating agent which only promotes the formation of small crystallites but not their growth.

Conclusions

Sodium and magnesium doped Fe_2O_3 thin films were synthesized by the sol-gel process and well characterized. First of all it must be outlined that it is possible to prepare high-purity coatings free from foreign ions by using metal alkoxides as precursors and pure reagents.

Sodium ion, as the dopant, is confined mainly in the surface region. Moreover, its presence delays the crystallization process so that the preparation of nanostructured $Fe_2O_3{:}Na^+$ coatings requires higher temperatures with respect to the pure coatings. On the other hand, no structural modifications are induced in the host matrix by the dopant since in both cases the observed crystalline phase corresponds to haematite.

In the Mg^{2+} doped coatings, the ion profile is homogeneous along the film thickness at each temperature thus suggesting a great tendency of this bivalent ion to follow the distribution of the trivalent one. However, the influence of the dopant on the microstructural and structural features of the iron (III) oxide matrix are not so evident as in the case of sodium.

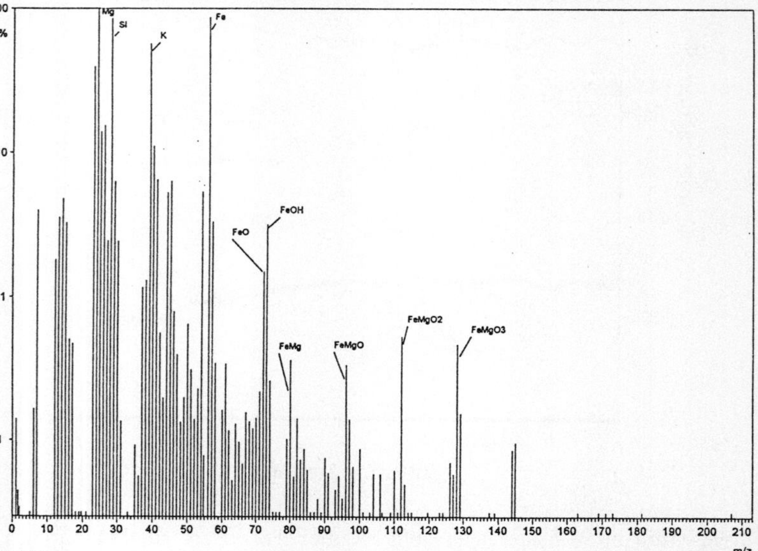

Figure 4 - SIMS spectrum of the Mg (50%) doped Fe2O3 thermally treated at 500°C.

We are indebted to Dr. P. Colombo for XRD measurements and to Dr. R. Bertoncello (University of Padova) for helpful discussions.

References

[1] L. Armelao, G. Granozzi, E. Tondello, P. Colombo, G. Principi, P. P. Lottici and G. Antonioli, Microstructural investigation of α-Fe_2O_3 thin films obtained via the sol-gel method, *J. Non-Crys. Solids* , in press

[2] C. J. Brinker and G. Scherrer, Sol-Gel Science: the Physics and Chemistry of Sol-Gel Processing, 1989, Academic Press, San Diego

[3] (a) J. Livage, M. Henry and C. Sanchez, *Prog. Solid State Chem.*, **18**, 1988, 258; (b) C. Sanchez and J. Livage, New J. Chem., **14**, 1994, 513

[4] T. Hashimoto, T. Yoko and S. Sakka, *J. Ceram. Soc. Japan*, **101**, 1993, 64

[5] L.Armelao, M.Bettinelli, M.Casarin, G.Granozzi, E.Tondello and A.Vittadini *J. Phys.Condens.Matter* , **7,** 1995, L299

[6] H. H. Kung, Transition Metal Oxides: Surface Chemistry and Catalysis, 1989, vol. 45, Elsevier, Amsterdam

[7] F. Orgaz and H. Rawson, *J. Non-Crys. Solids*, **82**, 1986, 378

[8] A. K. Atta, P. K. Biswas and D. Ganguli, *J. Non-Crys. Solids,* **125**, 1990, 202

[9] L. Armelao, Sintesi e Caratterizzazione di materiali Inorganici in Strato Sottile per Catalisi, Ottica ed Optoelettronica, Tesi di Dottorato in Scienze Chimiche, Università degli Studi di Padova, Novembre 1994

[10] D. A. Shirley, *Phys. Rev.* **55**, 1972, 4709

[11] manuscript in preparation

Materials Science Forum Vol. 195 (1995) pp. 61-66

Nanocrystalline Ferromagnetic Structures Produced by Devitrification of Amorphous Materials

F. Branda[1], A. Costantini[1], V. Iannotti[2], L. Lanotte[2] and P. Matteazzi[3]

[1] Dip.to di Ingegneria dei Materiali e della Produzione, Facoltà di Ingegneria, P.le V. Tecchio 80, I-80125 Napoli, Italia

[2] Dip.to di Scienze Fisiche, Unità INFM-GNSM, Univ. di Napoli „Federico II", P.le V. Tecchio 80, I-80125 Napoli, Italia

[3] Dip.to di Scienze e Tecnologie Chimiche, Facoltà di Ingegneria, Via Cotonificio 108, I-33100 Udine, Italia

Keywords: Nanocrystalline Materials, Devitrification, Soft Ferromagnets

Abstract.
Nanosized crystalline materials prepared by devitrification from the amorphous state are an interesting subject of physical research thanks to their attractive magnetic properties related to very low value of magnetostrition. This work is a comparison study of some devitrification tecniques to individuate both their advantages and disvantages in obtaining excellent nanocrystalline magnetic alloys. In particular we investigate differences between laser annealing, furnace heating both in an amorphous metal obtained by rapid quenching from the melt and in a CoP multilayer produced by electrochemical deposition.

Introduction.

The nanocrystalline materials produced by partial crystallization of Fe or Co based amorphous alloy exibit very soft magnetic behaviour due to their vanishing magnetocrystalline anisotropy and very small magnetostriction. The preparation methods may be summarized in the following main techniques: heating treatments in furnace[1,3], Joule heating[4,5], laser annealing[6]. The heating rate is very different in these methods and the structural trasformations thermally induced on the as cast ribbons are intrinsically connected to the heating behaviour. Therefore it is important to evaluate what treatment is more appropriate to reach peculiar purposes. In this paper we perform a comparative study of the effects induced by heating in programmable differential calorimeter or by laser treatment. The nanocrystallization process was particularly examined both in ribbons of $Fe_{73,5}Cu_1Nb_3Si_{13,5}B_9$, generally named FINEMET, produced by melt spinning and in ribbons of Co-P obtained by electrochemical deposition.

Experimental.

The classical FINEMET material was produced by melt spinning in the IEN institute (Torino-Italy) and kindly furnished by F.Vinai and P.Tiberto. Co-P amorphous alloy, also investigated in this study, was prepared by us using elettrochemical deposition[7]. The heat treatments were performed by means of a differential scanning calorimeter[8] (NETZSCH apparatus) and by CO_2 continuos Laser with programmable beam movement[9]. Magnetic properties were determined from quasi-static hysteresis loop measurement and saturation magnetostriction constant λ_s was obtained by cycle deformation under stress[10]. X-ray spectra were recorded by an INGL DIFFRACTOMETER using Co-Kα radiation and a position sensitive detector. Crystallite sizes were calculated from the line width using the Schener formula.

Influence of Heating Rate on the Structure of Devitrified Materials.

Fig.1 shows the temporal behaviour of temperature during a classical treatment in furnace (A) and a classical laser heating (B) of a FINEMET sample.
The most prominent differences between the two treatment consists in the interaction time and temporal change of temperature. During furnace annealing, the temperature is increased linearly with heating rates of 5-15 °C/min up to ≈ 540°C and subsequently, after ≈1 hour of isothermal annealing, the sample is furnace-cooled to room temperature with the same rate.
In laser annealing the process has whole evolution in ≈1 s and temperature/time ratio of 2000 °C/s are reached.
By changing the heating rate not only devitrification of the amorphous structure is stopped at a different degree of the crystallization process, but also different crystallite kinds can be produced. To clarify this feature we show in Fig.2 the differential scanning calorimetry spectra that are obtained

from FINEMET amorphous alloys by using different heating rate.

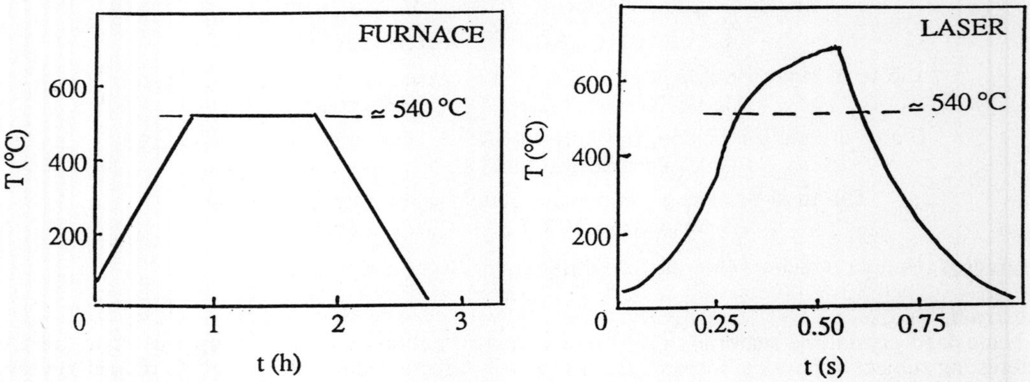

Fig.1

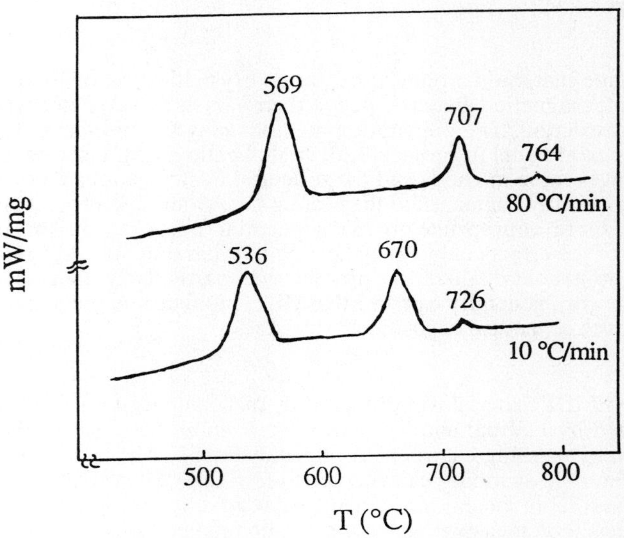

Fig.2

It is clear that when heating velocity increases the devitrification peaks occur at higher temperatures and at the same time, also sharpness and relative intensity are different. X-ray diffraction shows that the first devitrification peak corresponds to formation of α-Fe(Si) crystallites, the second peak is connected with the appearence of borides (Fe_3B and $Fe_{23}B_6$), and finally the third peak gives rise to Fe_3Si crystallites.

To prevent boride formation it is necessary to perform a 1 hour furnace heat treatment below 550 °C. Now we consider that if instantaneously the annealing temperature increases, the devitrification velocity of the first phase also increases, so that in very short time it is possible to perform crystallization of a large ammount of this phase. On the other hand if treatment time is very short the crystallization of the phases with higher transition temperature is stopped at its first stage. Laser heating to operates by means of this mechanism, obtaining both a great number of nanosized crystallites of the phase wich has low formation temperature (α-Fe) and a neglegible atomic fraction of the other phases. This may result in an advantage in comparison with classical treatment in

programmable furnace, because of both the increase of partial crystallization of soft magnetic phase and the obtaining of very small size crystallites.
The diffractometric spectra from $Fe_{73,5}Cu_1Nb_3Si_{13,5}B_9$ samples after heat treatment in furnace at 620 °C for 1 hour (A) and after laser annealing (B) at a maximum temperature of ≈ 700°C (following a temperature cycle as that in fig.1) are compared in Fig.3.

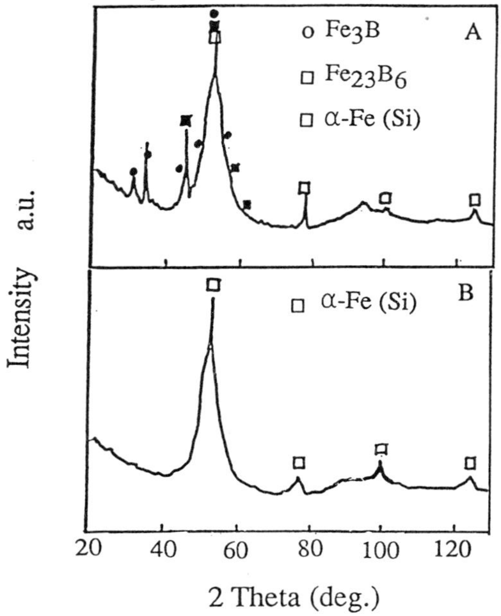

Fig.3

We can see that formation of Fe3B and Fe23B6 is not obtained by using laser treatment. This fact produces better soft magnetic properties as reported in the following **Table 1**:

Sample	Coercive Field (A/m)	λs (10^{-6})	Max.Permeability (10^4)	Remanence [T]	Crystallites Size [nm]
A	10	≈ 10	4	0.8	14
B	2	$-3 < \lambda s < 0$	8	0.6	15

Tab.1

New Results in Co-P Alloys Prepared by Electrochemical Deposition.
The consideration carried out in the preceding section are validated by the results obtained in a different amorphous alloy namely the CoP based multilayers produced by electrolysis[11]. The soft magnetism of this ribbons is already excellent as a consequence of the stratified structure magnetic layers ($Co_{.87}P_{.13}$) separated by non magnetic films ($Co_{.70}P_{.30}$).
Generally nanocrystallization of this alloys was not possible because of the large size of cobalt grains obtained by furnace heating (1 hour 310 °C; see figure subsequent A). Treatments in furnace at

higher temperature are not usefull since Co_2P intermetallic phase is also produced.
This phase, as well the formation of cubic cobalt, produces remarkable hardening of the magnetic behaviour[8]. The devitrification temperatures of $Co_{exagonal}$, Co_2P and Co_{cubic} crystalline phases are very near as one can see from Fig.4a.

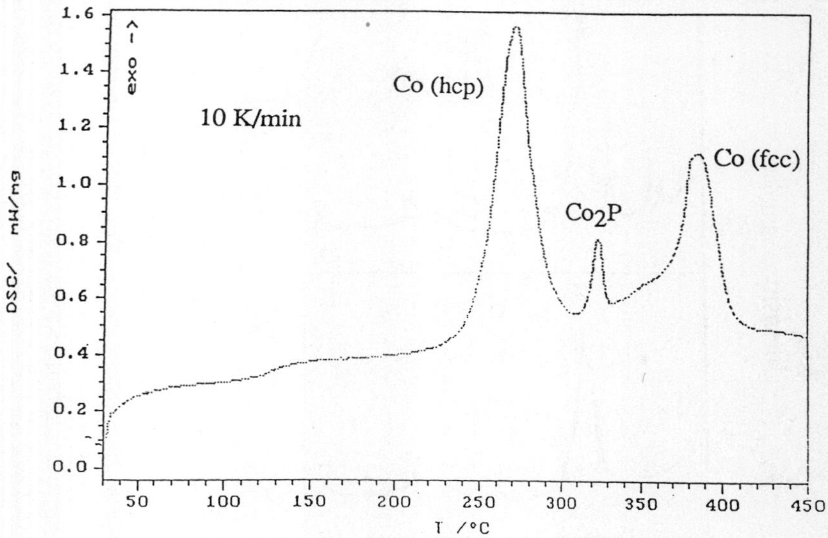

Fig.4a: Differential Scanning Calorimetry

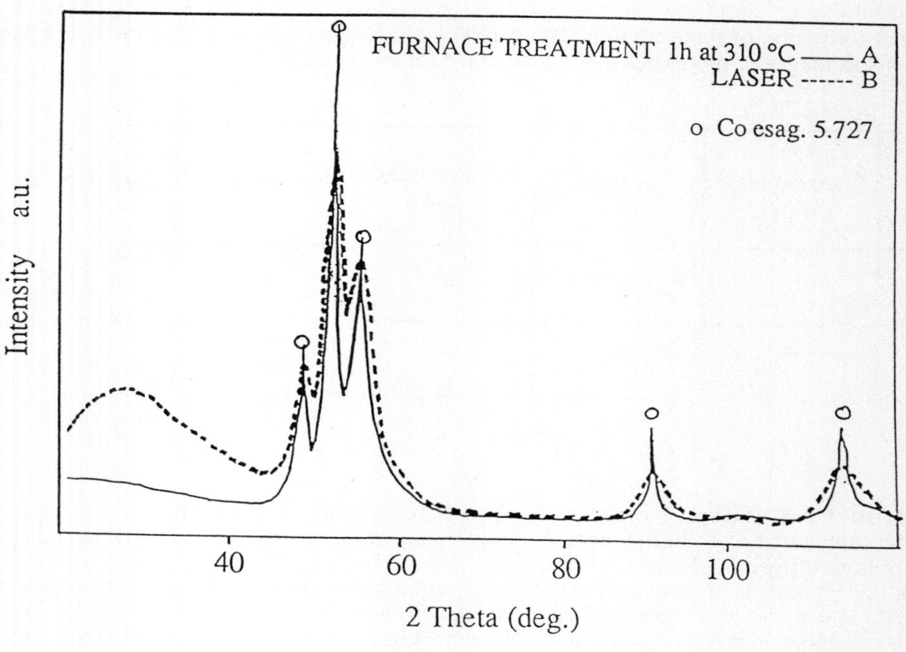

Fig.4b

In this case the effectiveness of laser annealing is more evident than in the case of FINEMET material; in fact conventional annealing cannot be used to produce the same laser effects.
Only a rapid treatment at a temperature so high to produce a large ammount of Co crystallites but not a big size of their grains can decrease magnetic anisotropy and magnetostriction as well.
In figure 4b the X-ray spectrum, after laser annealing, is reported. The used laser treatment cycle is shown in Fig.5.

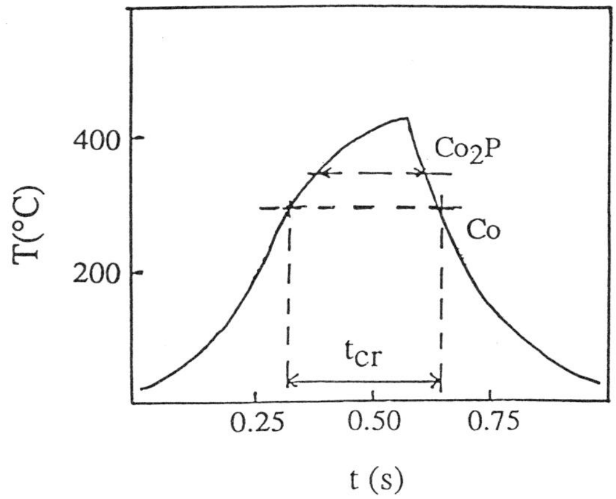

Fig.5

During laser heating, temperature values higher than the crystallization temperature of hexagonal cobalt phase were reached for an interaction time of 0.35 s while crystallization temperature of Co_2P phase was overcome only for 0.20 s.
This was sufficient to get 50% about of Co crystallites (average size 12 nm), without relevant presence of CoP grains. Conventional treatments aren't able to give similar results.
Saturation magnetostriction "λ_s", coercive field "H_c" and maximum permeability "μ" got improvements by the structure produced by laser annealing as it is shown in the following Table 2:

Sample	Coercive Field (A/m)	λ_s (10^{-6})	Max.Permeability (10^3)	Crystallites Size [nm]
A	10	5	7	≈ 64 nm
B	8	3	60	≈ 19 nm

A = heat treatment in furnace at 310 °C.
B = laser annealing as in Fig.5.

Tab.2

Extended investigation are in course to reveal the magnetoelastic feature of this partial nanocrystalline structure.

Final remarks

The devitrification behaviour of amorphous materials at different temperature rate is a fundamental investigation in order to clarify the possibility of preparing nanocrystalline structure.
Very slowly crystallization processes wich give rise to macrocrystalline grains can be performed at so high velocity that nanosized crystallites are produced. The best method in this case is laser annealing. On the other hand, if slowly devitrification can occurs with formation of a large number of crystallization nuclei, so that little size of grain is obtained, conventional heat tratments are to be preferred to laser annealing, since the difficulties involved by the last technique.

References

[1] Y. Yoshizawa, S. Oguna and K. Yamauchi, J. Appl. Phys. **64**, 6044 (1988).
[2] T. H. Noh, W. K. Pi, H. J. Kim and I. K. Kang, J. Appl. Phys. **69**, 5921 (1991).
[3] T. Kulik, R. Zuberek and A. Hernando, J. Magn. Magn. Mater.**140-144**, 433 (1995).
[4] P. Allia, M. Baricco, P. Tiberto and F. VIinai, Rev. Sci. Instru. **64**, 1053 (1993).
[5] E. Bonetti, L. Del Bianco, P. Tiberto, J. Magn. Magn. Mater. **140-144**, 477 (1995).
[6] L. Lanotte and V. Iannotti, J. Appl. Phys. (In course of pub., 1995).
[7] L. Lanotte, "Nanomagnetism" ed. A. Hernando, Kluwer Ac. Pub., **185** (1993).
[8] F. Branda, L. Lanotte, A. Costantini and P. Matteazzi, J. Mat. Sci. **28**, 1824 (1993).
[9] L. Lanotte, Z. Kackwoski, R. Pasquino, G. Palazzo and V. Tagliaferri, J. Magn. Magn. Mat. **135**, 215 (1994).
[10] L. Lanotte and L. Maritato, Elsevier Ser., Appl. Electr. Mater. **4**, 215 (1993).
[11] L. Lanotte, Z. Kackowski and L. Maritato, Europhys. Letters **9**, 717 (1992).

Materials Science Forum Vol. 195 (1995) pp. 67-72

X-Ray Absorption Study of Local Structure in Nanophase Palladium

S. De Panfilis[1], F. D'Acapito[2], V. Haas[3], H. Konrad[3], J. Weissmüller[3#], and F. Boscherini[1*]

[1] INFN, Laboratori Nazionali di Frascati, P.O. Box 13, I-00044 Frascati (Rome), Italy

[2] CNR, c/o GILDA CRG, ESRF,BP 220, F-38043 Grenoble Cedex, France

[3] Universität des Saarlandes, D-66121 Saarbrücken, Germany

Keywords: Nanophase Materials, X-Ray Absorption Spectroscopy, Grain Boundaries

Abstract

We present an X-ray absorption study of local order in nanophase Pd deposited by inert gas evaporation. Samples were subjected to various pressure and temperature treatments and were carefully characterized by X-ray diffraction. We find that there is a reduction in the average coordination numbers in the first few coordination shells; this reduction can be reproduced by considering the surface to bulk ratio and the distribution of dimensions of the nanometer-sized particles.

Introduction

The interest in the local structure of nanophase materials[1, 2] stems from the high density of grain boundaries present; in fact, one can attempt to study the structure of these defects by volume sensitive methods, which is impossible in coarse grained materials due to their low volume fraction. Fundamental to this is the question of what is the local structure of a grain boundary and whether it is different in a coarse grained or nanophase material. An x-ray diffraction study[3] reported a large diffuse scattering background which was interpreted as arising from a very disordered (in the limit "gas like") radial distribution function (RDF) in the grain boundaries. This result has been controversial with results from different techniques either supporting[4-6] or rejecting[7-10] this model.

X-ray absorption spectroscopy (XAFS)[11] has played an important role in this discussion[12]. In fact, all measurements reported to date[13-18] have shown a reduction of average coordination numbers in the first few coordination shells. A reduction in the average coordination number is expected on general grounds for an aggregate of nanometer sized clusters due to the reduced coordination of surface atoms and the high average surface-to-bulk ratio (SBR). A similar reduction is expected also for compacted nanophase materials in which free surfaces are replaced by incoherent grain boundaries. However, the reduction in coordination numbers reported by XAFS were greater than that compatible with a simple estimate of the SBR and most measurements were interpreted as compatible with the disordered grain boundary model[13-16] or with the presence of a high defect density [18].

A recent x-ray diffraction (XRD) study[19] of nanophase Pd (n-Pd) has reported reduced average coordination numbers in the first few shells for all samples studied. This reduction was satisfactorily explained by the SBR for all samples except for those investigated within 10 days of preparation; for the latter samples it was concluded that 10% of atoms are off lattice sites.

In the present paper we report XAFS measurements of n-Pd samples which have been carefully characterized as those studied by XRD by Löffler and Weissmüller[19]; of particular importance is the determination of the particle size distribution[20]. We note that XRD and XAFS are complementary in that they probe different parts of wave-vector space: 0 to 17 $Å^{-1}$ for XRD and 6

Present address: NIST, 223/B 152, Gaithersburg, MD 20899, USA.
* Corresponding author; e-mail: boscherini@lnf.infn.it

to 40 $Å^{-1}$ for XAFS. Results in qualitative agreement with those here reported have been already published[17], but no quantitative analysis or interpretation has been provided, to our knowledge.

Sample preparation and Experimental

Six samples of n-Pd were prepared by the inert gas condensation technique with different treatments following the deposition. One sample (n.1) was not consolidated, in order to serve as a reference for an ensemble of clusters with a negligible grain boundary density. All other samples were consolidated *in situ* under high vacuum for 5 to 10 minutes at 1 - 2 GPa; at this stage samples have a density from 84 to 87% of coarse grained Pd. Following room temperature consolidation one sample was kept at liquid nitrogen temperature (LNT) up to 1 week before measurement in order to reduce room temperature (RT) grain growth. Two samples were stored at RT for different periods of time and the remaining two samples were subjected to annealing and compaction procedures.

It is well known that nanophase materials are composed of grains whose size have a log-normal distribution function (LNDF)[21]. Volume and area weighted average grain size ($\langle D\rangle_V$ and $\langle D\rangle_A$ respectively) were determined by XRD using Kochendörfer's method and by a deconvolution of the experimental Bragg peaks[20], together with the rms strain; $\langle D\rangle_V$ and $\langle D\rangle_A$ completely describe the LNDF. The rms strain values lie between 0.03 and 0.3%; other sample characteristics are summarized in table 1.

Sample	Preparation /Treatment	$\langle D\rangle_V$ (nm)	$\langle D\rangle_A$ (nm)
1	Not compacted	12.1	8.4
2	Stored LNT	16.7	11.5
3	Stored RT for 1 month	18.6	16.7
4	Stored RT for 2 weeks	18.8	12.2
5	2 days 3.2 GPa, 100 °C	35.5	11.5
6	As n.5, + 1 h at 160 °C	>29.1	13.0

Table 1. Sample characteristics.

In order to perform XAFS experiments a thin (≈ 10 μm) homogeneous sample is required; inhomogeneity in the sample thickness distribution is a major source of experimental artifact[22, 23], which may reduce the measured coordination numbers; consequently much care was devoted to this aspect. Notice that all known experimental artifacts tend to reduce the measured coordination numbers, not to enhance them. The compacted samples were lightly ball milled at LNT in order to produce a powder; all samples were then dispersed in toluene using an ultra-sound bath. Only the finest particles (those forming a fine suspension at the top of the test tube) were then deposited on a poly-carbonate membrane forming a homogeneous sample. Homogeneity at the sub-micron level was checked optically and by measuring spectra on samples of different thicknesses, from 1 to 15 μm, which showed identical values of the coordination numbers. We found that samples prepared by using all particles produced by ball milling (without dispersion in toluene and subsequent selection of finest particles) exhibited artificially reduced coordination numbers, a fact which might explain the previously reported low coordination numbers, at least on materials prepared by inert gas evaporation.

XAFS measurements were performed on the GILDA CRG (D8) beamline of the European Synchrotron Radiation Facility, Grenoble, France. A Si(311) independent crystal monochromator with sagittal focussing was used; harmonics rejection was obtained by detuning the crystal to 1/3 of the rocking curve maximum using a closed loop feedback controller. Spectra were measured in the transmission mode with Ar filled ionization chambers in the energy range 24 - 26 keV; all spectra were measured at 77 K in order to increase the available k - space. For a few samples temperature dependent measurements were performed and no variation of the determined average coordination numbers was found, but only an increase of the thermal disorder factor in accordance to known vibrational behaviour[24]. A coarse grained Pd foil was measured as a reference sample.

Results and Discussion

In the present report we concentrate on the first shell environment, while a more complete determination will be published elsewhere. Spectra were analyzed according to standard procedures[11]. The raw XAFS functions, reaching 20 $Å^{-1}$, are reported in Fig. 1 and the good signal-to-noise is apparent. These functions were Fourier transformed in the range 3 to 20 $Å^{-1}$ and the result is displayed in Fig. 2. From the figures it is apparent that the spectra show different amplitudes and can be roughly grouped in three sets in order of decreasing amplitude: the coarse grained foil, samples 2 to 6 (i.e. all the compacted samples) and sample 1 (the un-compacted sample).

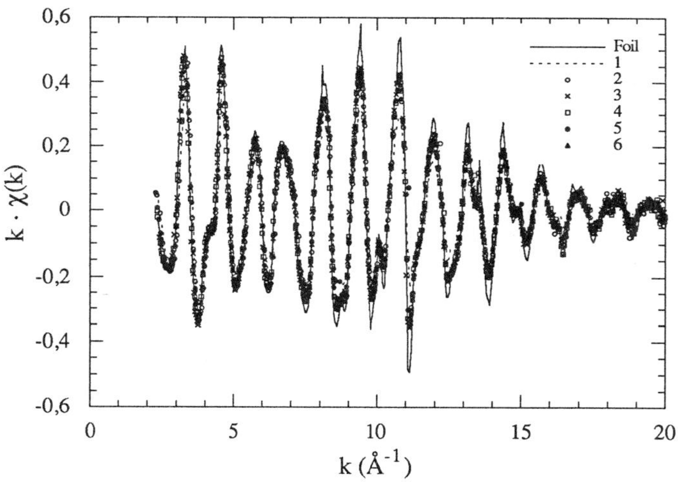

Fig. 1. Raw XAFS oscillations.

In order to quantitatively analyze the spectra an inverse Fourier transform was performed in the range 1.8 - 3.2 Å; the result was fitted to a model function using a non-linear least squares routine and experimental amplitudes and phases from the coarse grained Pd foil. Average coordination numbers, $\langle N \rangle_{EXP}$, and variations of the mean square relative atomic displacement (MSRD), $\Delta\sigma^2$, determined in this way are reported in Table 2; for both parameters the model is the coarse grained foil. The estimated errors are one standard deviations and include the effect of correlation between $\langle N \rangle$ and $\Delta\sigma^2$ which is, in fact, the major contribution to error bars; the errors on these two parameters should be interpreted in the following way: if $\langle N \rangle$ is increased (decreased) to its maximum (minimum) value then $\Delta\sigma^2$ must also be increased (decreased) in the same manner. The interatomic distances in all nanophase samples were found identical with those of the coarse grained reference; this is reasonable as size-dependent variations in interatomic distances set in at much smaller dimensions[25]. Finally, we found no evidence for anharmonic effects, justifying the use of Gaussian damping of the signal[26].

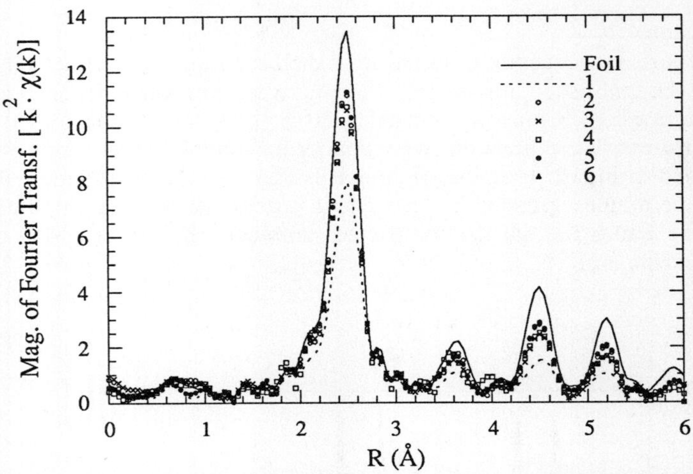

Fig. 2. Magnitude of Fourier Transform of the spectra reported in fig. 1

In order to interpret the data we have calculated the average coordination number expected for the samples considering only the SBR effect, in two independent ways. The average coordination of a single particle of given size D, N(D), was calculated for a cuboctahedron[27] and $\langle N\rangle_{LN}$ for the ensemble was then calculated as the volume weighted average of N(D):

$$\langle N\rangle_{LN} = \int_0^\infty N(D)\times LN(D)\times D^3 dD \Big/ \int_0^\infty LN(D)\times D^3 dD$$

where LN(D) is the LNDF of particle sizes. The parameters of LN(D) were taken from the XRD determination. As an alternative estimate we have used the relation[19]:

$$\langle N\rangle_\alpha = 12(1-\tfrac{1}{2}\alpha R_1)$$

where R_1 is the nearest neighbour distance and $\alpha = 3/\langle D\rangle_A$ (the SBR). Both estimates of $\langle N\rangle$, which are quite similar, are reported in table 2.

The experimental values reported in Table 2 are compatible with the rough division of spectra based on visual inspection referred to earlier: sample 1 shows a greatly reduced $\langle N\rangle_{EXP}$ while samples 2 to 6 have similar values for this parameter. We will discuss first the results on the compacted material and then sample 1.

Sample	$\Delta\sigma^2$	$\langle N\rangle_{EXP}$	$\langle N\rangle_{LN}$	$\langle N\rangle_{\alpha}$
	10^{-3} Å^2	atoms	atoms	atoms
	± 0.5	± 1.0		
1	1.2	9.7	11.33	11.41
2	0.8	11.7	11.50	11.57
3	0.9	11.5	11.65	11.70
4	1.0	12.0	11.53	11.59
5	0.7	11.8	11.52	11.57
6	0.8	12.0	11.56	11.62

Table 2: Experimental variation of MSRD with respect to Pd foil, average coordination number of the first coordination shell and two estimates of $\langle N\rangle$, as described in the text.

$\langle N\rangle_{EXP}$ values for samples 2 to 6 are all quite close to the estimates of $\langle N\rangle$ based on the SBR. The present measurements indicate, therefore, that the reduction of $\langle N\rangle$ for compacted materials can be simply explained by a size effect with no evidence of further reduction. This is the main conclusion of this paper, which contradicts nearly all previous reports. The slight enhancement of σ^2 with respect to the coarse grained foil indicates that the process of evaporation and compaction produces a slight distortion of the first shell environment which is insensitive to further hot pressing or annealing. It is interesting to compare the present result with the determination of disorder by XRD. In a real space analysis of XRD data[19] the slight increase of σ^2 was not found; however, we believe that the present results are more reliable as a greater portion of reciprocal space is probed by XAFS than by XRD. In a k-space analysis of XRD[9] an increase of 3×10^{-3} Å^2 in the Mean Square Displacement (MSD) was found, to be compared with the $\sim 0.8 \times 10^{-3}$ Å^2 we find for the increase in MSRD; in this case the two techniques measure different quantities, XAFS being sensitive to the relative atomic displacements while XRD measures the atomic displacements from the lattice positions. By comparing the numerical results of the two techniques we conclude that the increase in MSD is due in the most part to atomic displacements which are correlated in the first coordination shell.

Sample 1 (with a negligible grain boundary density) shows a greatly reduced $\langle N\rangle_{EXP}$, which is not reproduced by our estimate. There may be a number of reasons for this, including that our model of size distribution is not correct; it may well be that there are many more small diameter grains than presently estimated and that they coalesce during compaction. The slight increase in σ^2 may have both a structural and a dynamic[25] component, which cannot be decoupled from the present measurements alone. The small reduction of σ^2 upon compaction indicates that surface atoms have greater constraints of motion in grain boundaries than on a free cluster surfaces and/or reduced static disorder.

In conclusion we have presented an XAFS study of local order in n-Pd samples subjected to various treatments. By considering the distribution of particle sizes we have shown that the observed reduction in average coordination number can be explained as a size effect originating from the reduced coordination of the atoms present of the surface of the particles, with no evidence of further reduction. Additional information on the disorder was obtained and the results were compared to the local structure in an un-compacted sample.

Acknowledgments

We are very grateful to all the staff of the GILDA beamline (especially to S. Pascarelli and Prof. S. Mobilio) for making these measurements possible. We are also grateful to R. Celino, V. Rosato and G. D'Agostino for stimulating discussions on MD simulations of nanophase materials. XAFS measurements were performed under the public user program of the ESRF, BP 220, F-38043 Grenoble Cedex, France. Partial financial support was provided by the Deutsche Forschungsgemeinschaft (SFB No. 277).

References

1. H. Gleiter, in *Proc. Second Risø int. symp. on metallurgy and materials science*, eds. N. Hansen, T. Leffers and H. Lilholt (Røskilde, Denmark, 1981).
2. H. Gleiter, Progress Mat. Sci. **33**, 223 (1989).
3. X. Zhu, R. Birringer, U. Herr, and H. Gleiter, Phys. Rev. B **35**, 9085 (1987).
4. E. Jorra, H. Franz, J. Peisl, G. Wallner, W. Petry, R. Birringer, H. Gleiter, and T. Haubold, Phil. Mag. B **60**, 159 (1989).
5. U. Herr, J. Jing, U. Gonser, and H. Gleiter, Solid State Commun. **76**, 197 (1990).
6. H.E. Schaefer, in *Mechanical properties and deformation behavior of materials having ultra-fine microstructure* , eds. M. Nastasi, D.M. Parkin & H. Gleiter (Kluwer, Amsterdam, 1993).
7. C.A. Melendres, J. Mater. Res. **4**, 1246 (1989).
8. M.R. Fitzsimmons, J.A. Eastman, M. Müller-Stach, and G. Wallner, Phys. Rev. B **44**, 2452 (1991).
9. J.A. Eastman, M.R. Fitzsimmons, and L.J. Thompson, Phil. Mag. B **66**, 667 (1992).
10. P.G. Sanders, J.R. weertman, J.G. Barker, and R.W. Siegel, Scripta Metall. et Mater. **29**, 91 (1993).
11. T.M. Hayes and J.B. Boyce, Solid State Phys. **37**, 173 (1981).
12. F. Boscherini, in *Fundamental properties of nanostructured materials, eds. D. Fiorani and G. Sberveglieri* (World Scientific, Singapore 1994).
13. T. Haubold, R. Birringer, B. Lengeler, and H. Gleiter, Phys. Lett. A **135**, 461 (1989).
14. T. Haubold, W. Krauss, and H. Gleiter, Phil. Mag. Lett. **63**, 245 (1991).
15. T. Haubold, F. Boscherini, S. Pascarelli, S. Mobilio, and H. Gleiter, Phil. Mag. A **66**, 591 (1992).
16. T. Haubold, Acta Metall. Mater. **41**, 1769 (1993).
17. J.A. Eastman, M.R. Fitzsimmons, M. Müller-Stach, G. Wallner, and W.T. Elam, Nanostruct. Mater. **1**, 47 (1992).
18. A. Di Cicco, M. Berrettoni, S. Stizza, E. Bonetti, and G. Cocco, Phys. Rev. B **50**, (1994).
19. J. Löffler and J. Weissmüller, Phys. Rev. in press, (1995).
20. J. Weissmüller, J. Löffler, C.E. Krill, and R. Birringer, J. Mater Res. submitted, (1995).
21. C.G. Granqvist and R.A. Buhrman, J. Appl. Phys. **47**, 2200 (1976).
22. E. Stern and K. Kim, Phys. Rev. **B 23**, 1981 (1981).
23. J. Goulon, C. Goulon Ginet, R. Cortes, and M. Dubois, J. Phys. (Paris) **43**, 539 (1982).
24. G. Beni and P.M. Platzman, Phys. Rev. B **14**, 1514 (1976).
25. A. Balerna and S. Mobilio, Phys. Rev. B **34**, 2293 (1986).
26. P. Eisenberger and B. Lengeler, Phys. Rev. B **22**, 3551 (1980).
27. A. Khoutami, B. Legrand, C. Mottet, and G. Tréglia, Surf. Sci. **307-309**, 735 (1994).

Materials Science Forum Vol. 195 (1995) pp. 73-78

Formation of Nanophases by Crystallization of Amorphous Alloys

M. Baricco[1], P. Tiberto[2] and L. Battezzati[1]

[1] Dipartimento di Chimica Inorganica, Chimica Fisica e Chimica dei Materiali, Università di Torino, Via P. Giuria 9, I-10125 Torino, Italy

[2] Istituto Elettrotecnico Nazionale "Galileo Ferraris" C.so M. D'Azeglio 42, I-10125 Torino, Italy

Keywords: Nanophases, Crystallization, Nucleation Frequency, Growth Rate, Amorphous Alloys, Magnetic Materials

ABSTRACT

A comparison is made of crystallization mechanisms in Fe-based amorphous alloys ($Fe_{73.5}Cu_1Nb_3Si_{13.5}B_9$, $Fe_{90}Zr_7B_3$), which lead to a nanocrystalline bcc phase with peculiar soft magnetic properties through primary crystallization. In all samples, the nucleation takes place uniformely throughout the material. The presence of atoms with low diffusivity (Zr, Nb) reduces the grain growth rate so that a small size of crystals is maintained. On the contrary, even extremely fast heating and short heat treatment do not produce nanocrystals in $Fe_{40}Ni_{40}B_{20}$ although the crystallization mechanism is modified. However milling the ribbon leads to nanostructured crystalline phases.

The formation of nanophases by crystallization is discussed considering the temperature dependence of nucleation frequency and growth rate. The effects of various thermodynamic and kinetic parameters are analyzed and the most favourable conditions for nanocrystallization are outlined.

INTRODUCTION

The knowledge of crystallization mechanisms and kinetics has been recognized of primary importance in understanding the formation and stability of amorphous phases since the first appareance of metallic glasses [1]. More recently, the crystallization of amorphous alloys has been suggested as a means to produce materials with ultrafine microstructures. These materials could find interesting applications for their excellent mechanical [2] and soft magnetic [3] properties.

The formation of a crystalline phase with nanosized grains can be obtained, during the crystallization of an amorphous alloy, if the conditions of an high nucleation frequency (I_v) and a low growth rate (u_c) are simultaneously fullfilled. Considering the temperature dependence of I_v and u_c, these conditions are generally verified in small temperature ranges and particular annealing techniques (i.e. pulse and Joule heating) are necessary to access them [4]. In some amorphous alloys nanophases are obtained by crystallization in a wide temperature range by means of conventional annealing treatments [5], and by mechanical treatments at room temperature [6] (mechanical crystallization).

Aim of this paper is to report some results of nanophase fomation by crystallization in Fe-based metallic glasses with interesting soft magnetic properties. The primary crystallization mechanism will be analyzed in $Fe_{73.5}Cu_1Nb_3Si_{13.5}B_9$ and $Fe_{90}Zr_7B_3$. Different crystallization procedures, applied to amorphous $Fe_{40}Ni_{40}B_{20}$, will be compared. The conditions for nanophase formation by crystallization of metallic glasses will be discussed in terms of the temperature dependence of nucleation frequency and growth rate.

EXPERIMENTAL

Amorphous $Fe_{73.5}Cu_1Nb_3Si_{13.5}B_9$ and $Fe_{40}Ni_{40}B_{20}$ alloys have been prepared in ribbon form by planar flow casting in air. Amorphous $Fe_{90}Zr_7B_3$ ribbons have been kindly provided by ALPS electric Co. Ltd. (Japan). The amorphous-to-crystal transformation has been followed by means of thermal (DSC) and structural (XRD, TEM) analyses. DSC traces have been carried out in isothermal and tempering conditions and the calorimetric data have been used to obtain kinetic parameters. An high temperature DSC (Setaram HFDSC) has been used for thermal analysis at temperatures higher than 700°C. The lattice constant, a_0, of nanocrystalline phases during phase transformation has been determined by a careful analysis of XRD patterns. In order to minimize the experimental uncertainty, values from all the available diffraction peaks of the bcc phase were plotted vs $\cos\theta \cot g\theta$ and a_0 was obtained from the intercept of a linear best fit [7]. The effect of transformation mechanism on grain size distributions has been investigated on the basis of experimental XRD and TEM data.

RESULTS

The DSC trace of amorphous $Fe_{90}Zr_7B_3$ heated at 5 K/min is shown in fig. 1. In the inserts, XRD patterns obtained at various stages of annealing are shown. The amorphous structure transforms to a nanocrystalline bcc phase between 500°C and 580°C. The corresponding heat of crystallization is 3.4±0.4 kJ/mol. A second exothermal peak can be observed at 730°C, which corresponds to the formation of Fe_3Zr and Fe_2Zr intermetallic compounds. In this case, the heat of transformation is 0.8±0.1 kJ/mol. A kinetic analysis, performed by means of the Kissinger method, gave an activation energy for the first crystallization step of 360 kJ/mol. The crystallization sequence is in agreement with that previously reported for the same alloy [8]. Similarly, during the annealing of $Fe_{73.5}Cu_1Nb_3Si_{13.5}B_9$, the first crystallization step leads to a nanocrystalline bcc phase, and at higher temperatures a mixture of borides is produced [9]. A comparison of the DSC signal corresponding to the amorphous to nanocrystal transformation for the two alloys [5] showed a long tail on the high temperature side of the peak for $Fe_{73.5}Cu_1Nb_3Si_{13.5}B_9$, whereas a more symmetric trace was obtained for $Fe_{90}Zr_7B_3$. This was attributed to different solute partitioning during the transformation. Here the effect is confirmed by the analysis of the

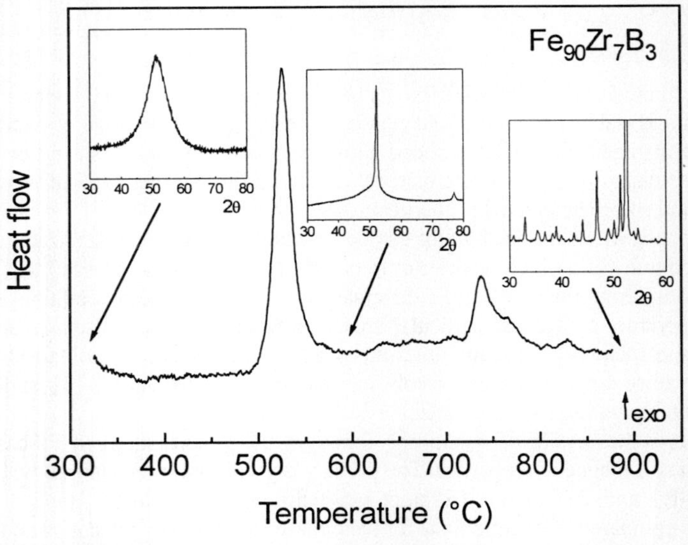

Figure 1 - DSC trace of $Fe_{90}Zr_7B_3$ heated at 5 K/min. The inserts show the XRD patterns (Co Kα) obtained after annealing up to the temperatures indicated by the arrows.

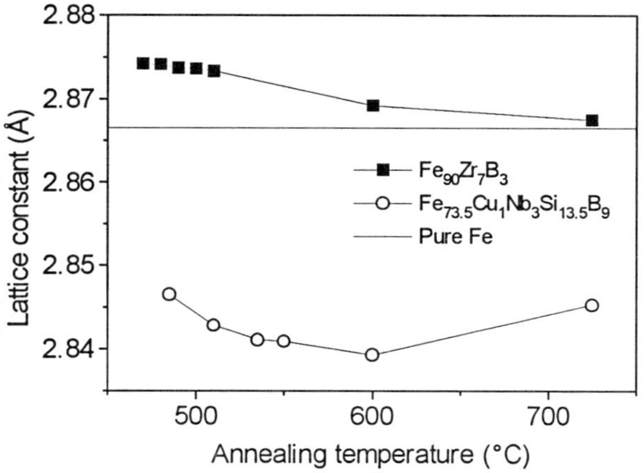

Figure 2 - Lattice constant of the nanocrystalline bcc phase as a function of annealing temperature. The annealing time was 20 minutes.

temperature dependence of the lattice constants of the nanocrystalline phases. The results are shown in fig. 2, evidencing significantly different values of a_0 for the two alloys. In $Fe_{90}Zr_7B_3$ the nanocrystalline phase has a lattice constant higher than that of pure iron, owing to the solution of Zr in bcc-Fe, whereas the solution of Si in bcc-Fe reduces the lattice parameter in $Fe_{73.5}Cu_1Nb_3Si_{13.5}B_9$. In addition, a_0 remains nearly constant in $Fe_{90}Zr_7B_3$ during the formation of nanocrystals, indicating little composition changes. On the contrary, a significant decrease of a_0 is observed in $Fe_{73.5}Cu_1Nb_3Si_{13.5}B_9$, indicating a significant rearrangement of the composition during the phase transformation [10].In both cases nanocrystals are in metastable local equilibrium with the matrix since major changes in a_0 occur when intermetallic compounds crystallize above 600°C.

The microstructure of the nanocrystalline phase has been analyzed by TEM. The crystals appear homogenously distributed in the volume with a spherical shape [9]. The particle size distributions are shown in fig. 3 for samples partially crystallized. The average size is slightly higher for $Fe_{90}Zr_7B_3$ with respect to $Fe_{73.5}Cu_1Nb_3Si_{13.5}B_9$ and agrees with those obtained from the Scherrer analysis of XRD patterns. In both cases, the distribution is wide suggesting the occurrence of nucleation and growth processes. Comparing the present results with an analysis of particle size distribution for various crystallization mechanisms [11], a nucleation process involving a transient time can be inferred in both cases. Analogous results are found for samples annealed in a furnace up to 550°C and also when the crystallization process is shifted up to an estimated temperature of 650°C by means of fast Joule heating, obtained applying to the sample an electrical current of 2.6 A for 60 s [10]. Details on the experimental apparatus can be found in ref. [12]. Crystal densities of about 10^{23} m^{-3} were found after all

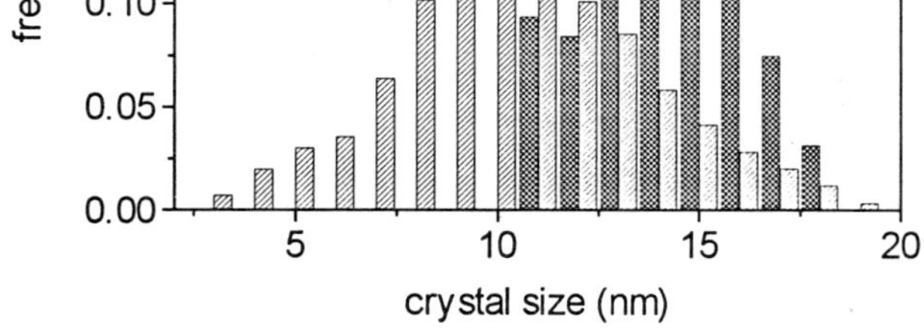

Figure 3 - Crystal size distribution of partially crystallized samples. Annealing conditions: 543°C, 1 hour for $Fe_{73.5}Cu_1Nb_3Si_{13.5}B_9$ and 480°C, 30 minutes for $Fe_{90}Zr_7B_3$.

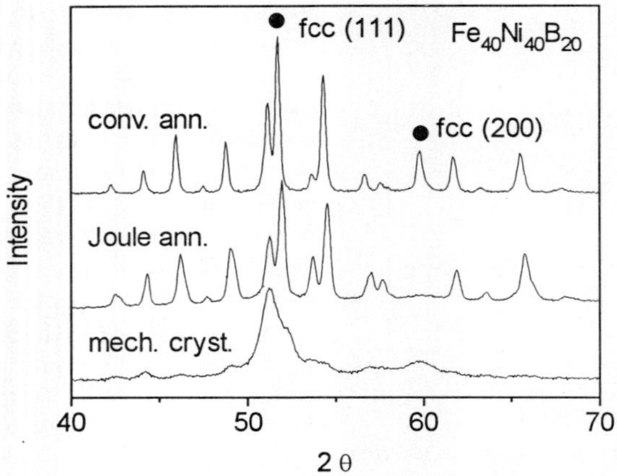

Figure 4 - X-ray diffraction pattern (Co Kα) of amorphous $Fe_{40}Ni_{40}B_{20}$ crystallized by different techniques.

annealings between 485°C and 650°C [10]. These evidences show that the nucleation frequency must be very high in a wide temperature range.

The conditions for nanocrystallization are not always met. No fine grained microstructure is obtained in $Fe_{40}Ni_{40}B_{20}$ either by conventional annealing or Joule heating. Conventional annealing has been performed by heating the sample in the DSC up to 600°C at a constant heating rate of 30 K/min. Joule heating has been obtained with an electrical current of 1.88 A for 15 s [13]. The XRD patterns of crystallized samples are shown in fig. 4. The diffraction peaks are sharp for both patterns. At low heating rates the crystallization of amorphous $Fe_{40}Ni_{40}B_{20}$ produces a eutectic mixture of fcc-(Fe,Ni) solid solution and orthorombic $(Fe,Ni)_3B$ [13]. The main effect of the strong increase of the heating rate by Joule heating, is to change the crystallization mechanism from eutectic to polymorphic, as evidenced in fig. 4 by the absence of the diffraction peaks of the fcc-solid solution [13]. Therefore either thermal nuclei are formed in limited amount or their growth during the transient time is not restrained by slow diffusion. Milling of the same alloy in a stainless-steel vial for 5 hours with a high-energy ball mill kept at room temperature causes crystallization. The corresponding XRD pattern of fig. 4 displays broad overlapped diffraction peaks, which can be still assigned to the same phase mixture as before. The reason for mechanically induced crystallization is yet unclear. Local thermal spikes during milling should form large crystals as detailed above. Therefore it is likely that crystallization is promoted by the effect of contaminants and of the huge increase in surface area. In fact, the presence of defects and impurities introduced during the milling process can play an important role in increasing the heterogenous nucleation rate [6], therefore promoting the formation of nanophases at room temperature.

DISCUSSION

The crystallization of amorphous alloys is generally described by nucleation and growth processes. These can be shown by a model calculation using classical expression for the kinetic quantities [11,15]. The homogeneous nucleation frequency is expressed by:

$$I_v = N_v \frac{D_n}{a^2} \exp(\frac{-\Delta G^*}{kT})$$

where ΔG^* is the activation barrier for the formation of a nucleus of crytical size, D_n a diffusivity for nucleation, a the size of the moving particle and N_v the number of atom per unit volume. ΔG^* is usually given by:

$$\Delta G^* = \frac{16\pi}{3}\frac{\sigma^3}{\Delta G_v^2}$$

where σ is the liquid(amorphous)-crystal interface energy and ΔG_v is the free energy difference per unit volume between liquid(amorphous) and crystal phases. For duplex or polymorphic growth, the rate is:

$$u_c = \alpha \frac{D_g}{\lambda}[1 - \exp(-\frac{\Delta G_v}{kT})]$$

where D_g is a diffusivity for growth, α a function of the elemental concentration in the matrix and the growing phases and λ a diffusion distance. The various thermodynamic and kinetic parameters can affect significantly the temperature dependence of both processes and different microstructure can be found after treatments at different temperatures.

The temperature dependence of the free energy difference between liquid and solid phases can be reasonably described considering the entropy of fusion, ΔS_m, and the heat capacity difference between liquid and solid phases, ΔC_p, at the melting point, T_m [16]. ΔG_v increases with increasing undercooling and is generally lower the higher the glass forming tendency of the alloy. For temperatures below the glass transition temperature, T_g, ΔG_v remains nearly constants. For good glass formers, a low value of ΔG_v reduces the nucleation frequency and the formation of nanophases should not occur.

The interfacial energy is usually assumed as proportional to the heat of fusion of the alloy [17]. The heat of fusion of a metastable phase is lower than that of stable ones, so that the corresponding value of σ should lower ΔG^*. On the other hand, ΔG_v of a metastable phase is lower with respect to that of stable ones and ΔG^* is increased. The final result on I_v is a balance between the two effects which act in opposite directions. Larger effects on the formation of nanophased microstructures have been evidenced via a reduction in interfacial energy by oxygen contaminations [18].

Most important in producing nanocrystalline phases is the behaviour of diffusivity, because different diffusivities can be involved in nucleation and growth processes. Bulk diffusivity is often related to viscosity through the Stokes-Einstein equation [15]. The temperature dependence of viscosity depends on the temperature range [11]. In particular, for temperatures lower than T_g, an Arrhenius behaviour is generally considered. Above T_g, the viscosity follows the Vogel-Fulcher-Tamman equation [11,15]. Although the maximum crystal density after crystallization is often obtained for temperatures above T_g [11], nanocrystallization can occur in both regimes. Bulk diffusivities are usually too low to explain the mobility needed for the formation of such high crystal densities, whereas are compatible with most crystallization processes, e.g. that of $Fe_{40}Ni_{40}B_{20}$. A discussion on diffusivity during growth has been given recently [19,20]. In order to produce nanophases, a possibility is that there is fast diffusion of a species at short range (nucleation) but motion through the bulk is difficult (growth). For high I_v it is necessary that diffusivity of at least an atomic species is high. This happens in the case of the $Fe_{73.5}Cu_1Nb_3Si_{13.5}B_9$ alloy, for which Cu acts as nucleating element and compositional adjustments of both precipitates and matrix, involving a slow species, Nb, are needed for coarsening of the precipitates. Wide compositional re-adjustments may not be necessary for the occurence of nanocrystals since it is possible in simpler alloys such as $Fe_{90}Zr_7B_3$. Here Fe motion promotes nucleation and the diffusion becomes slow in the amorphous matrix enriched in Zr, as evidenced in ref. [21] for similar alloys. It has recently been shown that interdiffusivity in glasses made of early and late transition metals depends on the lenght scale of the diffusion being fast for short distances and dropping markely for long range motion. This effect arises through the development of stress gradients which have been evidenced in multilayered films. Since the interlayer distance in films is of the same magnitude as the interparticle distance in devitrifying alloys, it has been suggested that these stress

effects may be relevant in the formation of fine microstructures by crystallization, at least at low temperatures [22].
The crystallization by fast annealing techniques is aimed at the increase of I_v at high temperatures. On the other hand, also the growth rate increases as a function of temperature and the possibility to produce nanophases depends on the balance between the two kinetic functions. In any case, the existence of an high I_v is a necessary conditions, as evidenced in the results obtained for $Fe_{40}Ni_{40}B_{20}$. and $Fe_{73.5}Cu_1Nb_3Si_{13.5}B_9$.

ACKOWLEDGMENTS
Dr. Makino (ALPS electric Co. Ltd., Japan) is kindly acknowledged for providing $Fe_{90}Zr_7B_3$ samples. Work performed for Progetto Strategico CNR " Materiali Innovativi" .

REFERENCES

[1] M.G.Scott, in "Amorphous Metallic Alloys", F.E.Luborsky ed., Butterworth, London (1983) p. 144
[2] A.Inoue, Mat.Sci.Eng. **A179/180**, 57 (1994)
[3] Y.Yoshizawa, S.Oguma, K.Yamamuchi, J.Appl.Phys. **64**, 6044 (1988)
[4] T.Kulik, D.Bucka, H.Matyja, J.Mat.Sci.Lett. **12**, 76 (1993)
[5] M.Baricco, L.Battezzati, Mat. Sci. Forum **179-181**, 597 (1995)
[6] M.L.Trudeau, L.Dignard-Bailey, R.Schulz, D.Dussault, A.Van Neste, Nanostr.Mater. **2**, 361 (1993)
[7] B.D.Cullity, in "Elements of x-ray diffraction", Addison-Wesley Pup., Reading, MA (1978)
[8] K.Suzuki, A.Makino, A.Inoue,T.Masumoto, J.Appl.Phys. **74**, 3316 (1993)
[9] M.Baricco, L.Battezzati, G.Riontino, Key Eng.Mat. **81-83**, 177 (1993)
[10] M.Baricco, C.Antonione, P.Allia, P.Tiberto, F.Vinai, Mat.Sci.Eng. **A179/180**, 572 (1994)
[11] U.Köster, U.Schünemann, in "Rapidly Solidified Alloys", H.H.Liebermann ed., M.Dekker, New York (1993) p. 303
[12] P.Allia, M.Baricco, P.Tiberto, F.Vinai, Rev.Sci.Instr. **64**, 1053 (1993)
[13] P.Allia, M.Baricco, P.Tiberto, F.Vinai, Phys.Rev. **B47**, 3118 (1993)
[14] E.Angelini, C.Antonione, M.Baricco, Phil.Mag. **B61**., 567 (1990)
[15] L.Battezzati, C.Antonione, G.Riontino, J.Non-Cryst.Sol. **89**, 114 (1987)
[16] K.S.Dubey, P.Ramachandrarao, Acta Met. **32**, 91 (1984)
[17] F.Spaepen, Acta Met. **23**, 729 (1975)
[18] U.Köster, J.Meinhardt, H.Alves, Mat. Sci. Forum **179-181**, 533 (1995)
[19] L.Battezzati, M.Baricco, Phil. Mag. B **68**, 813 (1993)
[20] L.Battezzati, in "Fundamental Properties of Nanostructured Materials", D.Fiorani and G.Sberveglieri eds., World Scientific Pub., Singapore (1994) p. 172
[21] A.Makino, Y.Yamamoto, Y.Hirotsu, A.Inoue, T.Masumoto, Mat.Sci.Eng. **A179/180**, 495 (1994)
[22] A.L.Greer, to appear in "Proc. of IV Int. Workshop on Non-Crystalline Solids", M.Vazquez and A.Hernando eds., World Scientific Pub., Singapore

Materials Science Forum Vol. 195 (1995) pp. 79-86

Si_3N_4/SiC Ceramic Nanocomposites

A. Bellosi

CNR-IRTEC, Via Granarolo 64, I-48018 Faenza, Italy

Keywords: Ceramic Nanocomposites, Si_3N_4/SiC, Nanocomposites Processing, Nanocomposites Properties

ABSTRACT

This study focuses on several aspects of Si_3N_4/SiC ceramic nanocomposites: powder treatments, processing and properties. The use of Si-C-N ultrafine amorphous powders was studied relatively to: - the effects of addition of 20 vol% to commercial Si_3N_4 powders, - the production of dense nanocomposites Si_3N_4/SiC. In both cases densification behaviour, microstructure and some mechanical properties were evalutated and compared.

INTRODUCTION

Silicon nitride - based ceramics have been well known as excellent candidate for high structural applications because of their physical and thermo-mechanical properties. However, because of their low toughness and susceptibility to flaws, introduced either during processing or in service, industrial applications have been limited. Research is currently being carried out into nanostructured Si_3N_4 - based composites because of the potential gains they offer campared to existing materials(1). Improvements in the following properties have been reported: strength and toughness, machining and polishing behaviour, wear resistance, high temperature properties (creep and fatigue resistance, strength).

As a second reinforcing phase, SiC is one of the most promising for the improvement of properties of silicon nitride: it is well known that the incorporation of SiC into Si_3N_4 matrices influences the microstructure and increases hardness, creep, toughness, oxidation resistance properties compared to those in pure Si_3N_4 materials (2-9).

The production of composite materials, where the different phases are homogeneously distributed, requieres relatively complex milling procedures applied to the starting powder, which however do not guarantee the homogeneous distribution of the phases and then the reproducibility of the porperties(2). The densification process of composite systems is difficult due to the different sintering behaviour of matrix and second phase. These problems may be overcome by the laser synthesis of composite powders from Si-C-N containing gaseous precursors, which appears the ideal method for growing homogeneous mixtures of pure and ultrafine silicon nitride-silicon-carbide particles(10, 12-16). Furthemore the laser synthesis process offers the possibility of controlling the composition of the material. This is especially interesting in view of the recently reported production of a superplastic composite after sintering of Si-C-N amorphous powder with a C-N atomic ratio near to 0.2 (13).

The purpose of this study is: i) to investigate the effects of an amount of ultrafine amorphous Si-C-N powder introduced into commercial submicron Si_3N_4 in order to modify the rate of α-Si_3N_4 (equiaxed) $\rightarrow$ β-Si_3N_4 (elongated) grains and to enhance densification; ii) to set up experimental methods suitable to treat ultrafine Si-C-N powders and to evaluate their densification behaviour. In both the cases, microstructural analyses and results of mechanical tests are compared with the results on materials produced by commercial submicron powders.

EXPERIMENTAL

a) Raw materials.
Nanosized Si-C-N powder was synthesized by CISE Spa (Milano, Italy), through CO_2 laser irradiation of SiH_4, NH_3 and C_2H_4 mixture. The main characteristics of this powder are: specific surface area 130 m^2/g, mean grain size 12 nm, mass chemical composition 62.4 wt% Si, 17.6 wt% N, 16.8 wt% C, and 3.2 wt% O.
Si_3N_4 powder (SN-E10) produced by UBE (Japan), was selected as submicron commercial powder: specific surface area: 10.6 m^2/g, particle size 0.1 - 0.3 μm, crystalline composition: 95 % α-Si_3N_4 and 5 % β-Si_3N_4, oxygen amount 1.4 %, overall metal impurities: 0.03 %. As sintering aids, La_20_3 (Merck, Germany), Y_20_3 (fine grade, H.C. Starck, Germany), Al_2O_3 (Alcoa A.16) have been used.
b) Materials fabrication.
i) Compositions Si_3N_4/ Si-C-N. The following batches were prepared:
A: Si_3N_4 + 3 wt% Y_2O_3, + 3 wt% La_2O_3.;
B: (Si_3N_4 + 20 vol% Si-N-C) + 3 wt% Y_2O_3 + 3wt% Y_2O_3.
The powders were mixed in hexane for 15 hours in plastic jair and silicon nitride balls, then were dryied, crushed and sieved. Densification was carried out by hot pressing at 1850°C for 40 min (mix A) and 20 min (mix B) using a pressure of 30 MPa, under vacuum (Table I);
ii) Compositions based on Si-C-N. Due to the high affinity to oxygen, special procedures had to be set up to treat these powders. Details on the experimental procedures are described elsewhere[17-18].
Several compositions were prepared following different procedures for the addition of the sintering aids:
- composition I: Si-C-N + 5 wt% Y_2O_3 + 2 wt% Al_2O_3 (chemical coprecipitation method starting from Y- and Al alkoxides);
- composition II: Si-C-N + AlN + YN (amount of metal cation like in composition I; mixing in hexane);
- composition III: Si-C-N + 6 wt% Y_2O_3 + 3 wt% Al_2O_3 + AlN* + YN* (* in an amount to incorporate the oxygen from Si-C-N powders into α-Sialon; mixing in isopropanol).
After drying and sieving the mixes were hot pressed under the experimental conditions listed in Table II, where data related to materials produced with commercial powders are also shown.
c) Characterization of dense samples
The densities of the samples were measured using the Archimedes' water-immersion method. The phase composition was determined by X-ray diffraction analysis. Microstructural characteristics were investigated by optical and scanning electron microscopy (SEM) on fracture and polished surfaces. The polished specimens for SEM observation were plasma etched using a gas mixture of CF_4 and O_2 for 20 seconds. The Young's modulus (E) was measured on samples with dimensions 0.8 mm × 8.0 mm × 28 mm (height by width by length) by the frequency resonance method. Flexural strength, both at room and high temperature, was measured on samples 3 mm × 3 mm × 30 mm in a 4-point bending fixture, with 26 mm outer span, 13 mm inner span, and a crosshead speed of 0.5 mm/min: The fracture thoughness (K_{Ic}) was evaluated by the direct crack measurement method with a load of 98.1 N in a hardness tester using the formula proposed by Anstis *et al.*[19]. On the same apparatus, the Vickers microhardness (*HV*) was measured on polished surfaces with a load of 4.91 N. The thermal expansion was evaluated up to 1300 °C, using a heating rate of 5°C/min.

RESULTS AND DISCUSSION

a) Compositions Si_3N_4 /Si-C-N
The comparison of the sintering behaviour of compositions A and B (fig.1 and Table I) evidence that the densification rate is higher in composition B which contain 20 vol% of amorphous Si-C-N powder. For a given composition and hot-pressing regime, good sintering behaviour for silicon nitride powder is characterized by the attainment of high density and by the absence of grain growth.

Table I - Main features of samples A and B hot pressed at 1850°C, 30 MPa and, respectively for 40 and 20 minutes.

a) Densification behaviour

Sample	ρ_o %	ρ_f%	$\dot{\rho}$ 10^{-3} g $cm^{-2}s^{-1}$			$m\Gamma^*$		
			ρ = 0.85	ρ = 0.90	ρ = 0.95	ρ = 0.85	ρ = 0.90	ρ = 0.95
A	48	~ 100	1.29	0.72	0.23	6.3	14.0	34.8
B	47	~ 100	2.59	2.14	41.57	2.2	4.4	10.8

b) Microstructural characteristics

Sample	Crystalline phases				I (200/I_{002}) β-Si_3N_4		Grain size		
							μm	μm	aspect ratio
	α-Si_3N_4	β-Si_3N_4	α-SiC	Si_2N_2O	Parall.	Perpend.	l	d	
A	3.4	96.6	-	-	2.8	30.9	2.2 ± 1.3	0.35 ± 0.2	7.3
B	29.8	62.7	4.9	2.6	2.7	16.3	2.6 ± 0.7	0.26 ±0.12	10.8

c) Mechanical properties

Sample	λ (20-1000°C)	E	HV	K_{IC}	σ (MPa)		
	10^{-1} °C	GPa	GPa	Mpa $m^{1/2}$	RT	1000 °C	1300 °C
A	3.49	320	18.6 ± 0.4	5.9 ± 0.1	1063 ± 43	807 ± 48	700 ± 52
B	3.75	323	17.0 ± 0.9	4.4 ± 0.1	622 ± 39	-	-

Table II - Caracteristics and properties of Si-C-N based compositions hot pressed at 30 MPa

Sample	H.P.		Density	Crystalline phases				H_v	K_{IC}
	T (°C)	t (min)	g cm^{-3}	α-Si_3N_4	β-Si_3N_4	Si_2N_2O	SiC	GPa	Mpa √m
I	1700	60	2.94	-	36	51	13	17.6 ± 1.1	4.8 ± 0.9
II	1790	30	2.68	37	50	-	13	21.1 ± 2.0	4.9 ± 0.7
III	1800	30	3.12	65■	25	-	10	21.0 ±1.3	3.8 ± 0.9
IV	1700	60	3.26	-	100	-	-	19.7 ± 0.6	5.2 ± 0.1
V	1750	60	3.26	-	80	-	20	20.7 ± 0.5	4.7 ± 0.1

I Si-C-N + 5% Y_2O_3 + 2% Al_2O_3
II Si-C-N + AlN + YN
III (Si-C-N + AlN + YN) + 3% Al_2O_3 + 6% Y_2O_3
IV Si_3N_4 (Ube SNE10) + 5% Y_2O_3 + 2% Al_2O_3
V comp. IV + 20 vol% SiC particles
■α-sialon

Pickup *et al.* [20] have described a method to evaluate the behaviour of powders in terms of the densification rate (ρ) and the grain coarsening rate (G).
The grain size dependence of densification rate is expressed by equation: $\dot{\rho} = k\ G^{-m}$ where G is a grain size, k and m are constants, m being a characteristic of the densification mechanism. From this equation the value of Γ*, regarded as a figure of merit of Si_3N_4 -based compositions and powders, can be obtained: $\Gamma^* = \frac{\dot{G}}{G} \cdot \frac{\rho}{\dot{\rho}} = -\frac{1}{m} \cdot \frac{\rho\,\ddot{\rho}}{\dot{\rho}^2}$

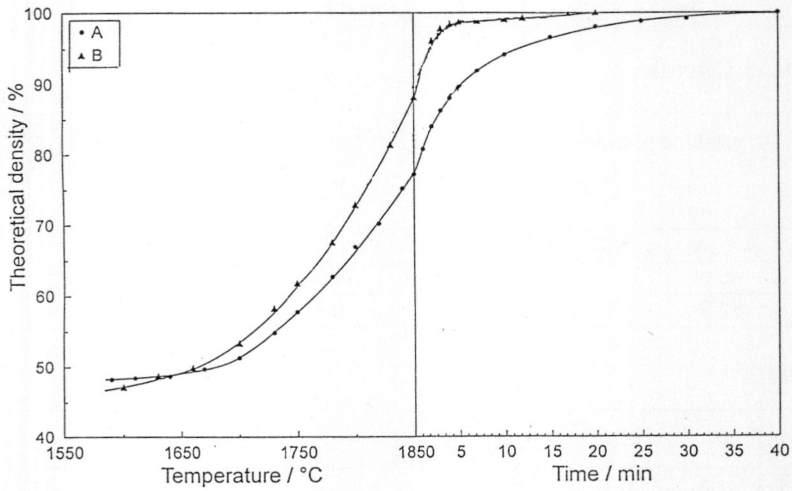

Fig. 1- Densification behaviour of mixes A and B.

A good Si_3N_4 material should exhibit high density and small grain size, therefore low values of Γ* are desirable, as it is proportional to the ratio of grain coarsening rate ($\dot{G}$) over densification rate ($\dot{\rho}$). Although the value of changes for different mechanism of densification (m=2 for lattice diffusion creep, m=3 for grain boundary diffusion creep), changes of mechanisms for two different matrix powders processed under identical conditions are less likely to occur than changes in rate for a given mechanism. It is also difficult to distinguish which of the mechanisms is dominant, therefore the $m\Gamma^*$ values were evaluated instead of Γ*. Table I shows the $m\Gamma^*$ values for the two compositions at various relative densities.
The results enphasize the good figure of merit of the silicon nitride mixture containing 20 vol% of nanosized and amorphous Si-C-N powder. Microstructural characteristic shown in fig. 2a,b and Table I evidence the presence of elongated grains with high aspect ratio (~10). Equiaxed grains are also present, mainly in sample B. These features give an example of the so called "in situ composite": this microstructure is well known to be the optimal one for the improvement of the mechanical properties. Moreower the introduction of nanosized amorphous Si-C-N phase into the mixture results in the enoforcement of the homogeneous nucleation and refinement of microstructure.
As crystalline phases, sample A contains a low amount of residual α-Si_3N_4 while sample B has an higher quantity of residual α-Si_3N_4 due to the thermal cycle (only 20 min at 1850°C) which did not allow a complete α → β- Si_3N_4 phase transformation. The addition of nanosized Si-C-N leads to the formation of ~ 5% of crystalline SiC and, owing to the oxygen content, induces the formation of low amounts of Si_2N_2O (~ 2.6%). A strong preferred orentation of β- Si_3N_4 grain is also evident.
The values of some physical and mechanical properties evidence that: - thermal expansion and Young's modulus (E) slightly increase in sample B owing to the presence of ~ 5% of SiC; - the

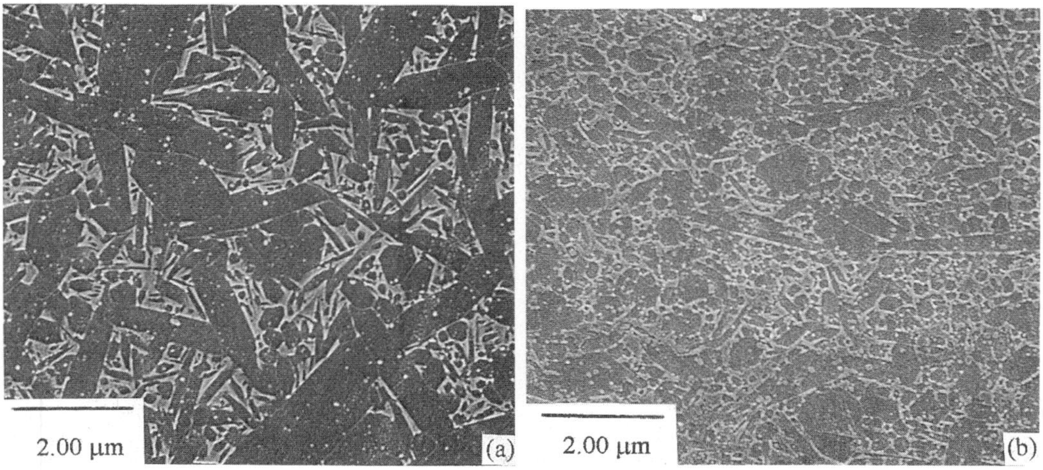

Fig. 2 - Microsturcture of hot pressed samles from composition A (a) and B (b).

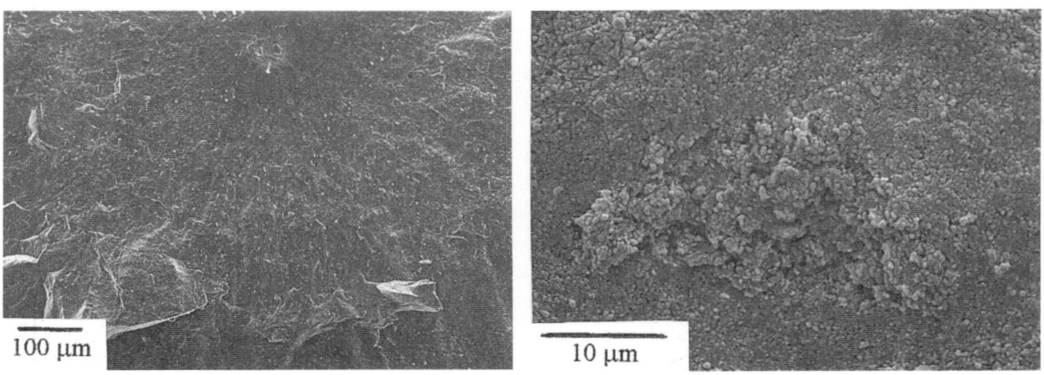

Fig. 3 - Crytical defect origin of fracture in hot pressed sample from composition B.

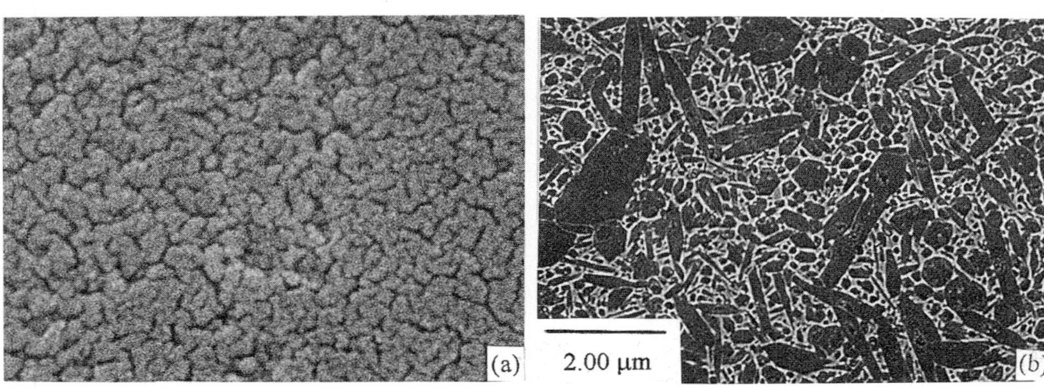

Fig. 4 - Microstructure of hot pressed sample produced with composition I from Si-C-N powders (a) and produced with the same composition from commercial Si_3N_4 powders (b).

different value of hardness (H_v) are to be related to different composition of the integranular phase and to the presence of some Si_2N_2O in sample B; - sample A is tougher than B owing to its coarser microstructure. Concerning flexural strength, sample A has excellent values also at high temperature due to its well designed composition and microstructure [21]. Material B is spite of its finer grained microstructure, has low strength which has to be associated to microstructural defects: porous areas which are due to scarce homogeneity of the mixed Si_3N_4/Si-C-N powders (fig.3).

b) **Compositions based on Si-C-N**

The comparison of the densification behaviour of Si-C-N-based composition I with samples produced by commercial powders: sample IV (Si_3N_4 UBE-SN10) and sample V (Si_3N_4 + 20%vol of SiC particles), containing the same dopants of composition I added by chemical coprecipitation[22], showed worse results for Si-C-N. The low sinterability of the amorphous Si-C-N powders can be mainly related to: - the low green density (26%) which inhibits high densfication; - the presence of C in the Si-C-N because carbon partially reacts with oxide additives and SiO_2 to form CO and reduce the Si_3N_4 or SiC, depending on the processing conditions.

In composition I, all the α-Si_3N_4 is transformed to β-Si_3N_4; contemporary Si_2N_2O (silicon oxynitride) is present. In composition II a certain amount of α-Si_3N_4 is still present, but no Si_2N_2O formed.

Material from composition III has 65% of α-sialon with a lattice parameters a = 0.78155 nm and c = 0.5698 nm, thus exhibiting a clear expansion of the α-Si_3N_4 unit cell (a = 0.7818 nm and c = 0.5591 nm [23]). These data correspond to the value of x = 0.46 in the formula $Y_{2x}(Si_{12-9x},Al_{9x})(O_{3x},N_{16-3x})$. No Si_2N_2O was detected in this sample.

The microstructure of the samples I, II and III, was very fine (grain size < 100 nm). In fig. 4 the microstructure of hot pressed samples deriving from commercial Si_3N_4 powder and from Si-C-N powder is compared.

Some mechanical properties of the Si-C-N materials are shown in Table II, where also values relative to of the hot-pressed samples produced from commercial powders with composition I are reported. Very high hardness value was determined for samples II and III. The lower hardness of I can be attributed to the lower density, to presence of microcrack and to the absence of α-Si_3N_4 . The values of fracture toughness are similar to those of materials produced from commercial powders. The low toughness of sample III can be ascribed to its high amount of glassy phase.

CONCLUSIONS

a) Si_3N_4-based materials were fabricated by hot pressing, evaluating the effects of an amount of ultrafine amorphous Si-C-N powder mixed to commercial Si_3N_4 on microstructure and properties. The presence of Si-C-N increases the densification rate and results in substantial refinement of the resulting microstructure. In both the cases, high aspect ratio (~ 10) of the β-Si_3N_4 grains have been observed: starting powder characteristics and the additive system were tailored to give "in situ composites" microstructures. The addition of 20% vol of Si-C-N reduces the flexural strength at room and high temperatures, due to defects which derive from difficulties in acheving homogenity of mixing and dispersing such refined powders. Powder treatment methods still needs to be improved when nanosized powders are added to submicron powders.

b) dense materials with very fine microstructure (~ 100 nm), high hardness (21 GPa) and good toughness (up to 5 MPa $\sqrt{m}$) were produced by hot pressing starting from Si-C-N with sintering aids. Oxygen (SiO_2) was incorporated into α-sialon by suitable additives. However the densification behaviour of ultrafine and amorphous Si-C-N powders was worse in comparison with commercial crystalline powders. Preliminary heat treatments which induce crystallization and simultaneus increase of the partice size is a necessary processing step before densification of these ultrafine powders.

ACKNOWLEDGMENT

The author would like to express her gratitude to Dr. Z. Lences (SAS, Institute of Inorganic Chemistry, Bratislava), for valuable help in the experimental activity and discussion.

REFERENCES

[1] K. Niihara, J. Ceramic Soc. Jap. Int. Ed. **99**, 945 (1992)
[2] R. Riedel, M. Seher, G. Becker, J. Eur. Ceram. Soc. **5**, 113 (1989)
[3] B. A. Bender, R. P. Ingel, W.J. McDonough, J.R. Spawn, Adv. Ceram. Mat. **1**, 137 (1986)
[4] K. Niihara, K. Suganuma. A. Nakahir, K. Izaki, J. Mater. Sci. Lett., **9**, 598 (1990)
[5] T. Rouxel, F. Wakai, M.E. Brito, A. Iwamoto, K. Izaki, J. Eur. Ceram. Soc., **11**, 431 (1993)
[6] K. Niihara, K. Izaki, T. Kawakahi, J. Mater. Sci. Lett. **10**, 112 (1990)
[7] A. Sawaguchi, K. Toda, K. Niihara, J. Am. Ceram. Soc., **74**, 1142 (1991)
[8] A. Keber, R. Wagner, in Third Euro-Ceramics, Vol. 3, Ed. by P. Duran, F. Fernandez, Faenza Editrice Iberico, p.595 (1993)
[9] M. Hermann, T. Millers, A. Rendtel, T. Reich, G. Boden, S. Schubert, ibid., p.603
[10] M. Suzuki, Y. Maniette, Y. Nakata, T. Okutani, J. Am. Ceram. Soc., **76**, 1195 (1993)
[11] E. Borsella, S. Botti, R. Fantoni, R. Alexandrescu, I. Morjan, T. Dikonimos-Makris, R. Giorgi, E. Enzo, J. Mater. Res., **6**, 1 (1991)
[12] M. Luce, O. Croix, C. Robert, M. Cauchetier, in Ceramic Powder Processing Science, February 4-7, San Diego, CA, (1990)
[13] F. Wakai, Y. Kodama, S. Sakaguchi, N. Murayama, K. Isaki, K. Niihara, Nature, **344,** 421 (1990)
[14] S. Futaki, Y. Shimizu, K. Shirashi, Y. Moryoshi, T. Sato, T. Sakai, J. Mater. Sci. , **22,** 4331 (!987)
[15] F. Allaire, R. Langlois, J. Mater. Sci., **27**, 1265 (1992)
[16] H.T. Sawhill, J.S. Haggerty, Comm. Am. Ceram. Soc. , **65**, C 131 (1982)
[17] Z. Lences, A. Bellosi, F. Monteverde, Mater. Chem. and Phys. **99**, (1995)
[18] K.J. Nilsen, S.C. Danforth, H. Wautier, in "Ceramic Powder Science, Advances in Ceramics", Vol.21, Ed. by G. Messing, K. Mazdiyasni, J. McCauley, R. Haber, The Am. Ceram. Soc. Inc., Westerville, Ohio, p.537 (1987)
[19] A.G. Evans, E.A. Chales, J. Am. Ceram. Soc., **59**, 371 (1976)
[20] H. Pickup, E. Gilbart, R. J. Brook, A. Leriche, F. Cambier, in "Ceramic Materials and Components for Engines", Ed. by W. Bunk, K. Hausner, Verlag DKG, p.63, (1986)
[21] A. Bellosi, G.N. Babini, " Production and Characterization on of high performance silicon nitride" in print on Advanced Materials
[22] A. Bellosi, V. Blasini, C. Galassi, Processing of Advanced Ceramics, **3**, 1153 (1993)
[23] Z. Panek " Standard Gibbs Energy of formation of β-Si_3N_4", submitted to J. Amer. Ceram. Soc.
[24] V. Biasini, S. Guicciardi, A. Bellosi, Refr. Mat. & Hard Mater., **11**, 213 (1992).

Materials Science Forum Vol. 195 (1995) pp. 87-92

TEM Characterization of Palladium and Silver Nanoclusters in Glass Matrix

M. Catalano[1], G. De[2,3], A. Licciulli[2] and L. Tapfer[2]

[1] Istituto IME-CNR, Università di Lecce, Via Arnesano, I-73100 Lecce, Italy

[2] PASTIS-CNRSM, ss. 7 Appia km. 712, I-72100 Brindisi, Italy

[3] permanent Address: Sol-Gel laboratory, C.G. & C.R.I., Calcutta 700 032, India

Keywords: Nanoclusters, Coatings, Sol-Gel, TEM, HREM, Glasses

Abstract

Palladium and silver nanoclusters doped silica gels and coatings were prepared by the sol-gel process and characterized by conventional transmission electron microscopy, high resolution electron microscopy and x-ray diffractometry. The Pd/Si and Ag/Si atomic ratios were 0.01 and 0.12, respectively. The Pd nanoclusters are very homogeneously distributed in the matrix on a micron scale having a very narrow size distribution with an average value of (4.5±0.5) nm. Ag clusters show a relatively broad size distribution. The average diameter distribution obtained from the analysis of several bright field images from a coating heat-treated at 550 °C exhibits two clear modes around 3.5 nm and 19 nm, respectively. The concentration of larger clusters is, however, very small in comparison to the smaller ones. Heat treatments were performed on both Pd and Ag samples, in order to have a deeper understanding of the clusters' structural evolution. No noticeable variations in the cluster diameter distribution are observed in the Pd samples after the heat treatment. The formation and growth of the Ag clusters depend significantly on the temperature and heat treatment time.

I. INTRODUCTION

In our earlier communications [1,2] we have reported on the preparation and characterization of Ag and Pd colloid doped bulk gels and coatings in pure and organically modified silica matrices, respectively. The above two noble metals have been chosen mainly for two different reasons. First, nanosized Pd colloids may give some interesting optical properties, since Pd is already being used in technical applications as coating (Schott calorex ®). Beside this Pd colloids doped in porous amorphous matrix could be used as catalysts or sensors. Second, silica coatings doped with a relatively high concentration of Ag colloids (Ag/Si atomic ratio = 0.12) could be interesting for non-linear optical applications. Recent literature survey shows that $\chi^{(3)}$ can be significantly increased by increasing the volume fraction of metal colloid in the glass matrix [3,4].

Due to the nanometer dimensions of the clusters, high spatial resolution techniques are necessary in order to characterize these materials. In particular, high resolution transmission electron microscopy (HREM) offers an unique opportunity of imaging the nanoclusters and, as a result, their structure, size, shape and distributions can be directly visualized. In the present study, the formation and growth of the Ag and Pd nanoclusters are investigated by conventional transmission electron microscopy (CTEM), HREM along with their selected area diffraction patterns (SADP) and electron nanodiffraction patterns (NDP) analyses. X-ray diffractometry (XRD) is used for additional supporting evidences.

2. EXPERIMENTAL

The Pd clusters studied in this work were obtained by a modified sol-gel process [2,5]. The Pd/Si atomic ratio of 0.01 was maintained in this preparation. To avoid the precipitation of Pd metal in the alcoholic solution, a Pd-DAMO [DAMO = $(NH_2CH_2CH_2NH(CH_2)_3Si(OMe)_3)$] complex solution (Pd/DAMO ratio = 1:2) was used as Pd source. A GPTS/TEOS (γ-glycidoxypropyltriethoxysilane/tetraethoxy silane) sol was mixed with Pd-DAMO solution. The solution was then distilled to partially remove the solvents and promote the hydrolysis and condensation reactions. The resulting concentrated sol was then cast into polythene moulds for gellation. The gels were then dried at 60 °C and heated at 200 °C in nitrogen atmosphere for cluster formation. Gels containing Pd clusters were heat treated up to 400 °C in N_2, in order to study the structural evolution of the clusters. Further details about the cluster preparation technique are reported elsewhere [2].

Ag clusters doped silica films with relatively high concentration of silver metal clusters (Ag/Si atomic ratio = 0.12) were deposited on silica glass substrates by the dip-coating method. The sol was prepared by mixing the requisite amount of silver nitrate (dissolved in minimum volume of water) to a previously hydrolyzed tetraethyl orthosilicate (TEOS) in presence of alcoholic solvents. The coated slides were first dried in air at 60 °C for 30 min. The dried gel films were then heat treated in air at temperatures ranging from 300 ° to 550 °C with different heat treatment times, in order to understand the cluster formation mechanisms and their structural evolution. The film baked at 550°C shows a strong surface plasmon peak at 410 nm. Further details about the cluster preparation technique are reported elsewhere [1].

The chemical compositions of the samples are analyzed by XPS (X-ray photoelectron spectra) and RBS (Rutherford back scattering) techniques. The results are in good agreement with the calculated values [1,2].

A JEOL 4000 EX II microscope, operating at 400kV, with an interpretable resolution limit of 0.16 nm was used for the HREM work. The analytical work was performed using a Philips CM30 TEM/STEM microscope. XRD spectra were obtained with a Philips PW 1880 diffractometer with a CuK_α source. TEM samples were obtained by scratching the Pd doped monoliths and the Ag doped coatings using a diamond tip and depositing the fragments onto thin carbon support films. A thin carbon film was then deposited onto the fragments to prevent the sample from charging under the electron beam during TEM observations.

III. RESULTS AND DISCUSSION

Figure 1 is a Bright Field (B.F.) image of Pd clusters in organically modified silica matrix. The clusters are very uniformly distributed in the matrix on a micron scale with very narrow diameter distribution peaked around (4.5 ± 0.5) nm. Figure 2 is a HREM image of some clusters in the as grown sample. Note the presence of some nanotwins and other structural defects in the clusters. Similar micrographs are obtained for the samples heated at 400 °C in N_2 atmosphere. However, in the latter case less defects are observed in the clusters. A stabilization of this type of dispersed nanophasic clusters with respect to heat-treatment time and temperature was also confirmed by XRD [2] analyses.

By applying Scherer's equation [7] the size of the clusters is evaluated by measuring the full width at half maximum of the Pd(111) XRD peak. The average diameter estimated by this method was about (3.0 ± 0.5) nm, for the samples heat treated within the temperature range 200 °C- 400 °C. This diameter value, however, does not agree with the value directly measured from TEM. In fact, the width of X-ray peaks is related to the crystallite diameter, their structural defects and to the cluster strain status [7].

TEM studies can provide independent information about these parameters. In this study, TEM analyses show no appreciable variation in the cluster diameter with increasing heat-treatment time or temperature. However, one noticeable difference between the as grown (200 °C /1hr) and heat-treated sample (400 °C) is the presence of nanotwins (Fig. 1 and 2) and other defects in the former case, whereas, these defects are absent in the latter one. Therefore, TEM offers a unique opportunity of imaging the nanoclusters, and their size, shape and other structural information can be visualized directly.

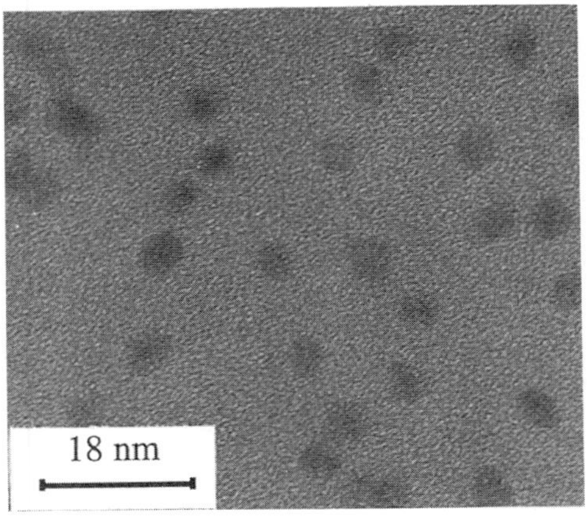

Fig.1: B.F. image of Pd clusters in the as grown sample.

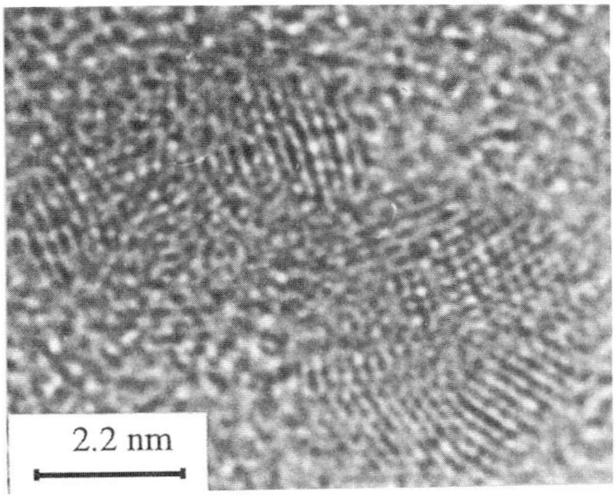

Fig. 2: HREM image of Pd clusters in the as grown sample.

Figures 3 and 4 are B.F. images of the silver doped coatings heat treated at 300 °C and 450 °C, respectively. The majority of the clusters in Fig. 3 and Fig. 4 have a very narrow diameter distribution around an average value of 1nm, although few large particles with diameters ranging between 10-20 nm are observed. Although NDP experiments on large clusters produce patterns that can be attributed to *fcc* Ag, no crystalline features are obtained from the SADP and HREM images from the smaller clusters. According to the other experimental results reported elsewhere [1], we suggest that these clusters are probably formed by a mixture of silver silicates and oxides.

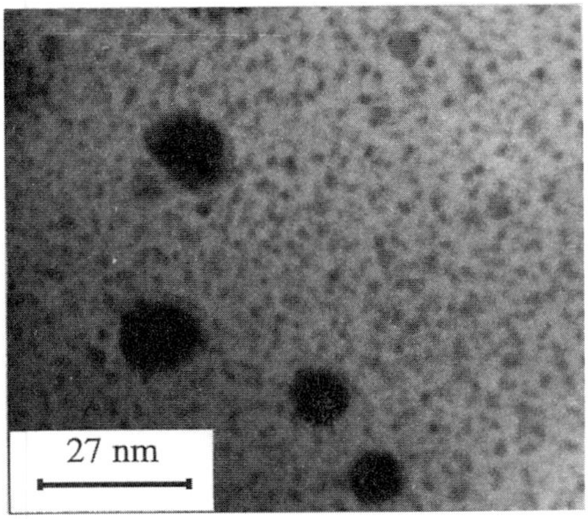

Fig. 3: B.F. image of Ag clusters in sample heat treated at 300°C for 100 min.

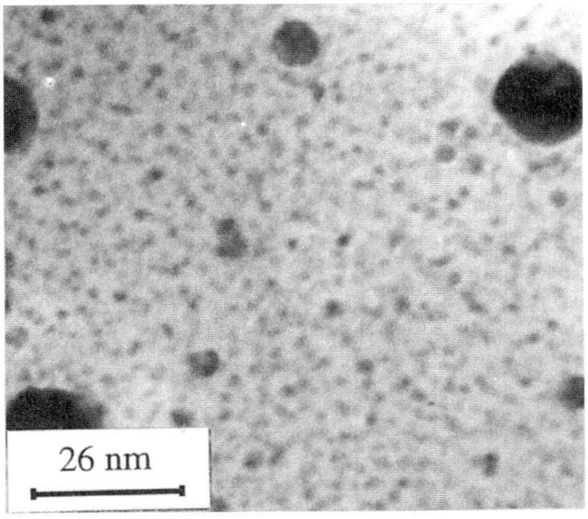

Fig. 4: B.F. image of Ag clusters in sample heat treated at 450°C for 100 min.

Figure 5 shows a B.F. image of the film heat treated at 550°C for 200 min. The image shows a broad size distribution of spherical silver nanocrystallites having diameters in the range from less than 1 nm to about 20 nm. The SADP patterns obtained from these crystallites confirmed a *fcc* crystallographic structure. The size distribution of the clusters obtained from several B.F. images is shown in Fig. 6. It can be seen that the percentage of large clusters is much lower than that of the smaller ones. In our previous work [1] optical absorption spectra from Ag coatings were obtained and fitted by using the Mie theory [8]. Input parameters for this fitting are the cluster diameters and their respective concentrations derived from the analysis of several TEM images. However, in order to correctly fit the experimental spectra, it is necessary to use a higher concentration of larger clusters than experimentally observed by TEM. This result indicates that the Mie theory may overestimate the distribution of larger clusters. Figure 7 shows the XRD pattern obtained from the same film examined by TEM. The pattern shows a series of peaks that are characteristic of crystalline silver. The sharp peaks are probably generated by the larger clusters.

The analysis of the experimental results outlined above shows that stabilization of the dispersed nanocrystalline phase is obtained in case of Pd after their formation. On the contrary the Ag system depends on the temperature and heat-treatment time. The most probable reason for this is the atomic ratio of the two systems, which differ significantly. The Pd system is very much diluted (Pd/Si atomic ratio = 0.01) and the possibility of cluster-cluster interaction is not feasible, whereas the Ag system is much concentrated (Ag/Si atomic ratio = 0.12) and a variation of clusters' size occurs following the interaction of the clusters through Ostwald ripening. Moreover, the use of DAMO in case of Pd as a complexing and cross linking agent ensures a uniform distribution of the dopant metal in the matrix [5].

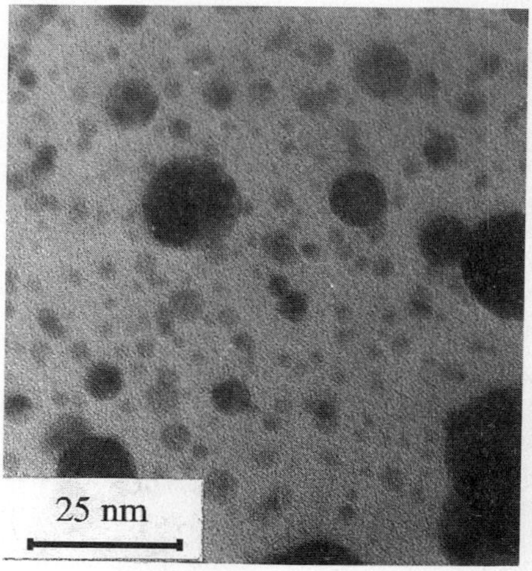

Fig. 5: B.F. image of Ag clusters in sample heat treated at 550°C for 200 min.

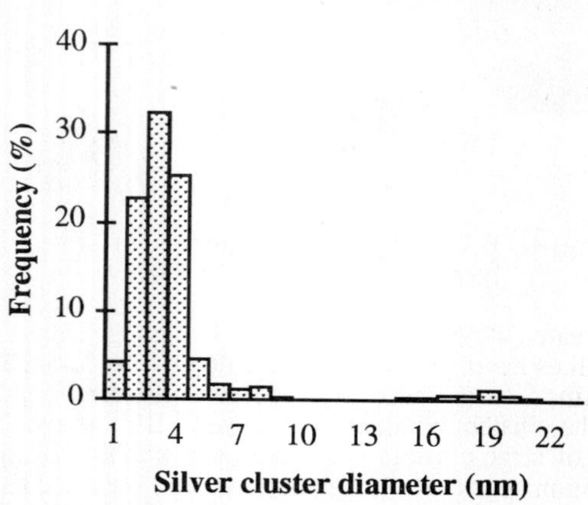

Fig. 6: Ag clusters size distribution for sample heat treated at 550°C for 200 min.

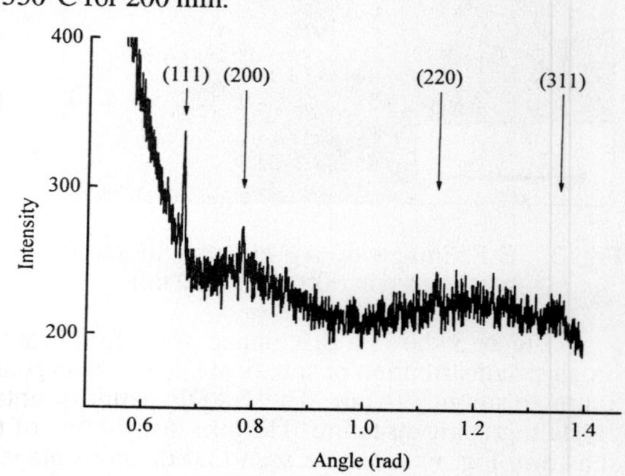

Fig. 7: XRD spectrum from coating heat treated at 550 °C for 200 min.

IV. CONCLUSIONS

In this we investigated nanocrystal doped glasses by using transmssion electron microscopy. Palladium clusters homogeneously nucleate in the bulk gel as a result of the growth process. The clusters have very narrow size distribution (4.5±0.5) nm and are very homogeneously distributed in the matrix on a micron scale. No significant variations in the size and shape of the clusters are observed for longer heat treatment times at 400°C in N_2 atmosphere, demonstrating that the cluster formation is complete at 200°C.

High concentration silver nanocrystal doped glassy coatings prepared by the sol-gel dip-coating method were characterized by TEM. The coatings heat treated up to 450°C consist of a mixture of silver metal and silver silicate/oxide clusters in a porous silica matrix. Stable and fully densified films were obtained at 550°C which contain only metal nanoclusters. The size distribution of the clusters shows the presence of mainly two different clusters, one in the range 1-5 nm in diameter and few of about 16-20 nm.

ACKNOWLEDGMENTS

The authors would like to thank M. Palmisano and L. De Caro for the skillful specimen preparation, E. Carlino for the stimulating discussions and M. A. Tagliente and A. Sacchetti for the XRD analyses. G. De acknowledges the financial support for a research fellowship from INFM (Istituto Nazionale per la Fisica della Materia), Genova.

REFERENCES

[1] G. De et al. 1995 (accepted for publication in J. Non-Cryst. Solids)
[2] A. Licciulli et al., Italian Crystal Growth 1995 (to be published in "Advanced Materials Research", Trans. Tech. Publications, Switzerland)
[3] K. Uchida et al. J. Opt. Soc. Am. **B 11**, 1236 (1994)
[4] K. Fukumi et. al, J. Appl. Phys. **75**, 3075 (1994)
[5] T. Burkhart et al., MRS Symp. Proc. (Better Ceramics through Chemistry VI), **346**, 779 (1994).
[6] M. Catalano et al 1995, Microscopy, Microanalysis, Microstructures, in press
[7] P. Klug et al, *X-ray Diffraction Procedures for Policrystalline and Amorphous Materials*, Wiley, New York, (1974)
[8] G. Mie, Ann. Phys. (Leipzig), **25**, 376 (1908)

Materials Science Forum Vol. 195 (1995) pp. 93-102

Nanoscale Platinum Particles by Clustering of Arene Solvated Pt Atoms

G. Vitulli[1], E. Pitzalis[1,2], R. Lazzaroni[1,2], P. Pertici[1], P. Salvadori[1,2], O. Salvetti[2], S. Coluccia[3] and G. Martra[3]

[1] Centro del CNR per le Macromolecole Stereordinate ed Otticamente Attive, Via Risorgimento 35, I-56126 Pisa, Italy

[2] Dipartimento di Chimica e Chimica Industriale, Università degli Studi di Pisa, Via Risorgimento 35, I-56126 Pisa, Italy

[3] Dipartimento di Chimica Inorganica, Chimica Fisica e Chimica dei Materiali, Via P. Giuria 7, I-10125 Torino, Italy

Keywords: Solvated Pt Atom, Clustering, Ultrafine Pt Powders

Abstract Nanoscale Pt particles (≈30-40 Å) have been generated by clustering of Pt atoms from arene solvated Pt atoms, obtained by reaction of Pt vapour and toluene or mesitylene. HRTEM and SEM analyses have been performed on the prepared Pt aggregates.

Results and discussion There is an increasing deal of attention in nanometer scale metal particles: they frequently exhibit hybrid properties, different from those of atomic or molecular species and bulk solids and are of interest in several fields, ranging from material science to biology [1]. Among the established methods of preparation, the technique which is variously known as metal atoms or metal vapour synthesis is beginning to take a relevant position [2]. Metal atoms (vapour) when deposited into low temperature weakly complexing organic media (such as toluene, tetrahydrofurane, pentane or similar organics) can allow the formation of metal atom solution stable at low temperature (solvated metal atoms). Such solutions, which contain solvent stabilized metal atoms as metal microclusters, can be used directly as precursors for a wide range of interesting and unusual homogeneous catalytic reactions, or, by further clustering at room temperature, can act as source of active metal aggregates which can be isolated as ultrafine metal powders or supported nanometer particles [3] (Scheme 1).

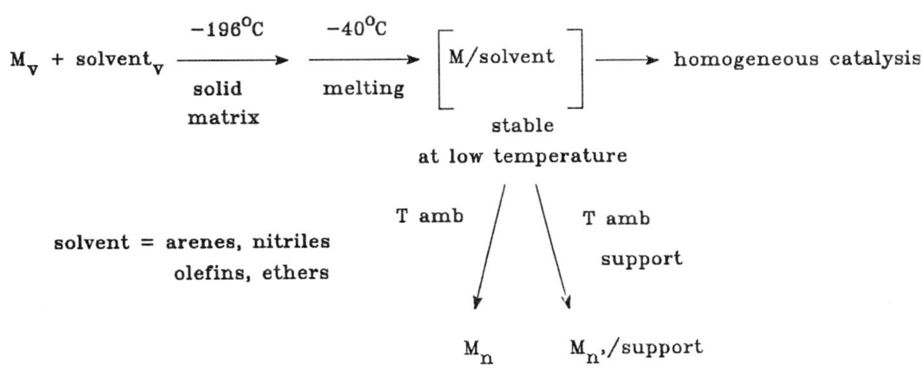

Scheme 1

We have recently shown that mesitylene solvated metal atoms, obtained by reaction of Pt vapour with mesitylene, can be conveniently used as a homogeneous catalytic precursor or as starting material for the deposition of active Pt particle on organic and inorganic support [4].We report here that in the absence of supports the clustering of Pt atoms in aromatic solvents affords nano-scale Pt aggregates of very narrow size distribution (Scheme 2).

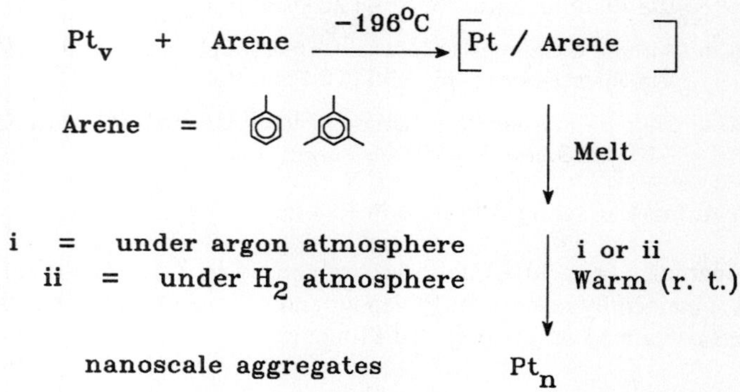

Scheme 2

The codeposition of platinum vapour with arenes (toluene, mesitylene) at liquid N_2 temperature, in a glass reactors described in the experimental part, affords red-brown matrices, which give on melting brown solutions stable at low temperature (≈-70°C with toluene, -20°C with mesitylene). Warming to room temperature results in precipitation of Pt as powder. The clustering processes are quite slow under an inert atmosphere but are more rapid in the presence of H_2. The samples so obtained have been characterized by HRTEM at different magnifications and by SEM analysis. The results are summarized in table 1.

Table 1

POWDER	MORPHOLOGICAL CHARACTERISTICS	AGGREGATES SIZE (μm)	PARTICLES DIAMETER (nm)
Pt_n from Pt/mesitylene	spongy aggregates	<1	3.8
Pt_n from Pt/mesitylene +H_2	spherical compact aggregates	0.15 ÷ 0.3	2.8
Pt_n from Pt/toluene	aggregates grouped in chains	<0.5	4.4
Pt_n from Pt/toluene + H_2	compact aggregates grouped in chains	<0.5	3.0

The Pt particles derived by Pt/mesitylene co-condensates, by clustering in inert atmosphere, appear as spongy agglomerates of irregular form, smaller than 1μm (Fig. 1).

Fig. 1 HRTEM micrograph of Pt powder resulting from Pt/mesitylene co-condensate upon warming up to 300 K under argon.

Higher magnification (Fig. 2) shows that such agglomerates are formed by Pt nanoparticles, with an average size of 3.8 nm. They are surrounded by an amorphous layer of variable thickness (≤ 1nm), which is probably a carbonaceous deposit formed by decomposition on warming at room temperature of the organic molecules bonded to the surface of the metal particles.

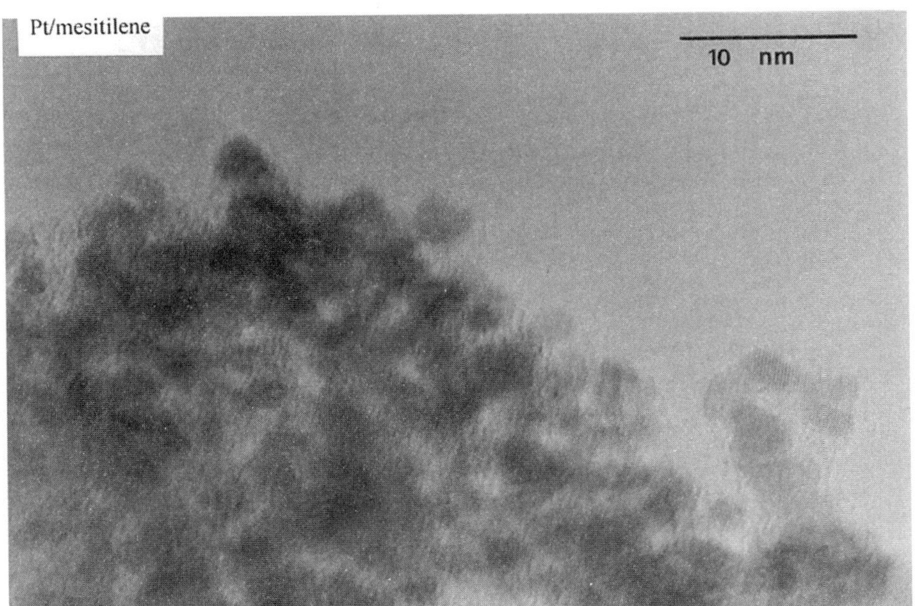

Fig. 2 HRTEM micrograph of Pt powder resulting from Pt/mesitylene co-condensate upon warming up to 300 K under argon, higher magnification.

The precipitation of Pt particles in the presence of H_2 produces powders with a different morphology consisting of spherical agglomerates, more compact than in the previous phase, whose diameter is in the range 150-300 nm, as shown by HRTEM and SEM (Fig.3, Fig.4).

Fig. 3 HRTEM micrograph of Pt powder resulting from Pt/mesitylene co-condensate upon warming up to 300 K under hydrogen.

Fig. 4 SEM micrograph of Pt powder resulting from Pt/mesitylene co-condensate upon warming up to 300 K under hydrogen.

Such particles are polycrystalline in nature, being formed by closely packed Pt nanoparticles (Fig. 5), with an average size (2.8 nm) slightly smaller than in the previous sample.

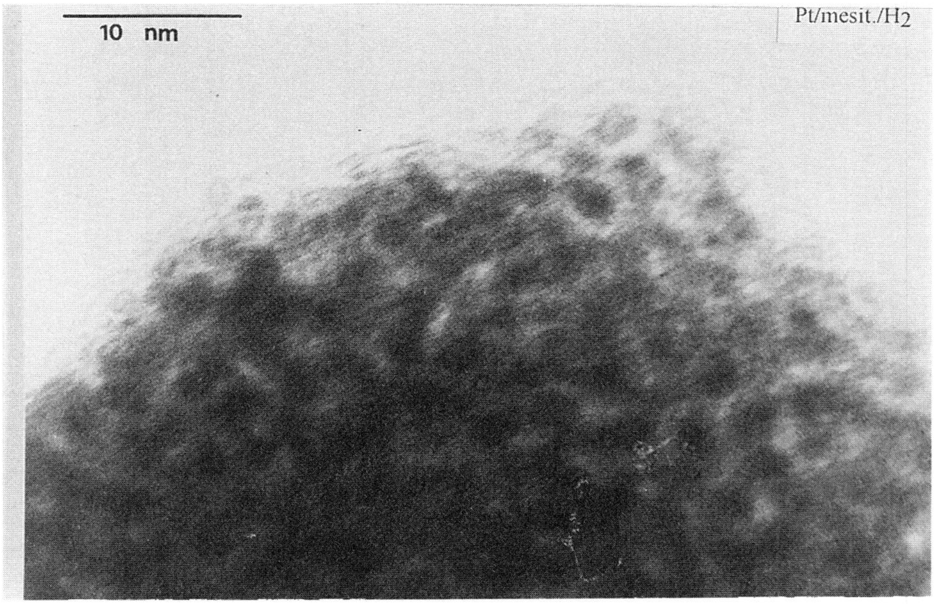

Fig. 5 HRTEM micrograph of Pt powder resulting from Pt/mesitylene co-condensate upon warming up to 300 K under hydrogen, higher magnification.

Similar results were obtained for the powders precipitated from toluene solutions (micrographs are not reported for the sake of brevity). However, it must be noted that these samples contain bigger Pt nanoparticles, having average sizes of 4.4 nm (powders precipitated on warming up under argon) and of 3.0 nm (powders precipitated in the presence of H_2). These differences may be related to the lower stabilization effect of toluene as ligand [5], allowing the growth of bigger metal particles.

Preliminary experiments on the preparation of bimetallic aggregates have been made. The vaporization of Pt and Fe using as stabilizing ligands mesitylene and/or cycloolefins allows to isolate at low temperature red brown solvated metal atoms: from these solutions by treatment with hydrogen at atmospheric pressure and warming up to room temperature metal powder is gently deposited. Platinum/iron molar ratio of 5/1 has been used. The powder so obtained shows, at atomic absorption and EDS microanalysis, Pt/Fe ratio of 4.8-4.9. SEM and HRTEM micrographs (Fig. 6, Fig. 7) showed homogeneous aggregates, whose diameter is 200 nm, formed by particles of 3 nm. These preliminary results are of interest and are promising, indicating in the metal vapour synthesis a convenient tool for the preparation of bimetallic aggregates of well defined chemical composition and homogeneous particles size.

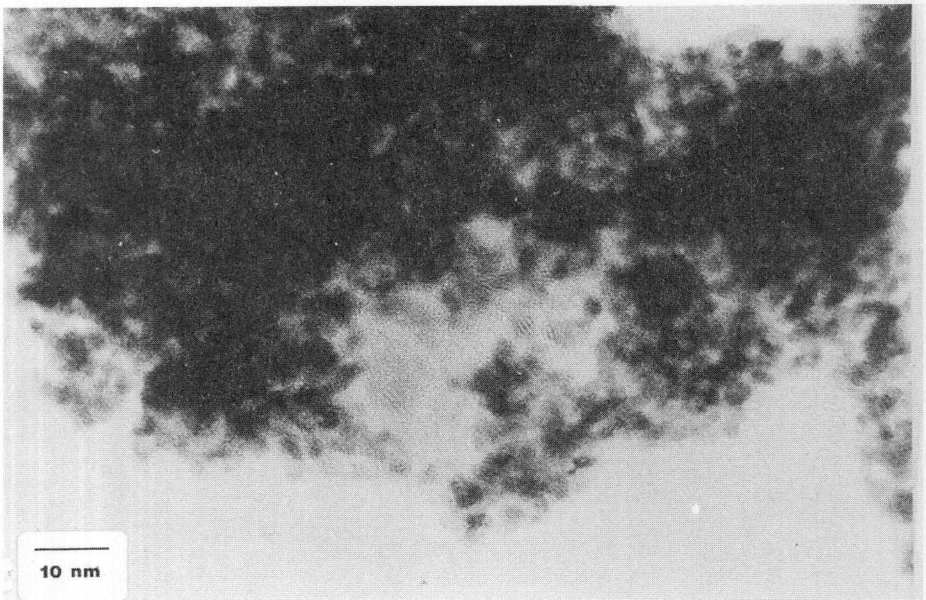

Fig. 6 HRTEM micrograph of Pt/Fe powder resulting from Pt/mesitylene and Fe/cicloolefins upon warming up to 300 K under hydrogen.

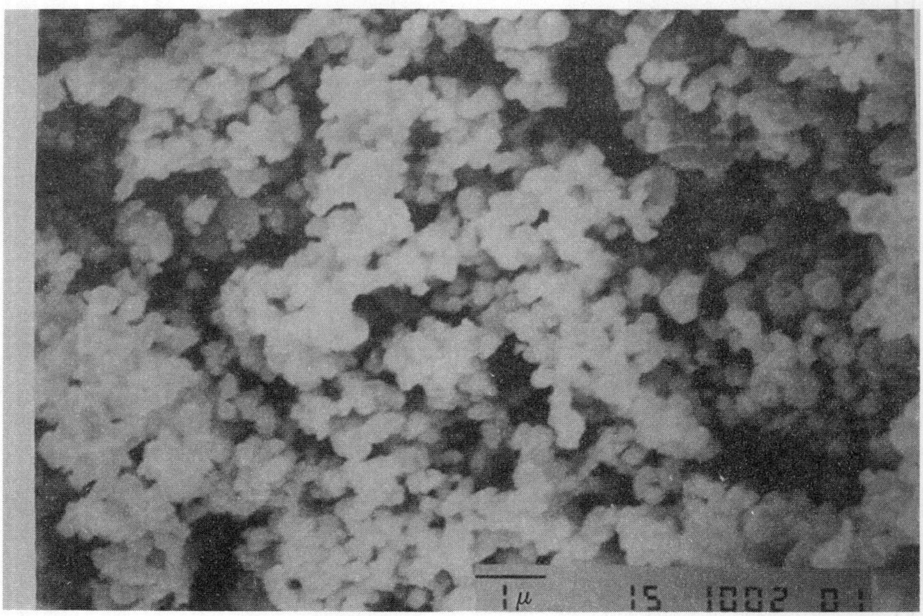

Fig. 7 SEM micrograph of Pt/Fe powder resulting from Pt/mesitylene and Fe/cicloolefins upon warming up to 300 K under hydrogen.

Experimental The prototype MVS apparatus has been constructed by Rial company, as indicated by CSMSOA (Centro di Studio del CNR per le Macromolecole Stereordinate ed Otticamente Attive) research group, following literature reports [6]. The general assembling of the various parts is shown in Fig. 8.

Fig. 8 MVS apparatus.

The pumping unit contains a Turbomolecular ALCATEL Mod. 2008 A of 10m^3/h capacity pump. A liquid nitrogen trap is fitted above the turbo pump, giving an ultimate vacuum of 10^{-5}-10^{-6} torr, and onto this is bolted an electromagnetic valve. The thermocouple RIAL mod. DTH controls the primary vacuum and allows the turbomolecular pump to start when the vacuum is better than 10^{-2} torr.

The liquid nitrogen trap serves as protection both for the pumping unit against substrate vapours, and for the reaction vessel against eventual diffusion of oil from the pumps. The electric panels consists of a general control panel, a penning gauge, the turbo control switches, and the two Variac of the power supplies with two amperometers to measure the current. The reaction vessel consists of a thick walled 5 l round bottomed flask of 140 mm diameter and 5 mm thickness. A stainless steel dewar of 10 l capacity is lifted by a motor along an endless screw. The flask is attached to the head of the reactor by a flange LV 1509 with an "O" ring which make the joint vacuum tight.

The reaction vessel and the evaporating system is schematically shown in Fig. 9.

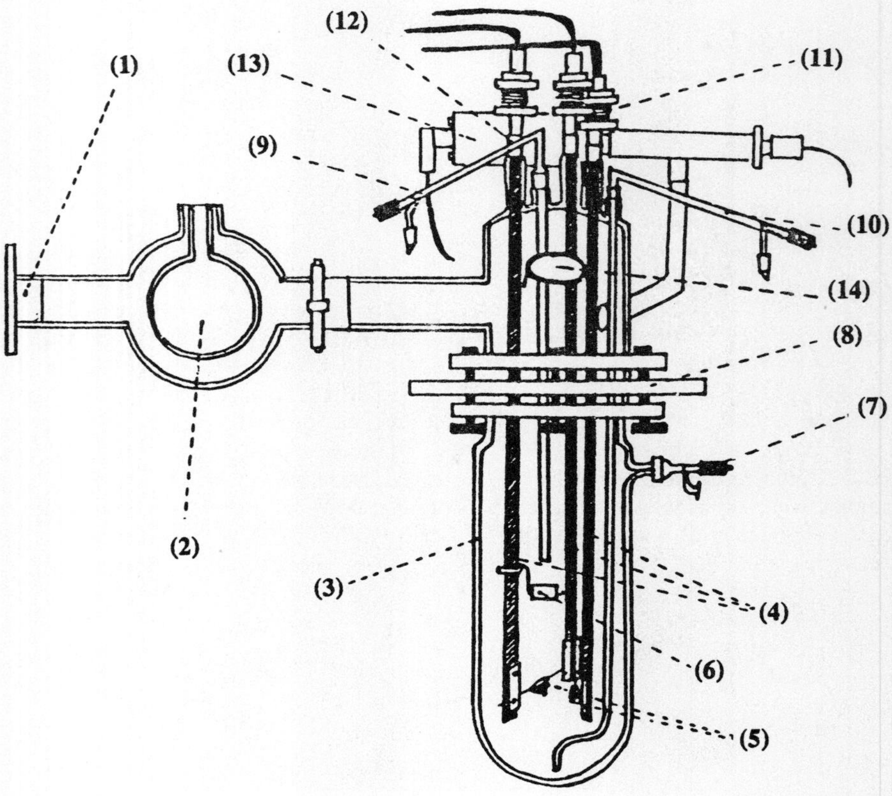

(1) metal-glass seal for the connection with the pumping unit; (2) liquid nitrogen trap; (3) thick glass reaction vessel; (4) copper electrodes; (5) alumina-tungsten crucibles; (6) glass ring; (7) rotaflo tap and ball-and-socket joint for the connection to the vacuum manifold; (8) alumina flanges connecting the reaction vessel to his head; (9) glass tube for the introduction of the co-reactant vapours; (10) glass syphon for the products recovery; (11) steel bellows regulating the slope and height of the electrodes; (12) stainless steel flanges with Viton "O" ring; (13) Pirani and Penning Gauges; (14) optical pyrex window.

Fig. 9 MVS apparatus, scheme of the reaction vessel and of the evaporating system.

Three copper self-cooling electrodes of 15 mm diameter are screwed in metal joints which are sealed with glass and stainless steel flanges with Viton "O" ring (12). Steel bellows (11) allow to regulate the slope of the electrodes and their height. Two different crucibles (5) can be fixed between each couple of electrodes with steel plates screwed on their ends. By means of a 4 ways transformer it is possible to have three different maximum currents passing through each crucibles: 0-400 with 5 V, 0-200 with 10 V, 0-100 with 20 V, for a maximum power of 2 Kw.

A glass ring (6), suspended under the ligand inlet tube and above the crucibles, prevents the co-reactant to drop onto the hot crucibles and allows the operator to see the beginning of the vaporization of the metal from the darkening of the glass. The ligand inlet system consists of a glass tube (9), fitted with a high vacuum rotaflo tap and the joint for the connection to the ligand flask. The products outlet system is a glass syphon which touches the bottom of the reaction vessel and it is fitted through a Rotaflo tap to a joint for the connection to the Schlenk tube in which the reaction product is recovered.

At the back the vessel is connected by a glass tube to the Pirani and Penning gauges (13) for the measurement of the pressure up to 10^{-3} mbar to 10^{-6} mbar, respectively. The reaction vessel is connected through a teflon "O" ring, a Rotaflo high refilling of the vessel at the end of the experiment.

An optical pyrex disk (14) of 50 mm diameter has been sealed on the front wall of the reaction vessel to allow a pyrometer to be focused on the crucibles, and their temperature to be measured. In a typical experiment Pt vapour, generated by resistive heating of a W wire surface coated with electrodeposited Pt (100-300 mg), was cocondensed at liquid nitrogen temperature with toluene or mesitylene (40 ml) in the glass reaction chamber, in $\approx$20 minutes. After the reaction, the reactor chamber was warmed at the melting point of the solid matrix and the resulting yellow-brown solution siphoned at low temperature in a Schlenk tube and allowed to decompose at room temperature, under argon or hydrogen, to give Pt metal. The electron micrographs were obtained with a Jeol 2000EX high resolution, transmission electron microscope, equipped with a top entry stage. The samples were ultrasonically dispersed in organic solvent (mesitylene or toluene) and then deposited on copper grids covered by an holey carbon film. SEM analyses have been performed using a Cambridge-Oxford 440S electron microscope.

Acknowledgement

This work was supported by the Commission of the European Community under the contract Brite-Euram n. BREU.CT91-0406 and by the Consiglio Nazionale delle Ricerche, "Progetto Strategico: Tecnologie Chimiche Innovative".

References

[1] C. Hayashi, Physics Today **40**, 44 (1987); G. Schmid, Chem. Rev. **92**, 1709 (1992); G. Schmid, H. Harms, J.-O. Malm, J.-O. Bovin, J. van Ruitenbeck, H.W. Zandbergen and W.T. Fu, J. Am. Chem. Soc. **115**, 2046 (1993) and references therein; C. Amiens, D. De Caro, B. Chaudret, J.S. Bradley, R. Mazel and C. Roncan, J. Am. Chem. Soc. **115**, 11638 (1993) and references therein.

[2] M.P. Andrews, G.A. Ozin, Chem. Mater. **1**, 174 (1989); K.J. Klabunde, Y.Y.X. Li, B.J. Tan, Chem. Mater. **3**, 30 (1991).

[3] See as examples: B.A. Scott, R.M. Plecenik, G.S. Cargill III, T.R. Mc Guire and S.R. Herd, Inorg. Chem. **19**, 1252 (1980) and references therein; C.F. Kernizan, K.J. Klabunde, C.M. Sorensen and G.C. Hadjipanayis, Chem. Mat. **2**, 70 (1990) and references therein; S. Bertozzi, N. Campigli, G. Vitulli, R. Lazzaroni and P. Salvadori, J. Organomet. Chem. **487**, 41 (1995) and references therein.

[4] C. Polizzi, A.M. Caporusso, G. Vitulli and P. Salvadori, J. Organomet. Chem. **451**, C4 (1993); G. Vitulli, R. Falorni, P. Salvadori, A. Parmaliana, F. Frusteri and N. Giordano, Catal. Lett. **17**, 151 (1993); G. Vitulli, E. Pitzalis, P. Salvadori, G. Capannelli, O. Monticelli, A. Servida and A. Julbe, Catal. Today, in press.

[5] P.H. Maitlis, Chem. Soc. Rev. **10**, 1 (1981).

[6] J.R. Blackborows and D. Young, Metal Vapour Synthesis in Organometallic Chemistry, Ed. Springler-Verlag, Berlin (1979).

Materials Science Forum Vol. 195 (1995) pp. 103-110

Engineering Ordered Mesoporous Solids in the Nineties

A. De Stefanis, S. Foglia and A.A.G. Tomlinson

Institute of Materials Chemistry, Rome Research Area, C.N.R., CP 10 Monterotondo Staz., I-00016 Rome, Italy

Keywords: Mesoporous Solids, PILCs, MCMs, Surfactants, Liquid Crystals, Separations, Catalysis

Recent advances in preparing ordered mesoporous solids, especially pillared clays (PILCs) and other layered materials and MCM-type zeolites are described, including prospects for 'engineering' such solids with its relevance for new molecular sieve and catalytic materials for industry. Some problems of correlating structure and porosity in mesoporous materials are discussed, including overcoming well-known limitations of the classical BET equation, *via* catalytic shape selection indexes and scanning probe microscopies.

The explosion in interest in porous materials over the last five years has been particularly centred on methods for inducing ordered porosity in inorganic solids. Classical alkylammonium templates for producing mid-pore zeolites have been extended to liquid crystal systems, giving rise to new mesoporous, MCM, zeolites [1]. Simultaneous advances in restructuring smectite clays, especially using pillaring techniques [2] are leading to more rationally engineered clays. This short overview puts some of the issues raised into focus, and especially their impact on perceptions of porosity.

Mesoporous Zeolites

Mobil Oil workers exploited liquid crystal ordering by long-chain surfactants $C_nH_{2n+1}NR_3^+$ (n = 8-16, R = Me) [3], to produce mesoporous zeolites in which the pores can be tailored to have diameters between 16 - >100 Å and BET (N_2) surface areas > 700 m^2g^{-1}, as shown schematically in Fig. 1.

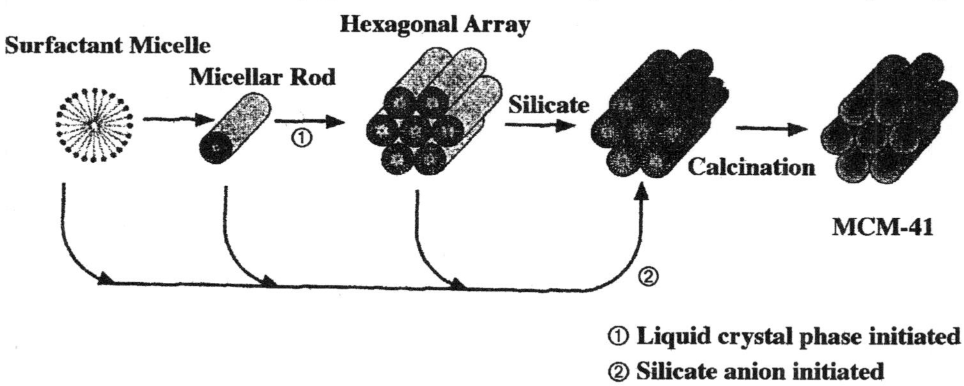

Fig. 1. Suggested templating mechanism for mesoporous MCM zeolites.

Very mild hydrothermal preparation conditions (70-150°C) and short reaction times for MCM-41 and MS41S zeolites, has meant that sorption, catalysis and structural studies followed rapidly [4] (although industrial applications lag). An exception is MCM-22, instead prepared from hexamethylenimine and triethanolamine as templates, and requiring long (280 days) hydrothermal templating [5], which provides a good example of interplay between probing of porosity and structure for such materials. The pore properties were originally scoped *via* a catalytic shape selection index: the decane test [6], which suggested the presence of a double-channel structure. For the ill-defined XRD of MCM-22, established Rietveld methods can be ambiguous, so the structure was solved by HREM and model building with verification *via* high resolution synchrotron radiation XRD [5]. As seen in Fig. 2, the structure indeed consists of a double channel arrangement, giving

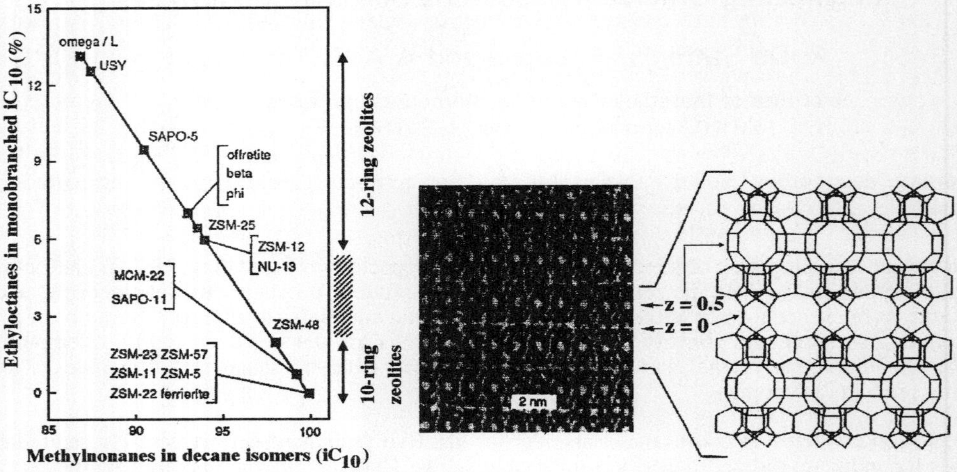

Decane test probing (at 5% conversion) [6]

HREM *hOl* lattice image with model structure [5]

Fig.2. The 2-Channel structure of MCM-22.

rise to a pore topology reminiscent of that in dodecasil-1H. MCMs are rapidly being derivatised, titanium and iron having already been substituted into the lattice and comparisons with classical xylene transformations are well advanced [4]. The hexagonal transformation *via* metastable kanemite structures suggests that analogous non-aluminosilicate M41S are accessible.

Pillared Layered Structures

Pillared clays (PILCs) were originally investigated as porous catalysts stable to high temperature steam treatment in FCC catalysis after the substitution of acidified clays by zeolites in the petrochemicals industry [2].

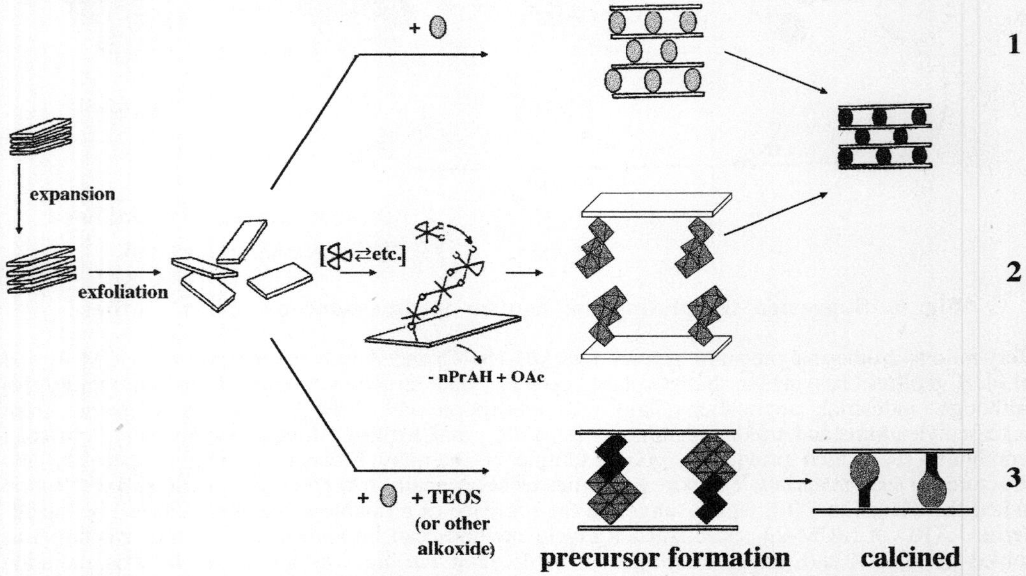

Fig. 3. Current Methods for producing Oxide-Pillared Layered Solids.

They are formed by insertion of polyhydroxyoxy aluminium cationic species $[Al_{13}O_4(H_2O)_{12}(OH)_{24}]^{7+}$ (Keggin ion), inside the layers of a smectite, followed by calcination to give cross-linking, as shown very schematically in Fig. 3. Subsequently, alumina-pillared rectorite PILCs were indeed found to be excellent FCC catalysts, although low availability of rectorite has impeded industrial use to date [7], so a need still exists for a substitute. The aqueous chemistry of $[Al_{13}O_4(H_2O)_{12}(OH)_{24}]^{7+}$ (a very simplistic formulation of the species present) was then clarified *via* mas nmr spectroscopy [8], and a start made in following complex changes on surfaces, although much remains to be done in this field. Instead, the current literature boom in these materials is due to the realisation that the pore system can be at least partially engineered, and utilised for much softer reactions. (PLS have long been considered to have supermesh ordering, i.e. packet supermolecular structures with variable pore shapes and accessibilities [9], a consequence of incontrollable ordering of colloidal platelets [10]). Recent advances now mean that - though not yet as tailorable as MCMs - PLS do in fact contain well-defined pore systems, as in an industrial application exploiting pore acidity: the production of glycol ethers from ethylene oxide + ethanol on BP-PILC (an alumina-pillared montmorillonite) :

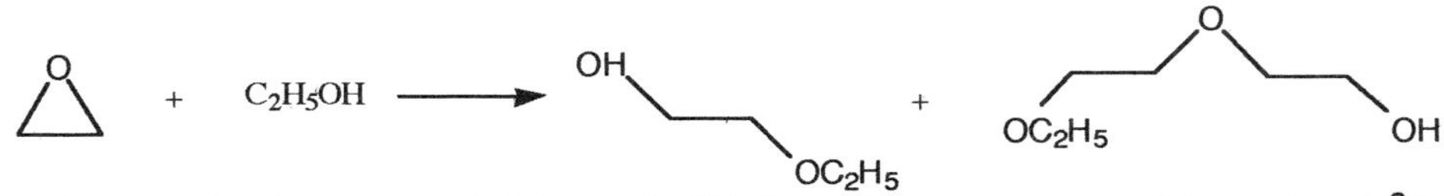

Although conversion is 100% at 90-93°C on the H^+ form, it decreases to < 65% on the Mg^{2+}-form [11].
Sorption-separation and catalysis on other montmorillonite PILCs and mixed montmorillonite-beidellite alumina pillared analogues, denominated AZA and FAZA (derived from Greek Zenith clays) confirm the presence of ordering even though the materials are derived from traditional path 1 of Fig. 3. For example, air-gas separations are feasible and change systematically with simple pore-blocking *via* cation-exchange (see Fig. 4) [12]. Even large hydrocarbons and complex five-component mixtures can be separated with efficiencies approaching commercial organo-smectites (bentone 34) (smectites as such do not permit such separations) [13].

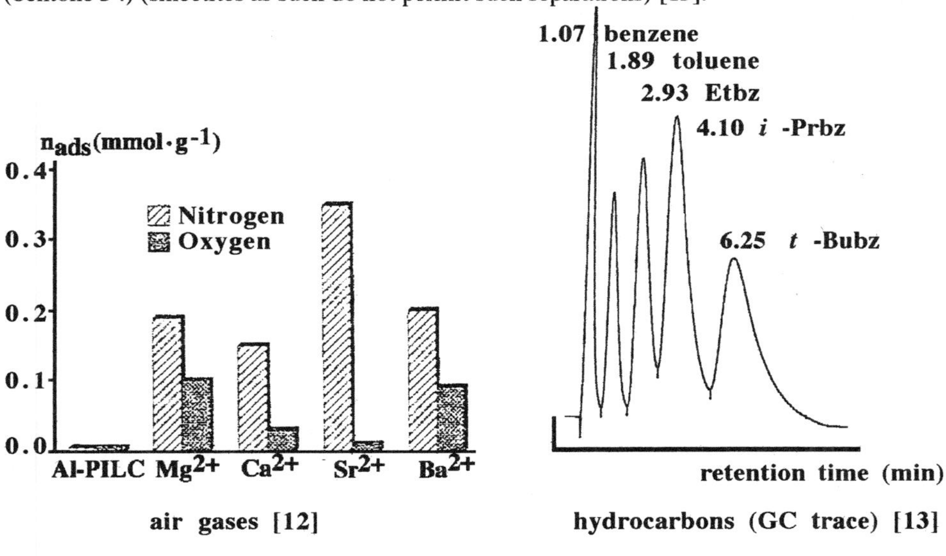

Fig. 4. Some separations on PILCs.

Further evidence for ordered pillaring is provided by the role of AZA and FAZA as selective Lewis acid reaction vessels in terpene transformations [14]. However, the clearest evidence to date is that observed in the dehydrogenation of ethylbenzene to styrene, which occurs with high selectivity on cation-exchanged FAZA (Co^{2+}, Cu^{2+}, Ni^{2+}, Ca^{2+}, K^+), K^+-FAZA being a particularly selective catalyst for this reaction, showing virtually no benzene by-product [15]. Also shown in Fig. 5 is a possible active site for K^+-FAZA, i.e. formation of $KFeO_2$ on the surface of "pancake" Fe^{2+}/Fe^{3+}-alumina pillars present in this PILC (and which may be analogous to the active site suggested to be present in commercial K^+-α-Fe_2O_3/alumina catalysts).

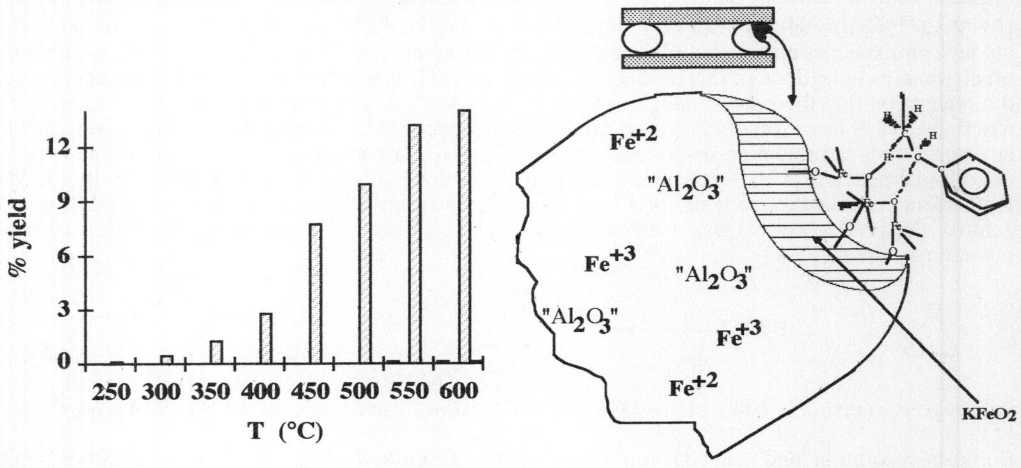

Fig. 5. Ethylbenzene ---> styrene conversion on K^+-FAZA (suggested active site at right)

Surface hydrolysis of a metal acetate at controlled pH - the second method of Fig. 3 - was first utilised for preparing chromia-pillared materials of group IV layered α-phosphates (α-$M(HPO_4)_2 \cdot nH_2O$, M = Zr, Sn, colloidalised by dispersion *via* *n*-$PrNH_2$) [16]. It has subsequently been utilised generally, not only for phosphates and smectites but also for layered oxides such as niobates, etc. The original chromia-pillared phosphate materials provided the first examples of pillared structures (via EXAFS spectroscopy) demonstrating that the technique provides more ribbon-like pillars, and formation of the amorphous phase (at 800°C) occurs with gross rearrangement to corner-shared octahedra (very different from chromia itself in both cases) [17], as seen in Fig. 6.

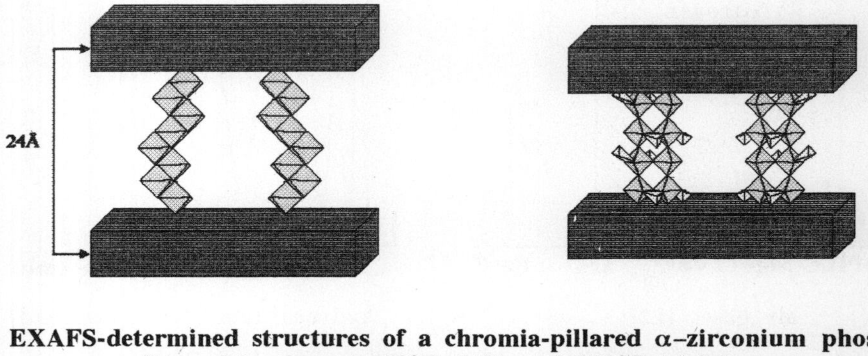

Fig. 6. EXAFS-determined structures of a chromia-pillared α-zirconium phosphate, after calcination at 400°C (left) and 800°C (right).

A newer extension, combining wet pillaring methods and sol-gel, path 3 in Fig. 3, has provided mixed Al/Si oxide pillared montmorillonites. The inorganic chemical literature on alkoxides is so vast that this method will be of wide utility in the near future. Already, variants of this method are leading to emergence of homologies with MCM topologies. For example, using micelles from long-chain amine surfactants (in place of n-$PrNH_2$) in method 2 can also transmit ordering *via* templating. This has been done for silica pillars [18], but should be also applicable for a variety of micelle structures. Less developed are PLS derived from oxides and sulphides, mainly Keggin ion solid precursors and their surface dynamics *via* mas nmr, have been reported to date. However, as shown in Fig.7, neutral MoS_2 can indeed give alumina precursors *via* path 1, although it appears that pH-control is required to obtain more ordered solids [10].

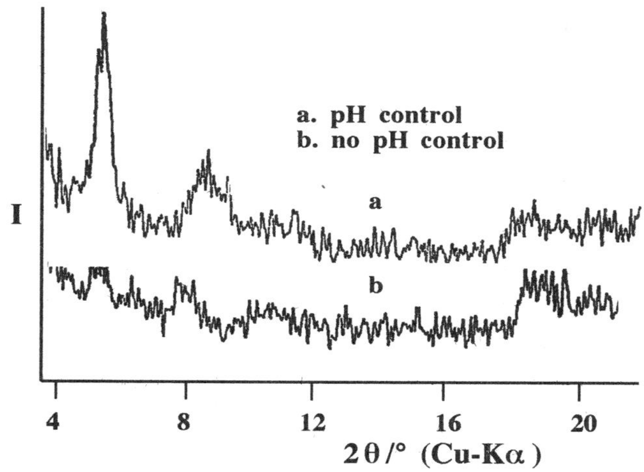

Fig. 7. XRD of Al_{13}-Keggin ion MoS_2 pillar precursor

Porosity Assessment

The above advances have highlighted problems in assessing porosity, which are acute for PILCs (and pillared layered structures - PLS - in general) not only in the mesoporosity range, but also as regards microporosity (i.e. < 20Å). In his magisterial overview, reformulating multilayer physical adsorption in fractal terms for application to rough surfaces, Fripiat underlined the limitations of the classical "α" form of the BET equation [19] : i.e. neglect of lateral interactions and assumption that the first adsorbate layer alone has an adsorption energy different from the heat of condensation of the liquid adsorbate. Recall that in the basic BET equation :

$$Ns/Nm = C\sum_{i=1}^{n} ix^i \Big/ 1 + \sum_{i=1}^{n} x^i$$

*N*s = no. of adsorbed molecules,
*N*m = no. of molecules necessary to cover surface with a monolayer

n = α, giving :

$$Ns/Nm = Cx/(1 - x)[1 + (C-1)x]$$

Further, the original EB proposal rests on space requirements based on properties of 3-dimensional condensed N_2 phases *via*

$$\alpha = \phi(M/\rho N)^{2/3} \quad (\alpha = \text{area of } N_2 \text{ molecule})$$

where : ρ = density of 3-dimensional reference phase
M = molecular weight
N= Avogadro's No.

The calculation was performed using the density of liq. N_2, i.e. f = 1.091, giving α = (N_2) = 16.2 $Å^2$, i.e. α depends on the density of liq. N_2 and ultimately on the coordination number. Subsequently, in a landmark paper, Jelinek *et al* experimentally demonstrated that the assumption of a straight herringbone arrangement for N_2 is flawed [20]. Utilising nanochemical techniques, these workers found a dense layer is obtained on silica, giving α = 13.5 $Å^2$, i.e. the coordination number is lower. Further impact on solving porosity problems in this field will probably also derive from nanochemical techniques, e.g. scanning probe methods - especially AFM - backed up by density function theory and simulation. The feasibility of using AFM for adsorption was shown some time ago in studies of dye sorption on mica surfaces and extended to other oxidic surfaces [21]. Further, it has already been demonstrated that pore sizes in pillared silicons can be directly measured by AFM [22], an example being shown in Fig. 8.
It seems only a short step to observing - in real time - the interaction of chemical moieties in such pores.

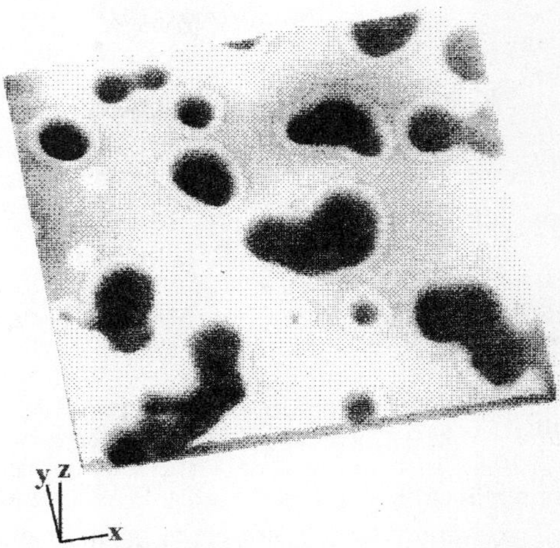

Fig. 8. Measurement of pores *via* AFM (⊥ cut of a pillared Si(111) film).

Acknowledgements
We thank the E.C. for financial support through BRE2-CT93-0450, and Drs.G.Perez and M.P.Atkins for encouragement.

References

[1] C.T. Kresge, M.E. Leonowicz, W.J. Roth, and J.C. Vartuli, USP, No. 5,098,684 (1992); 5,102,643 (1992).

[2] M.A. Occelli and H. Robson, *Expanded Clays and Other Microporous Solids,* Academic Press, New York, (1992); *Pillared Layered Structures, Current Trends and Applications,* I.V. Mitchell ed., Elsevier, Amsterdam, (1990).

[3] C.T. Kresge, M.E. Leonowicz, W.J. Roth, J.C. Vartuli, and R. Beck, *Nature*, **359**, 410 (1992); J.C. Vartuli, K.D. Schmitt, C.T. Kresge, W.J. Roth, M.E. Leonowicz, et al., *Chem. Mater.*, **6**, 2317 (1994), and refs therein.

[4] e.g. see : A. Corma, *Chem. Revs.*, **95**, 559 (1995); A. Sayari, C. Danumah and I.L. Moudrakovski, *Chem. Mater.*, **7**, 814, (1995).

[5] M.E. Leonowicz, J.A. Lawton, S.L. Lawton, and M.K. Rubin, *Science*, **264**, 1910 (1994).

[6] W. Souverijns, W. Verrelst, G. Vanbutsele, J.A. Martens and P.A. Jacobs, *Chem. Commun.*, 1671 (1994).

[7] M.A. Occelli, in *Expanded Clays and Other Microporous Solids,* M.A. Occelli and H. Robson eds., Academic Press, New York, (1992) .

[8] R.A. Schoonheydt, private communication.

[9] F. Bergaya, L. Gatineau, H. Van Damme, in *Multifunctional Mesoporous Inorganic Solids,* NATO ASI Ser. C, p. 19-26, Vol 400 (1993).

[10] S. Mulley, S. Foglia, A. Sironi, and A.A.G. Tomlinson, *J. Mater. Chem.*, in press, (1995); S. Foglia and A.A.G. Tomlinson, in preparation.

[11] M.P. Atkins, in *Pillared Layered Structures, Current Trends and Applications,* I.V. Mitchell ed., Elsevier, Amsterdam, p. 159 (1990).

[12] N. Maes, A. Molinard, K.K. Peeters, H.Y. Zhu, I. Heylen, H. Zaïdani, E.F. Vansant, *European Symposium (CEA-PLS) on Adsorption, Separation and Environmental Applications of Pillared Layered Substrates*, Antwerp (B) 4-5th June 1993.

[13] A. De Stefanis, G. Perez, and A.A.G.Tomlinson, *J. Mater. Chem.*, **4**, 959 (1994).

[14] G. Perez, A. De Stefanis, and A.A.G. Tomlinson, *Appl. Catal.*, in press, (1995).

[15] G. Perez, M.R. Raimondo, A. De Stefanis, and A.A.G. Tomlinson, *J. Appl. Pyrolysis and Rel. Phenom.*, in press, (1995); G. Perez, A. De Stefanis, and A.A.G. Tomlinson, *Eur. Patent*, applied for, (1995); G. Perez, A. De Stefanis, and A.A.G. Tomlinson, *J. Chem. Soc. Faraday Trans.*, submitted.

[16] P.Maireles-Torres, P. Olivera-Pastor, E. Rodriguez-Castellon, A. Jimenez-Lopez, and A.A.G. Tomlinson, *J. Mater. Chem.*, **1**, 739 (1991).

[17] D.J. Jones, J. Roziere, A. Jimenez-Lopez, P.Maireles-Torres, P. Olivera-Pastor, E. Rodriguez-Castellon, and A.A.G. Tomlinson, *Inorg. Chem.*, in press, (1995).

[18] A. Gil, G. Guiu, P. Grange and M. Montes, *J. Phys. Chem.*, **99**, 301 (1995); A. Galarneau, A. Barodawalla and T.J. Pinnavaia, *Nature*, **374**, 529, (1995).

[19] J.J. Fripiat, in *The Fractal Approach to Heterogeneous Chemistry,* ed. D. Avnir, J. Wiley p. 331, (1989), and refs therein; J.J. Fripiat, L. Gatineau, and H. Van Damme, *Langmuir*, **2**, 562 (1986).

[20] L. Jelinek and Sz. Kovats, *Langmuir*, **10**, 4226 (1994).

[21] H. Cai, A.C. Hillier, K.R. Franklin, C.C. Nunn, and M.D. Ward, *Science*, **266**, 1551 (1994).

[22] S.K. Lazarouk and A.A.G. Tomlinson, *Eur. Patent*, applied for, (1995); S.K. Lazarouk and A.A.G. Tomlinson, *J. Mater. Chem.*, submitted.

Materials Science Forum Vol. 195 (1995) pp. 111-116

The Crystallization of Al-Sm Amorphous Alloys

P. Rizzi[1], M. Baricco[1], L. Battezzati[1], P. Schumacher[2] and A.L. Greer[2]

[1] Dipartimento di Chimica Inorganica, Chimica Fisica e Chimica dei Materiali, Università di Torino, Via P. Giuria 9, I-10125 Torino, Italy

[2] Department of Metallurgy and Materials Science, University of Cambridge, Pembroke Street, CB2 3QZ Cambridge (UK)

Keywords: Amorphous Alloys, Primary Crystallization, Aluminium Nanocrystals

ABSTRACT

Amorphous alloys with composition $Al_{100-x}Sm_x$ ($8 \leq x \leq 14$) have been prepared by melt spinning in controlled atmosphere. The quenching conditions significantly influence the microstructure of as-quenched samples and their crystallization mechanisms. The crystallization was followed by means of DSC, XRD, TEM and SEM. Different mechanisms are observed as a function of composition. For low Sm content ($Al_{92}Sm_8$), a primary crystallization occurs at about 180°C, with the formation of Al nanocrystals with a grain size of about 15 nm. A polymorphic crystallization is found in $Al_{90}Sm_{10}$, with the formation of a metastable intermetallic phase at about 220°C. For higher Sm content ($Al_{88}Sm_{12}$ and $Al_{86}Sm_{14}$), a eutectic crystallization at about 230°C gives a mixture of stable (Al and $Al_{11}Sm_3$) and metastable phases.

INTRODUCTION

The synthesis of Al-rare earth amorphous alloys by rapid solidification is leading to the development of new materials with interesting mechanical properties, which can be further improved by the presence of nanosized Al precipitates [1]. Materials with such microstructures can reach a fracture strain significantly higher with respect to conventional Al-base alloys produced by precipitation hardening. The conditions for glass formation in these systems are still under discussion, as well as the description of the crystallization mechanisms by which nanophases are produced. The glass-forming ability has been reported for binary Al-RE (RE ≡ Ce, Dy, Er, Gd, Ho, La, Nd, Pr, Sm, Tb, Y, Yb) and several multicomponent systems [2-5]. Among them most remarkable is $Al_{88}Ni_9Ce_2Fe_1$ the strength of which is around three times that of crystalline precipitation-hardened alloys when its microstructure is made of fine Al particles (size up to 10 nm) embedded in an amorphous matrix [1].

In this work we examine the crystallization behaviour of the best glass-former among binary system, Al-Sm. It was earlier reported [6] that in the narrow composition range from 8 at. % Sm to 14 at. % Sm primary polymorphic and eutectic transformations occurs. These findings are reviewed and extended with consideration of the relative effect of nucleation frequency and growth rate.

EXPERIMENTAL

Master alloys of compositions $Al_{100-x}Sm_x$ ($8 \le x \le 14$) have been prepared from pure elements by arc melting in purified argon. Amorphous ribbons have been produced by melt spinning in controlled atmosphere of pure argon in two different laboratories. The thickness of these ribbons is in the range of 20-50 mm. The surface of as-quenched samples was observed by scanning electron microscopy (SEM). Identification of the structure of the samples has been made by X-ray diffraction using Co Kα wavelength (λ = 1.7902 Å). The crystallization processes were examined by differential scanning calorimeter (DSC) using both isothermal and continuous heating conditions. Transmission electron microscopy (TEM) was used to study the crystal size and distribution in partially transformed samples.

RESULTS AND DISCUSSION.

Both amorphous ($Al_{90}Sm_{10}$, $Al_{86}Sm_{14}$) and partially crystalline ($Al_{88}Sm_{12}$, $Al_{92}Sm_8$) ribbons were obtained with careful control of the operating conditions (wheel speed, crucibles, atmosphere). However, Al-Sm is a marginal glass-former since crystalline phases tend to nucleate rather easily from the melt. We have taken advantage of this to study phase formation in as-quenched samples and compare them with devitrified alloys.

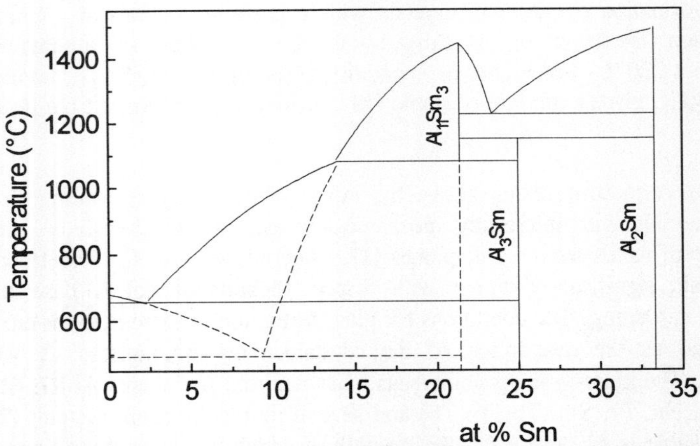

Figure 1 - Al-rich part of the Al-Sm stable phase diagram (continuous line) and metastable Al-$Al_{11}Sm_3$ eutectic (dashed lines).

XRD patterns of as quenched $Al_{88}Sm_{12}$ ribbons shows the presence of a crystalline fraction of $Al_{11}Sm_3$. As can be seen in the phase diagram of fig. 1, this intermetallic compound is stable only at high temperature but it appears to nucleate more easily than the equilibrium phase Al_3Sm. A SEM micrograph of the surface of an as-quenched $Al_{88}Sm_{12}$ sample is shown in fig. 2. This revealed the presence of $Al_{11}Sm_3$ dendrites on the air side. The crystals are non-uniformly distributed along the ribbon. They appear to have nucleated periodically at every turn of the spinning wheel possibly because of local heterogeneities. Another example of the tendency to nucleate the intermetallic $Al_{11}Sm_3$ can be found studying the $Al_{86}Sm_{14}$ amorphous alloy. In this case TEM observations showed the presence of quenched-in

Figure 2 - SEM micrograph of an $Al_{88}Sm_{12}$ as-quenched ribbon. $Al_{11}Sm_3$ dendrites are apparent on the air side.

$Al_{11}Sm_3$ nuclei. Crystallization occurs by growth of Al and $Al_{11}Sm_3$ as a eutectic in both alloys (fig. 3) [6]. Despite that as-quenched $Al_{90}Sm_{10}$ ribbons were fully amorphous to XRD, TEM analysis showed the presence of few isolated unidentified crystals. The crystallization was followed by continuous heating in DSC. The thermogram of fig. 4 shows three exothermal peaks: the first is relevant to a polymorphic transformation leading to a metastable phase that later Guo et al. [7] reported to be cubic with a=1.9154 nm. An isothermal DSC trace for the same transformation is shown in fig. 5. After reaching the isothermal condition there is a sloping trend of the signal which is attributed to surface formation of the metastable phase. Then the exothermal peak for bulk crystallization is well evident. An Arrhenius plot of the time corresponding to minima of several traces provides an activation energy value of 253 kJ/mol.

The most interesting alloy is $Al_{92}Sm_8$. XRD patterns of the as-quenched ribbons are shown in fig. 6 (curve a and b). They exhibit a broad halo with a low-intensity reflection due to a fraction of Al crystals on the air side. Continuous heating (fig. 4b) produces three peaks. The first exotherm occurs at about 180°C. Ribbons heated up to 215°C (fig. 6c) gave broad α-Al diffraction peaks indicating a small crystal size. The application of Scherrer analysis provided an average value of 15 nm. TEM analyses have been performed on samples heated up to 171°C for one hour. A dark field image (fig. 7) shows roughly spherical nanocrystals

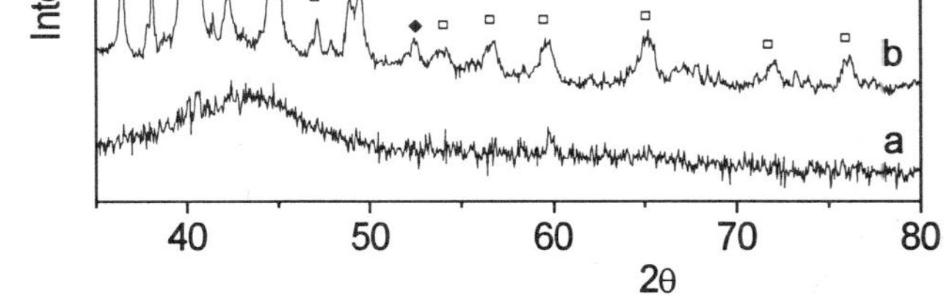

Figure 3 - XRD patterns of $Al_{86}Sm_{14}$ ribbons: (a) as-quenched; (b) heated up to 280°C.

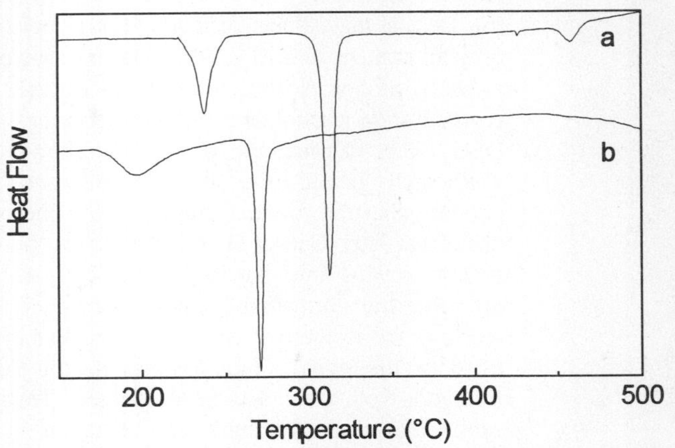

Figure 4 - Continuous heating DSC traces (40 K/min) given by rapidly solidified alloys: (a) $Al_{90}Sm_{10}$; (b) $Al_{92}Sm_8$.

disperded in the matrix. The average size is 10 nm, confirming the values obtained with the Scherrer analysis.

The first skewed calorimetric signal (fig. 4) is due to a primary crystallization of Al from the amorphous matrix. However if we consider the stable phase diagram, Al is not the primary phase at 8 at % Sm because the equilibrium eutectic composition is at about 3% at Sm; but a metastable eutectic Al-$Al_{11}Sm_3$ can be envisaged from extrapolation of liquidus lines at higher Sm content if the formation of Al_3Sm is neglected. So for sufficient undercoolings, there is a driving force for primary crystallization of Al in $Al_{92}Sm_8$ and for $Al_{11}Sm_3$ in $Al_{86}Sm_{14}$ and $Al_{88}Sm_{12}$.

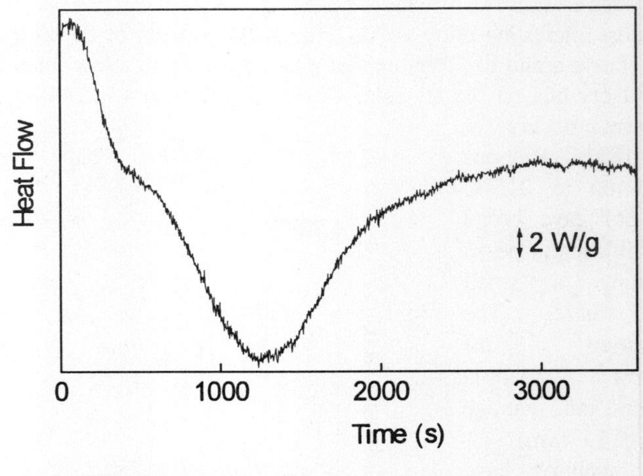

Figure 5 - Isothermal DSC traces obtained at 193°C for $Al_{90}Sm_{10}$.

A Kissinger plot for calorimetric peaks obtained at various heating rates gives an activation energy of 232 kJ/mol. This figure is close to that reported above for $Al_{90}Sm_{10}$ indicating similar kinetics for crystallization although different phases are produced. Isothermal DSC runs performed for the same transformation at several temperatures give only a sloping trace with no peak (fig. 8). This behaviour was already found for a Fe-based amorphous alloy [8] and may imply continuous

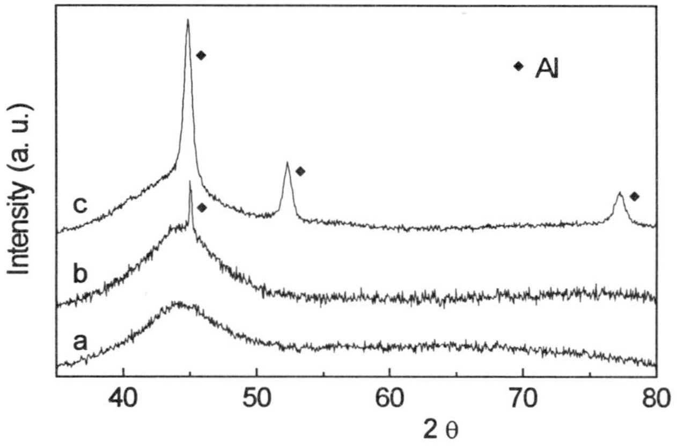

Figure 6 - XRD patterns taken on both sides of an Al92Sm8 rapidly solidified ribbon (curves a and b), showing the occurrence of Al crystals on the air side. Curve c shows the primary crystallization of Al after the first peak.

compositional readjustment between crystals and matrix or within the matrix.
Although these findings are preliminary, after a fast nucleation of Al-nanocrystals, there appear to be a hindrance to their subsequent growth. Moreover, the fact that several intermetallics both stable and metastable can nucleate in the remaining matrix may be relevant for explaining the slow crystallization kinetics.

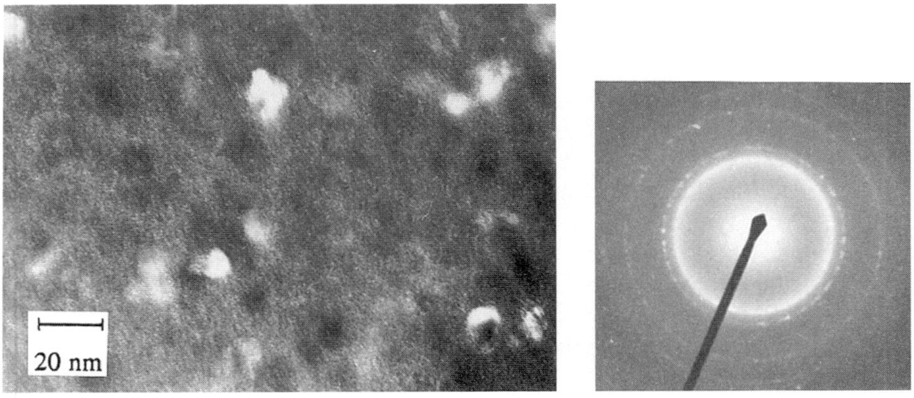

Figure 7 - TEM micrograph (dark field) taken with an $Al_{92}Sm_8$ ribbon heated at 171 °c for 1 hour. Al nanocrystals of the average size of 10 nm can be seen. The diffraction pattern shows rings due to fcc Al. The first ring is broad because of the remaining amorphous phase.

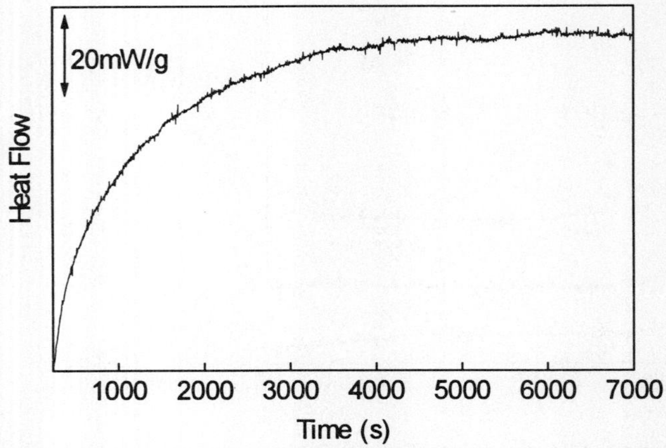

Figure 8 - Isothermal DSC traces obtained at 171 °C for $Al_{92}Sm_8$. No peak is obtained but only a sloping trace at every temperature.

ACKNOWLEDGMENTS
This work was performed within the agreement of the British Council and MURST for the British-Italian collaboration in research and higher education.

REFERENCES

[1] A.Inoue, *Mater. Sci. Eng. A*, 179/180 (1994) 57

[2] Y. He, G.M. Dougherty, G.J. Shiflet and S.J. Poon, *Acta Metall. Mater.*, 41 (1993) 337

[3] L. Battezzati, M. Baricco and C. Antonione, *J. Alloys and Comp.*, 209 (1994) 341

[4] R.F. Cochrane, P. Schumacher and A.L. Greer, *Mater. Sci. Eng. A*, 133 (1991) 367

[5] K. Ohtera, K. Kita, H. Nagahawa, A. Inoue, T. Masumoto, *Mater. Sci. Eng. A*, 179/180 (1994) 592

[6] L. Battezzati, M. Baricco, P. Schumacher, W.C. Shih and A.L. Greer, *Mater. Sci. Eng. A*, 179/180 (1994) 600

[7] J.Q. Guo, K. Ohtera, K. Kita, J. Nagahora, N.S. Kazama, A. Inoue and T. Masumoto, *The 4th International Conference on Aluminum Alloys, Atlanta, USA*, T.H. Sanders, Jr. and E.A. Starke, Jr., 1993, vol. 2, 753

[8] L.Battezzati, M.Baricco, *Phil. Mag.* 68 (1993) 813

Materials Science Forum Vol. 195 (1995) pp. 117-122

Morphological Characterization of α-Fe Nanostructures in Eutectic Spherulites

A. Deriu[1], F. Malizia[2], F. Ronconi[2] and M. Vittori-Antisari[3]

[1] INFM-Dipartimento di Fisica, Universtità di Parma, Viale delle Scienze, I-43100 Parma, Italy

[2] INFM-Dipartimento di Fisica, Università di Ferrara, Via Paradiso 12, I-44100 Ferrara, Italy

[3] ENEA, Settori Nuovi Materiali, C.P. 2400, I-00100 Roma, Italy

Keywords: Crystallization, Morphological Characterization, Small Angle Neutron Scattering, Transmission Electron Microscopy

Abstract

The evolution of the microstructure of the crystalline phases formed at the various stages of the $Fe_{80}B_{20}$ amorphous alloy crystallization process has been investigated by small angle neutron scattering and transmission electron microscopy. Transmission electron microscopy measurements indicate that the crystallization of $Fe_{80}B_{20}$ amorphous ribbons occurs by nucleation and growth of Fe_3B spherulites which include small finely divided α-Fe acicular crystallites. It is shown that the small angle neutron scattering signal is due to the contrast between α-Fe particles and the Fe_3B matrix in which they are embedded and can be interpreted in terms of two families of roughly cylindrical homogeneous particles with different characteristic dimensions. These families can be associated to individual α-Fe dendritic structure and to the clusters formed by associations of such dendrites.

Introduction

Iron-based metallic glasses, produced by melt spinning technique have been the object of much scientific interest in recent years. This interest is mostly due to the fact that their soft magnetic properties (such as high permeability and low coercitive force) are important for technological applications. These properties are to a large extent modified during even a partial crystallization of the metallic glasses; for this reason, the crystallization process of these systems has been studied in detail. For several years we have been involved in a systematic study of soft magnetic amorphous metal-metalloid (Fe-B) alloy with the aim of investigating the isothermal crystallization process in this glass. To reach this purpose we have used several techniques such as differential scanning calorimetry (DSC) [1,2], Mössbauer spectroscopy and SQUID magnetometry [3,7], transmission electron microscopy (TEM) and small angle neutron scattering (SANS) [4,5].

It is well established that the crystallization of $Fe_{80}B_{20}$ glass occurs by an eutectic reaction and gives rise to two crystalline phases: α-Fe and the metastable boride Fe_3B [6].

The kinetic of the crystallization of $Fe_{80}B_{20}$ has been examined by differential scanning calorimetry (DSC); this investigation has revealed that full isothermal crystallization analysis of amorphous $Fe_{80}B_{20}$ alloy must include not only a crystal nucleation-and-growth process but also the coarsening process, and that these two processes are well separated in time during isothermal annealing.

The first morphological characterization of the obtained crystalline phases has been carried out by us using TEM on a series of samples with different degree of crystallization obtained submitting the amorphous ribbons to isothermal treatments at 693 K for different times. Our results reveal that the crystallization occurs by nucleation and growth of spherulites of Fe_3B which include finely divided small α-Fe crystallites. The spherulites have an ellipsoidal shape and are randomly packed; their average dimension is about 0.7 μm and they show an internal columnar structure due to the presence of α-Fe crystalline phase.

We have used the Small Angle Neutron Scattering technique in order to characterize from the structural point of view the obtained $Fe_{80}B_{20}$ samples with different degree of crystallization, and to get detailed informations on the size, shape and packing of the α-Fe acicular particles embedded in the Fe_3B spherulites.

Experimental

The $Fe_{80}B_{20}$ metallic glass samples were kindly supplied by Dr. Franco Vinai from Istituto Elettrotecnico Nazionale G. Ferraris, Torino (Italy). They have been prepared by melt-spinning technique with constant quenching rate; they were ribbons with thickness of about 20 μm and width of 5 mm. In order to obtain specimens with different degrees of crystallization, amorphous samples of $Fe_{80}B_{20}$ were subjected to isothermal treatments at a temperature of 693 K for different times ranging from 120 to 2100 s.

The thermal treatments have been optimized on the basis of previous accurate studies of the kinetics of isothermal crystallization in the same materials [1,2]. The isothermal annealings were carried out using a Perkin-Elmer Differential Scanning Calorimeter DSC-7.

TEM observations on partially crystallized samples were carried out using a Jeol 4000 FX electron microscope with a maximum accelerating voltage of 400 KV. However, for this work, the microscope was operated at a rather lower voltage to avoid any possible radiation damage of the specimens.

The SANS measurements were carried out on the LOQ diffractometer at the ISIS pulsed neutron source (Rutherford Appleton Laboratory, Didcot U.K.) and on the PACE diffractometer at the Laboratoire Lèon Brillouin (CEA-Saclay, France).

LOQ utilizes a white beam of neutrons with wavelength from 2 to 10 Å to provide a simultaneous momentum transfer, Q, interval from $7.5 \cdot 10^{-3}$ Å^{-1} up to $2.0 \cdot 10^{-1}$ Å^{-1} .

The PACE diffractometer is a reactor based instrument and was operated at two wavelenghts, 13.9 Å and 6.28 Å, and at two sample-to-detector distances, 4.6 m and 2.0 m, to provide a combinated Q-range from $4.0 \cdot 10^{-3}$ Å^{-1} to $1.5 \cdot 10^{-1}$ Å^{-1} .

All the measurements were performed at room temperature. The samples were contained in thin glass cells which have a very high transmission ($\geq$ 95 %) at all the neutron wavelengths employed. The scattering data were corrected for background scattering and the sample transmissions using various instrument-specific procedures and converted to absolute units using a calibrated standard scatterer consisting of a solid blend of hydrogenous and perdeuterated polystyrenes.

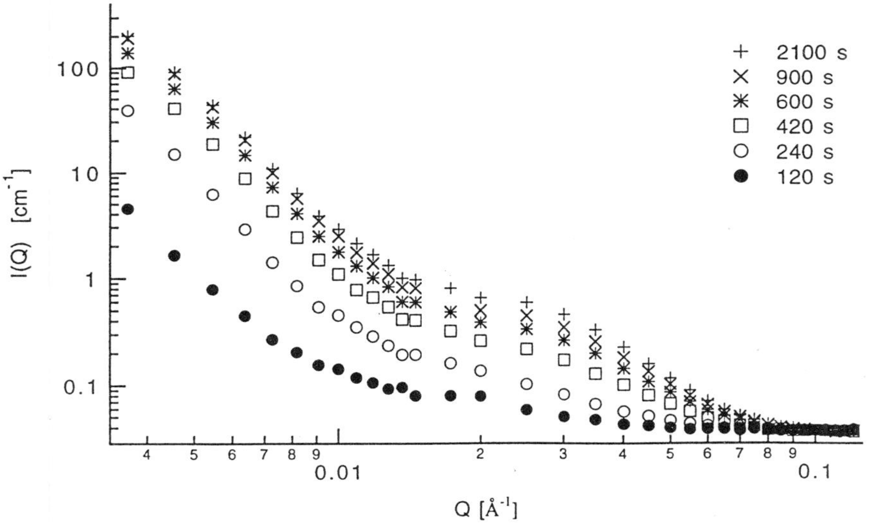

Fig. 1 : Scattering intensities $I(Q)$ for $Fe_{80}B_{20}$ samples annealed for various time at 693 K.

Results and discussion

Fig. 1 shows the radially averaged scattering intensities of the samples annealed at 693 K for different time ranging from 120 to 2100 s. A marked increase of the intensity with increasing the annealing time is evident in the low-Q region. This effect can be attributed to the progressive crystallization of the amorphous sample and to the nucleation and growth of eutetic spherulites.

A quantitative analysis of the obtained SANS profiles has been carried out by fitting the experimental intensities to a model derived from the morphological informations provided by TEM observations. The considered model includes: i) an incoherent scattering contribution I_0, ii) a form factor $F(Q)$ for elongated α-Fe particles, iii) a structure factor $S(Q)$ for interparticle correlations.

The SANS data has been fitted to above model with a standard χ^2 minimization algorithm to yield the parameters characterizing the scattering objects.

In Fig. 2 the experimental data for the sample annealed for 900 s (full circle) and the best fit of our model (line) are shown. All the experimental data could be fitted with the same model considering two families of cylindrical shaped α-Fe particles with different characteristic dimensions.

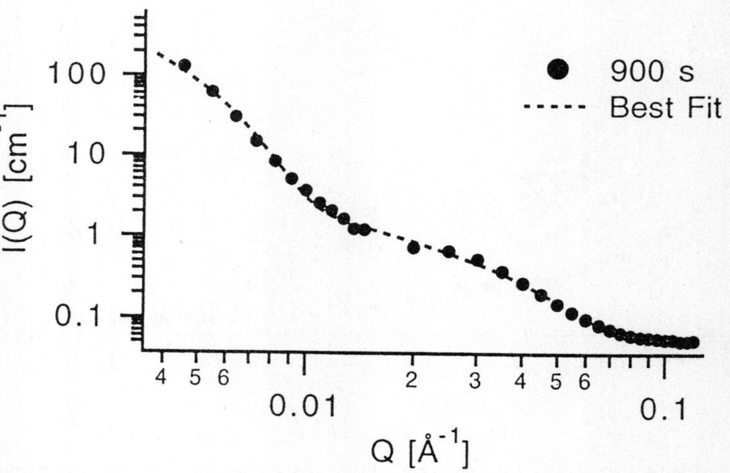

Fig. 2: Experimental SANS data (full circle) and the result of the fitting (line) for $Fe_{80}B_{20}$ sample treated at 693 K for 900 s.

Table 1: R_1 and R_2 indicate the section-radii of the cylindrical particles while α_1 and α_2 are the coefficient proportional to the number density of the object belonging to these different families.

Annealing time [s]	α_1 [$Å^{-2}$]	R_1 [Å]	α_2 [$Å^{-2}$]	R_2 [Å]
2100	0.02	48.5	1.35	450.9
900	0.02	46.5	1.34	442.0
600	0.017	46.3	1.15	444.3
420	0.014	47.0	1.00	455.4
240	0.0098	53.1	0.694	470.1

In Table 1 the section-radii of the cylinder, R_i, and the coefficients proportional to the population of the two families of cylindrical particles, α_i, as a function of annealing times are listed. We can note that both R_1 and R_2 are almost independent on the annealing time, while the coefficients α_i increase with increasing the annealing time. From the obtained values and considering our assumptions is was possible to calculate the ratio between the number of scattering objects belonging to each family per unit volume N_1/N_2 . We found a costant value of N_1/N_2 for all the annealing times. The constancy of R_1, R_2 and N_1/N_2 can be explained as a conseguence of a very high growth velocity of nucleated particles as expected for eutetic reaction; thus, even at the shortest annealing time, the nucleated spherulites are already full grown and have an internal structure very close to that of the fully grown ones. This conclusion is supported from TEM micrographies of the same samples [5].

Conclusions

In conclusion our study of partially crystallized $Fe_{80}B_{20}$ samples shows that the SANS signal is due to the contrast between α-Fe elongated particles and the Fe_3B matrix in which they are embedded. SANS data can be interpreted in terms of two families of roughly cylindrical homogeneous particles with different section-radii R_i and lenghts $h_i >> R_i$. These families can be associated to individual α-Fe dendritic structure and to clusters formed by associations of such dendrites.

Acknowledgements

The authors acknowledge financial support from the European Union throught its Human Capital and Mobility Programme for Large Scale Facility.

References

[1] F.Malizia and F.Ronconi, Phil. Mag. B **68**, 869 (1993).

[2] F.Malizia and F.Ronconi, J. Appl. Phys. **74**, 6988 (1993).

[3] F.Malizia and F.Ronconi, Mat. Res. Soc. Sym. Proc. **321**, 355 (1994).

[4] A.Deriu, F.Malizia, F.Ronconi, S.M.King, (in preparation).

[5] A.Deriu, F.Malizia, F.Ronconi, M.Vittori-Antisari, S.M.King, (submitted to J. Appl. Phys.).

[6] A.L. Greer, Acta Metall. **30**, 171 (1982).

[7] D.Fiorani, F.Malizia, F.Ronconi, A.M.Testa, J. Magn. Magn. Magn. **133** 295 (1994).

Materials Science Forum Vol. 195 (1995) pp. 123-126

Copper-Selenide Nanoparticles and Cu_2Se Bulk Phase Studied by X-Ray Photoemission Spectroscopy

D. van der Putten[1], D. Olevano[1], R. Zanoni[1], H. Krautscheid[2] and D. Fenske[2]

[1] Dipartimento di Chimica, Università degli Studi di Roma „La Sapienza", p.le A. Moro 5, I-00185 Rome, Italy

[2] Institut für Anorganische Chemie der Universität Karlsruhe, Engesserstrasse, D-76128 Karlsruhe, Germany

Keywords: XPS, Photoemission, Nanoparticles, Molecular Cluster Compounds, Cu_2Se

Abstract: XPS spectra of core levels, valence bands and Auger transitions are reported on the molecular cluster compound $Cu_{146}Se_{73}(PPh_3)_{30}$ and on Cu_2Se bulk phase obtained from the cluster after heat treatment. The photoemission features of both species are reported and commented in the light of a cluster-to-bulk transition.

Keywords: XPS, photoemission, nanoparticles, molecular cluster compounds, Cu_2Se

1. INTRODUCTION

Molecular cluster compounds [1, 2] are very well suited to study the chemical and physical properties of nano-sized particles. The compounds do not exhibit a size distribution and the insulating nature of the organic ligands prevents that experimental observations are obscured by unwanted and uncontrolled interactions with a substrate. There are, however, a number of unresolved problems associated with this relatively new class of small metal particles. First, the long-term stability and sensitivity to air exposure is largely unexplored. Second, the nature of the chemical bond of the core atoms directly bound to the organic ligands is a matter of dispute. In this paper we present XPS experimental data showing the transition from well defined molecular clusters to a bulk phase. This transition does not alter the 2:1 Cu/Se stoichiometry, as deduced from XPS relative quantitative data, thus resulting in an ideal case to study the electronic structure modification related to a nucleation process.

2. EXPERIMENTAL RESULTS

The compound $Cu_{146}Se_{73}(PPh_3)_{30}$ (to be named hereafter Cu_{146}) consists of molecules with a trigonal prismatic shape with an edge length of about 40 Å and a 20 Å thickness. The copper-selenide core has a maximum width of 26.7 Å and a thickness of 8.7 Å. The 73 Se atoms are organized into 3 planar layers, with 21, 31 and 21 atoms. Out of the 146 Cu atoms, 120 are located between the three Se layers, 12 Cu in the middle Se layer and 14 Cu atoms outside the Se layers. For a detailed description of Cu_{146}, see [3].

The XPS experiments were performed with a Scienta-300 spectrometer equipped with an Al Kα radiation source, installed at the Centre de Spectromicroscopie of the IPA-EPFL in

Lausanne. The energy resolution was 0.35 eV. Cu_{146} was spread onto a graphite substrate under an argon atmosphere.

The room-temperature (β) phase of Cu_2Se was obtained by heating the Cu_{146} sample in vacuo to 200 °C for about two hours. The reaction of Cu_{146} to Cu_2Se was monitored to completion by measuring the disappearance of P $2p$ and ligands-related C $1s$ peak component. The chemical composition of Cu_{146} was confirmed by comparing the relative intensities of the Cu $2p$, Se $3d$ and P $2p$ core lines, which were found in good agreement with the nominal ratio, after normalization for the respective theoretical cross-sections.

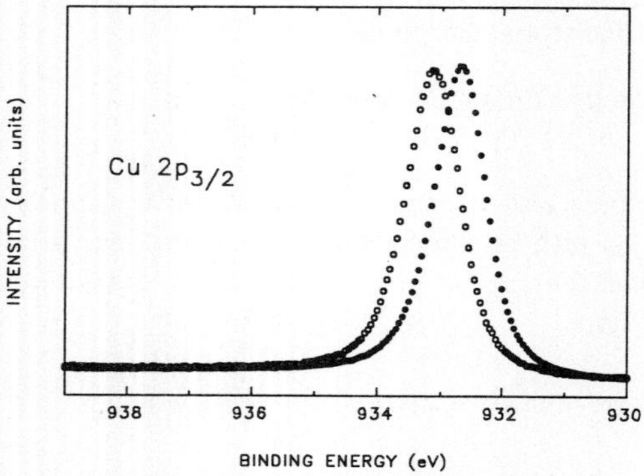

Figure 1. Cu $2p_{3/2}$ spectra of Cu_{146} (open circles) and β-Cu_2Se (filled circles).

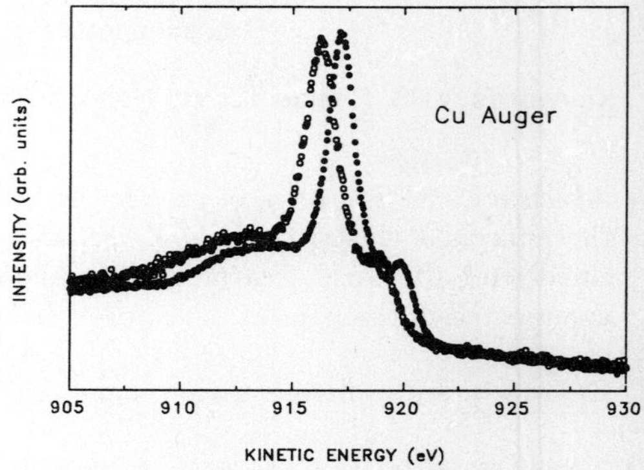

Figure 2. Cu Auger spectra of Cu_{146} (open circles) and β-Cu_2Se (filled circles).

Figs. 1 and 2 report the Cu $2p_{3/2}$ and $CuL_3M_{4,5}M_{4,5}$ spectra of both Cu_{146} and Cu_2Se. We note that the Cu $2p_{3/2}$ and Cu Auger position and shape of Cu_{146} and β-Cu_2Se are indicative of the monovalency of the Cu atoms. No satellite structures are observed in the Cu $2p$ region.

The valence band of Cu_2Se (Fig. 3) consists of a relatively narrow Cu $3d$ band at about 3.1 eV and two Se $4p$ bands at about 5 eV and 1 eV [4, 5].

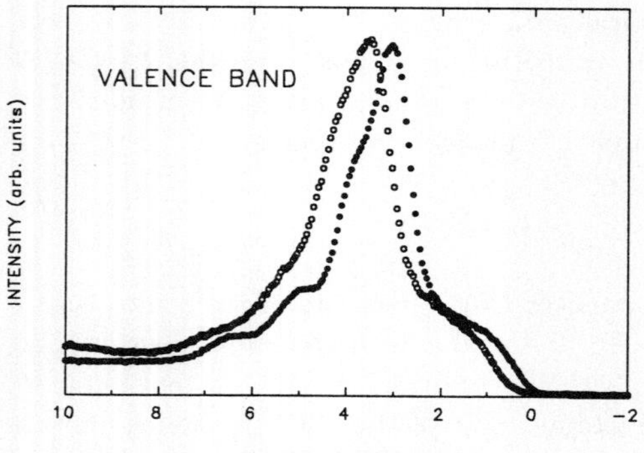

Figure 3. Valence-band spectra of Cu_{146} (open circles) and β-Cu_2Se (filled circles).

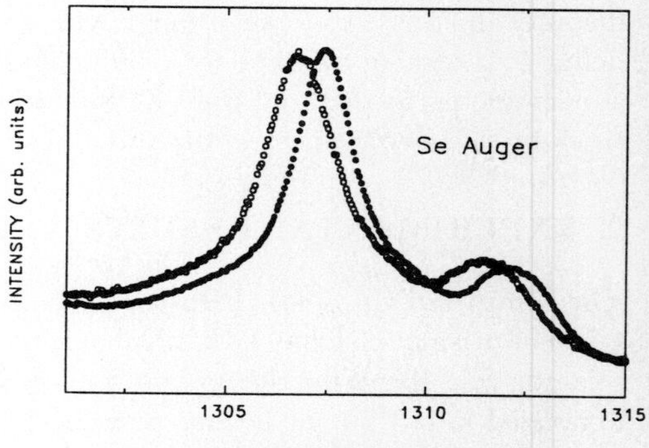

Figure 4. Se Auger spectra of Cu_{146} (open circles) and β-Cu_2Se (filled circles).

The valence band of Cu_2Se shows a density of states at the Fermi level, in accordance with the conductive properties of β-Cu_2Se [6].

The Cu $2p_{3/2}$ and Se $3d$ spectra of Cu_{146} are shifted to a higher BE as compared to bulk Cu_2Se by 0.49 eV and 0.29 eV, respectively. The Cu $3d$ levels of the valence band are shifted by 0.50 eV. The Se $4p$ levels, which constitute the top of the valence band, has a shift of 0.35 eV.

The similarity of the Cu $2p_{3/2}$ line shapes of Cu_{146} and bulk Cu_2Se is surprising since 30 Cu atoms are directly bound to PPh_3. Se Auger ($L_3M_{4,5}M_{4,5}$) spectra are shown in Fig. 4. The observed shifts on Cu and Se Auger lines are 0.91 and 0.61 eV, respectively, close to twice the BE shift of Se $3d$, Se $4p$, Cu $2p$ and Cu $3d$.

3. DISCUSSION

We observed a rigid BE shift of the entire XPS spectrum of Cu by 0.5 eV and of Se by 0.3 eV, which we interpret in favor of a Coulomb charge on the surface of the cluster. The absolute values of the shifts are relatively small compared to a unit charge on the Cu_{146} core, which would be about 0.8 eV, consistently with the reduced ability of the limited number of conduction electrons in small metal particles to screen an electrostatic charge. However, a difference in BE shift of Cu and Se is incompatible with a surface charge only. The response of the cluster to the creation of a core hole is a combined mechanism of limited screening, leading to a positive surface charge and a positive BE shift, and local polarization of the surrounding, resulting in a negative BE shift. The BE shifts of Cu and Se are different because the energy required to polarize the surroundings of a charged Cu or Se atom will not be the same.

Additional evidence for the reduced electron ability to screen an electrostatic field is obtained from the BE shifts of the Auger lines of Cu and Se. The expected Auger line shift is $\Delta E_{kin} = 3\Delta E_B$. ΔE_B is the final-state Coulomb shift of the primary core hole. We observed $\Delta E_{kin} \approx 2\Delta E_B$ (Table I) since the limited number of conduction electrons, which is not able to effectively screen a single core hole, is unable to screen two core holes to the same extent.

The appearence of a Coulomb charge on the surface of the cluster implies a delocalization of the $4p$ conduction electrons near the Fermi level. However, the degree of delocalization is reduced, leading to a limited conduction electron screening of a core hole. Core holes are localized and thus able to induce a local polarization of the environment as an additional extra-atomic relaxation energy. The fact that the Se $4p$ shift near E_F is not much different from the Se $3d$ core-level shift (0.3 eV) leads us to the conclusion that holes in the $4p$ band near E_F are localized because delocalized holes cannot induce a local polarization and the BE shift would then be the Coulomb shift [7], which is about 0.8 eV for a particle the size of Cu_{146}. Evidence for a localization of valence holes was also obtained by us in an XPS experiment on the molecular metal cluster compound Au_{55} [8].

In conclusion, we have shown that the binding-energy shifts of the copper-selenide cluster are caused by two mechanisms. A Coulomb charge on the cluster originating from a limited screening response of the conduction electrons to a core hole in combination with polarization of the surroundings of the core hole. Evidence is presented for hole-state localization near E_F.

Acknowledgments.

The invaluable support of Carlo Coluzza at the IPA-EPFL is gratefully acknowledged. This research was supported by the EU through HCM contract # ERBCHRXCT-940532.

References

[1] L. J. de Jongh (editor), Physics and Chemistry of Metal Cluster Compounds (Kluwer Academic Publishers, Dordrecht, 1994).

[2] G. Schmid (editor), Clusters and Colloids (VCH, New York, 1994).

[3] H. Krautscheid, D. Fenske, G. Baum and M. Semmelman, Angew. Chem. Int. Ed. Engl. 32 (1993) 1303.

[4] E. J. D. Garba and R. L. Jacobs, Physica B 138 (1986) 253.

[5] V. A. Terekhov, V. M. Kashkarov, V. V. Gorbachev, Y. A. Teterin and É. P. Domashevskaya, Sov. Phys. Solid State 25 (1983) 1424.

[6] Z. Vučić, O. Milat, V Horvatić and Z. Ogorolec, Phys. Rev. B 24 (1981) 5398.

[7] G. K. Wertheim, S. B. DiCenzo and D. N. E. Buchanan, Phys. Rev. B 33 (1986) 5384.

[8] D. van der Putten and R. Zanoni, J. El. Spectr. Rel. Phen., in press (1995); D. van der Putten, D. Olevano, R. Zanoni, H. Krautscheid and D. Fenske, J. El. Spectr. Rel. Phen., in press (1995).

Materials Science Forum Vol. 195 (1995) pp. 127-132

Probing Local Structure by Multiple-Scattering EXAFS Data-Analysis: Application to Nanocrystalline Iron

A. Di Cicco

Unità di ricerca INFM, Dipartimento di Matematica e Fisica
Università degli Studi di Camerino, Via Madonna delle Carceri, I-62032 Camerino (MC), Italy

Keywords: EXAFS, GNXAS, Multiple-Scattering, Defect, Structure

A multiple-scattering (MS) EXAFS study of nanocrystalline iron prepared by high-energy ball milling is presented. The use of MS signals has allowed an accurate investigation of the local structure. The dramatic decrease of the coordination numbers, followed as a function of the milling time, is assigned to the presence of a large number of ill-coordinated atoms. The stronger reduction of the three-atom multiple-scattering signal confirms this interpretation. An "in-situ" annealing (up to about 1100 ° C) is shown to increase the local coordination.

INTRODUCTION

X-ray absorption spectroscopy [1] has been used quite often for investigating local structure of nanocrystalline materials and/or clusters produced using different techniques. Generally speaking, a marked reduction of the first-neighbor coordination number (CN) has been observed for nanocrystallites of sizes in the same range of those of the present work (see for example ref. [2-5]). For smaller cluster sizes, variations of the average interatomic distances R and of bond variances σ_R^2 have been observed (see for example ref. [5]). The existence of anharmonic vibrations, leading to non-Gaussian shapes of the first-peak of the pair distribution function, has been also pointed out. [7]
In most cases, CN values have been found much lower than expected looking at the average size of the nanocrystallites. Even taking into account realistic distribution of diameters and systematic errors in the determination of the average size (usually evaluated by XRD), the reduction of the CN values is difficult to explain. The lack of coordination appears to be a real phenomenon, observed in many systems and related mainly to the preparation of nanophase systems. A *frozen-gas-like* model for grain-boundaries was also proposed to explain both diffraction and EXAFS results. However, this simplified picture has been subject of debate and appears to be not adequately justified. In particular, the complete lack of short-range order has no clear physical explanation. Recently, computer simulations [8] have shown that grain boundaries in nanocrystallites do not differ substantially from those of standard polycrystalline material.
In this paper an EXAFS (Extended X-ray Absorption Fine Structure) study of nanocrystalline iron prepared by high-energy ball milling is presented. [9] Iron powders show different average grain sizes in the range 70-200 Å.
Modern EXAFS data-analysis is based on the use of *ab-initio* multiple-scattering codes. The inclusion of higher order scattering terms is necessary for a quantitative evaluation of the structural signal and allows to extract important information on higher-order local distribu-

tion functions. This unique feature of the EXAFS technique can be very useful for structural determination in ill-ordered materials such as nanophase systems. As also reported in a recent paper [10], the quantitative analysis of Fe K-edge EXAFS spectra performed by using *ab-initio* MS codes for data analysis (GNXAS) [10-12] is able to give information on the local two-atom and three-atom distributions. The inclusion of the three-atom terms allows to select between different models or local structure.

RESULTS AND DISCUSSION

In Fig. 1 the $k^2\chi(k)$ EXAFS spectra of the 2h, 8h, 16h and 32h (milling time in hours) nanocrystalline iron are reported. In Fig. 2 the Fourier Transform (FT) spectra performed in the range 2.8-15 Å^{-1} of the EXAFS spectra are shown. Spectra of nanocrystalline samples are compared with the pure crystalline iron one, recorded at LURE (D42 beamline) during the same run. Measurements of Fig. 1 have been performed taking particular care for the preparation of samples. [10]

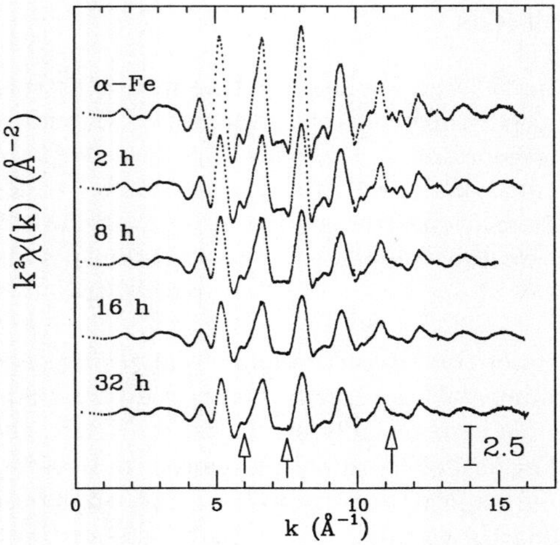

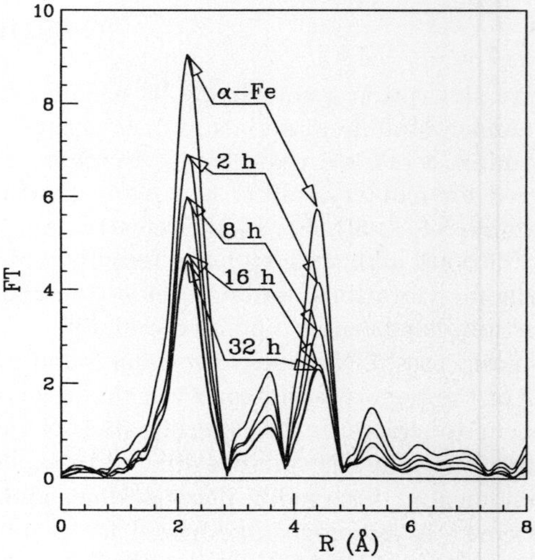

Fig. 1
Experimental Fe K-edge EXAFS $k^2\chi(k)$ spectra (a) of iron powders prepared by high-energy ball-milling at different milling times (2h, 8h, 16h, and 32h) compared with the bulk iron spectrum.
Fig. 2
Fourier Transform (FT) of the EXAFS $k^2\chi(k)$ spectra (reported in Fig. 1) of bulk iron and nanocrystalline samples.

Looking at Fig. 1 and Fig. 2 it is clear that the bcc local structure is preserved. There is also clear evidence for a progressive reduction of the first-neighbor CN (and consequently of the CNs of the farther shells) as testified by the decrease of the first peak in the Fourier Transform.
Of course, the strong reduction of the first-shell CN can not be explained by the estimated sizes of the nanocrystallites because the reduction due to the surface effect should be limited to a few percent for first-neighbors.
The lack of coordination indicates the existence of a large fraction of ill-coordinated Fe atoms. This means that the number of "surface" atoms is much larger than expected. This "surface" atoms can be regarded either as true surface atoms or as atoms in the proximity of a defect (dislocations, point defects). The Fe atoms could be also bonded to some other light atoms, but the EXAFS data-analysis rules out any appreciable contribution due to those atoms, with the exception of hydrogen.
It is worth to say that the increasing importance of defects as a function of milling time is also recognized by X-ray diffraction analysis. [13-15]
EXAFS spectra contain important contributions due to two-atom and three-atom configurations limited within a few Å. In particular the third FT peak around 4.5 Å is mainly due to multiple-scattering. [10-12]
The reduction of the FT peaks located at higher distance values is more pronounced, due to the different effect of the lack of neighbors on farther atoms and triangular configurations. In Fig. 1 and Fig. 2 (arrows) it is possible to see the decrease of the high-frequency (R) content of the EXAFS spectra.
Quantitative analysis of the high-frequency features, never attempted before, gives useful information about local atomic arrangement.
The GNXAS method [11,12] has been applied to the above mentioned EXAFS spectra in order to get information on the reduction of the CNs. Data-analysis is based on the GNXAS multiple-scattering analysis of bulk iron.[10,12] No significative change of the interatomic distances and bond variances has been found. The variance of the first-neighbor distances was found to be the same of that typical of bulk iron within the statistical error. No evidence for asymmetric distribution of first-neighbors has been detected. The amplitude modification of the EXAFS spectra of nanophase iron can be fully explained by the reduction of CNs and multiplicities of the triangular configurations. Fits of the nanocrystalline spectra show the same level of agreement obtained for bulk iron.[10-12] In particular, first-neighbors coordination numbers are found to be 5.7 (3), 5.0 (3), 4.1 (3) and 4.0 (3) for 2h, 8h, 16h and 32h samples respectively (8 is the "bulk" value). More interesting is the increased reduction of the degeneracy of the collinear configurations (4 per atom, "bulk" value) that are found to be 2.4(2), 1.8 (2), 1.5 (2) and 1.4 (2). These values have to be expected in presence of a large number of ill-coordinated atoms because the probability of finding a specific three-body configuration is in this case obviously lower than finding a two-body one. The degeneracy of the collinear configurations found in nanocrystalline irons is in agreement with a very simplified structural model that includes the presence of lattice defects. [10]
This very simple model is easily produced by using an iterative procedure acting on a perfect bcc lattice. The procedure begins with a central atom for which the first-neighbors are chosen randomly with a fixed probability. Connectivity is ensured by imposing a minimum number of first-neighbors. The procedure is repeated on any new atom generated at the previous step. A maximum distance from the center is allowed, limiting the actual cluster size. The

total number of atoms to be inserted is fixed according to a selected density. This procedure allows to build a highly connected network in presence of a large number of atomic defects which greatly increase the surface-to-volume ratio. The final clusters have an approximate spherical shape with very irregular surfaces and presence of voids and point defects. In Fig. 3 a model cluster of about 90 Å diameter is shown (left) and compared with bulk iron (right). The cluster is not spherical due the particular graphic perspective chosen. Parameters of the calculation were the generation probability p of the first-neighbors (in this case p=1/6) and the minimum number N_m of neighbors allowed (N_m=2). The final density is in this case 0.5. Obviously, the determination of a realistic model reproducing the atomic arrangement (and motion) in nanocrystalline materials including grain boundaries is beyond the scope of the present work. Using realistic interatomic potentials, such a low-density and low-coordinated model would be certainly not stable. In order to shed some light on the problem of a correct modelling of nanophase systems, it could be interesting to develop specific applications of molecular dynamics or other advanced computer simulation techniques in such systems showing large excess free-volume.
However, the EXAFS results indicate that the lack of coordination is real and must be taken into account. The evident decrease of the high-frequency multiple scattering signal is a strong signature of the presence of defects.

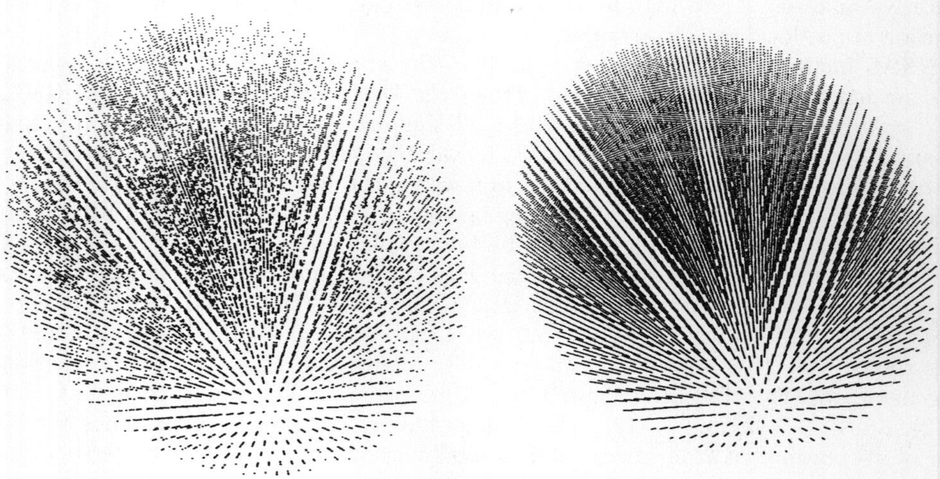

Fig. 3
Model Fe (bcc) cluster (diameter of about 90 Å) calculated using the iterative procedure described in the text (left) and compared with bulk iron (right)

In Fig. 4 the comparison between different best-fit FT spectra obtained using different structural models is presented (8h). Model A is obtained using an overall reduction factor for all of the MS signals (*frozen-gas-like* model). The relative intensity of the first-neighbor and collinear three-atom configuration peaks is not correctly reproduced. Model B is calculated using as coordination numbers those derived from clusters of about 10 Å diameter. In this case, the peak related to the collinear configuration is too much damped. The first-neighbor coordination numbers and the degeneracy of the collinear configuration derived from the above-mentioned model (density of about 0.8-0.9) are in good agreement with the experimental values found in the 8h case (upper curves).

In a further study, we have performed an "in-situ" high-temperature annealing of the 32h sample. This sample has been heated up to about 1100 °C for about half an hour and then cooled down to room temperature. The total duration of the thermal treatment was about 120 minutes.

At about 915±5 °C we have clearly observed the phase transition from α-iron (bcc structure) to γ-iron (fcc structure, $T_m \approx 910$ °C).

In Fig. 5 the $k^2\chi(k)$ EXAFS (a) and the Fourier Transform (b) spectra of the 32h sample before (upper curve) and after (lower curve) annealing are presented. Both experimental (dots) and calculated (solid curves) spectra are reported. The FT curves clearly show an increase of the amplitude of the EXAFS signal. EXAFS data-analysis shows unambiguously that a high-temperature treatment reduces the number of ill-coordinated atoms. In the case of the 32h annealed sample the first-neighbor CN is increased of about 35% and the EXAFS spectrum becomes very similar to that recorded for the 2h case.

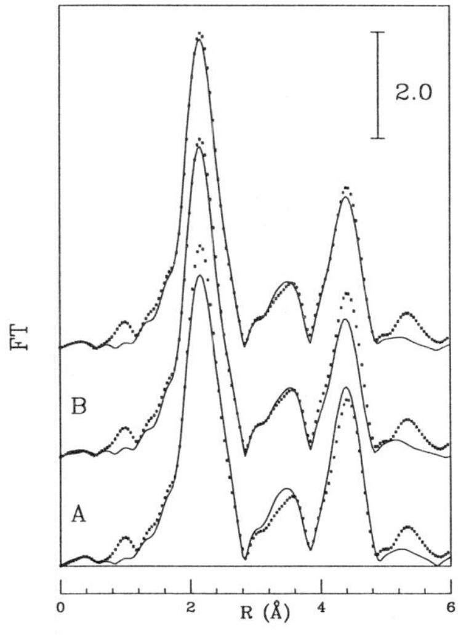

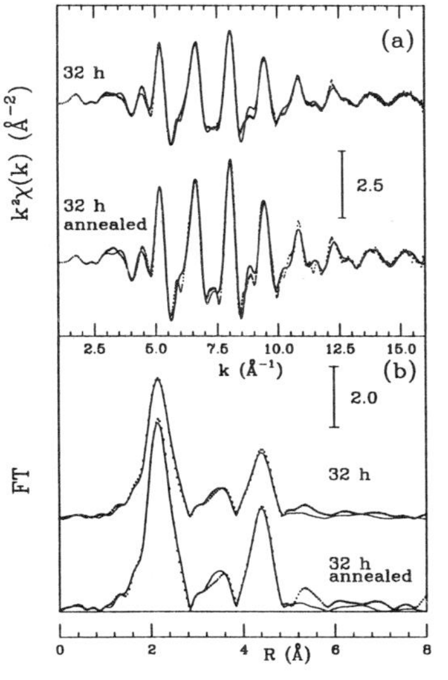

Fig. 4 Comparison between different best-fit Fourier Transform (FT) spectra (8 hours milling time) obtained using different structural models (see text).

Fig. 5 Upper panel (a): comparison between experimental $k^2\chi(k)$ spectra (dots) of the 32h nanocrystalline iron sample before and after annealing. The best-fit calculated signals obtained using multiple-scattering contributions are shown as solid curves. Lower panel (b): Fourier Transform (FT) of the $k^2\chi(k)$ reported in the upper panel.

CONCLUSIONS

In this paper the application to nanocrystalline iron of an advanced EXAFS data-analysis technique including multiple-scattering terms is presented. Experimental spectra show a progressive reduction of the local coordination. A more pronounced reduction has been found for the intense multiple-scattering signal related to the main three-atom collinear configuration in the bcc structure. The reduction of coordination numbers can not be explained looking at the estimated sizes of the crystallites under consideration. The presence of a large number of ill-coordinated atoms is necessary to explain the measured EXAFS spectra. A naive model accounting for atomic defects has been shown to be able to explain the observed differences in the decrease of the first-neighbor coordination numbers and of the degeneracy of the main collinear configuration. Development of more realistic models accounting for the observed reduction of local coordination are strongly stimulated. Further EXAFS investigation aimed to study local atomic structure beyond the pair distribution in nanophase systems are strongly encouraged by this research.

REFERENCES

[1] "X-ray Absorption: Principles, Applications, Techniques of EXAFS, SEXAFS and XANES" eds. D. C. Koningsberger and R. Prins, Wiley (1988).
[2] T. Haubold, R. Birringer, B. Lengeler, and H. Gleiter, Phys. Lett. A **135**, 461 (1989).
[3] M. Jaouen, P. Bouillard, T. Girardeau, P. Chartier, J. Mimault, and G. Tourillon, J. Phys.: Condens. Matter **2**, 8113 (1990).
[4] T. Haubold, F. Boscherini, S. Pascarelli, S. Mobilio, and H. Gleiter, Phyl. Mag. A **66**, 591 (1992).
[5] G. Cocco, S. Enzo, N. T. Barrett, and K. J. Roberts, Phys. Rev. B **45**, 7066 (1992).
[6] A. Balerna, E. Bernieri, P. Picozzi, A. Reale, S. Santucci, E. Burattini, and S. Mobilio **31**, 5058 (1985).
[7] B. S. Clausen, L. Gråbæk, H. Topsœ, L. B. Hansen, P. Stoltze, J. K. Nørskov, and O. H. Nielsen, J. of Catal. **141**, 368 (1993).
[8] D. Wolf, J. Wang, S. R. Phillpot, and H. Gleiter, Phys. Rev. Lett. **74**, 4686 (1995).
[9] E. Bonetti, G. Scipione, G. Valdré, G. Cocco, R. Frattini, and P. P. Macrí, J. Appl. Phys. **74**, 2053 (1993).
[10] A. Di Cicco, M. Berrettoni, S. Stizza, E. Bonetti, G. Cocco, Phys. Rev. B **50**, 12386 (1994).
[11] A. Filipponi, A. Di Cicco, T. A. Tyson, and C. R. Natoli, Sol. State Comm. **78**, 265 (1991).
[12] A. Di Cicco, Physica B **208 & 209**, 125 (1995).
[13] E. Hellstern, H. J. Fecht, Z. Fu, and W. L. Johnson, J. Appl. Phys. **65**, 585 (1989).
[14] M. S. Boldrick, E. Yang, and C.N.J. Wagner, J. of Non-Cryst. Sol. **150**, 478 (1992).
[15] C.N.J. Wagner, E. Yang, and M. S. Boldrick, Adv. X-ray Anal. **35**, 585 (1992).

Materials Science Forum Vol. 195 (1995) pp. 133-138

Investigation of the Oxidative-Pyrolysis Processes of $RuCl_3 \cdot xH_2O$ and $IrCl_3 \cdot xH_2O$

L. Nanni[1], A. De Battisti[1], S. Daolio[2], B. Facchin[2] and C. Paguera[2]

[1] Dipartimento di Chimica dell'Università, Via L. Borsari 46, I-44100 Ferrara, Italy

[2] Istituto di Polarografia ed Elettrochimica Preparativa, Consiglio Nazionale delle Ricerche, Corso Stati Uniti 4, I-35020 Padova, Italy

Keywords: Electrocatalysis, Oxide Films, SIMS

Abstract

In this paper the reactions of oxidative-thermic decomposition of the precursor salts $RuCl_3 \cdot xH_2O$ and $IrCl_3 \cdot xH_2O$ used in the DSA® preparation have been studied. The pyrolytic processes have been followed by secondary ion mass spectrometry (SIMS), by which it has been also possible to study the deposited film structure on the basis of the depth-profiles versus storage thickness. The mass spectra show, for both precursors, a progressive oxidation leading eventually to the formation of the respective oxides (RuO_2, IrO_2); by depth-profiles, on the other hand, it has also been verified that the films are not homogeneous but some components are segregated in different regions.

1. Introduction

Electrodes based on noble-metal oxides, prepared by thermal decomposition, played a very important role both in the field of industrial applications and in the studies of the electrochemical reactions of anodic oxidation [1-4].The flexibility by which amount and nature of the components can be changed, allow to modulate their catalytic properties and the selectivity, for certain electrolytic processes, in a wide range.

The dimensionally stable anodes (DSA®), consisting of supported films of metallic oxides belonging to VIII group B stabilised by other components, as TiO_2, Ta_2O_5, ZrO_2, are certainly the most studied. For example, the anodes formed by mixtures of RuO_2-TiO_2 deposited on Ti show good performances for the chlorine evolution, while those formed by IrO_2-Ta_2O_5 show a better performance for the oxygen discharge [1,2].

The electrical, catalytic and mechanical properties of these devices, are strongly influenced by the parameters of their preparation, as pyrolysis temperature, chemical nature of the precursor salts and their solutions etc. [5].

Previous studies, made by thermal analysis coupled with mass spectroscopy on the evolved gases during the pyrolysis [6] and RBS [7], show that the final result of a preparative method is influenced by the history of the chemical transformation of precursors. Along these lines the pyrolytic process of the $RuCl_3 \cdot xH_2O$ and $IrCl_3 \cdot xH_2O$ salts (precursors of the electrocatalytic components of DSA®) has been followed by SIMS.

2. Experimental procedure

2.1. Samples preparation

For both precursor salts, $RuCl_3 \cdot xH_2O$ and $IrCl_3 \cdot xH_2O$, samples have been heated at the following temperatures: 80°C, 150°C, 250°C, 350°C, 400°C, 450°C, 500°C; these values have been chosen on the basis of physical-chemical transformations showed by DTA curves for the same precursors [6,8]; samples at room temperature have also been prepared.

The solutions used were 1% of $RuCl_3 \cdot xH_2O$ (Fluka puriss.) and 1% of $IrCl_3 \cdot xH_2O$ (Fluka puriss.) in 2-Propanol (Backer Analysed HPLC); the film were prepared dropping the solution on amorphous silica plates (1 cm x 1 cm), evaporating the solvent at room temperature and pyrolyzing for 20 minutes at the fixed temperature in the oxygen current.

2.2. SIMS analysis

The SIMS equipment is custom-built and has been described elsewhere. A mass-filtered duoplasmatron ion gun (model DP50B, VG Fisons, Manchester, UK), generating a monochromatic (1-10 KeV), highly collimated (up to 5 µm), O_2^+, O^+ ion beam, and an inert-gas gun with a traditional electron-impact source (model IQE/12-38, by Leybold-Spacs GmbH, Berlin), were used as bombarding sources. The secondary-ion optics are based on a three-lens design with central stop and are interfaced to a Balzers (Balzers A.G., Liechtenstein) model QMA 400 quadrupole mass analyser (mass range extended up to 2048 u). A secondary electron multiplier (90° off-axis, SEM) is used for negative and positive ion detection in counting mode. The lens potentials, the quadrupole electronic control units and detection system are controlled by Balzer QMS 421 unit. Two personal computers, connected in a local area network, allow control of the instrument and data collection. The samples were exposed to ion beams (O_2^+) with different energies and current densities.

3. Results and discussion

3.1. Mass spectra

In the tables 1 and 2 are reported secondary ion fragments for the samples $RuCl_3 \cdot xH_2O$ and $IrCl_3 \cdot xH_2O$.

Table 1

Pyrolysis Temperatures	Ion clusters
Room temperature	Ru, RuC, RuO, RuOH, $RuOH_2$, RuCO, RuCl, Ru_2, Ru_2O, Ru_2OH, Ru_2Cl
80°C	Ru, RuC, RuO, RuOH, RuCl, Ru_2, Ru_2O, Ru_2Cl, Ru_2OH
150°C	Ru, RuC, RuCO, RuCl, Ru_2, Ru_2O, Ru_2OH, Ru_2Cl_2
250°C	Ru, RuC, RuCO, RuCl, Ru_2, Ru_2O, Ru_2OH, Ru_2Cl
350°C	Ru, RuO, RuOH, RuCl*, Ru_2, Ru_2O, Ru_2OH, Ru_2O_2
400°C	Ru, RuOH, RuO, Ru_2, Ru_2O
450°C	Ru, RuO, RuOH*, Ru_2, Ru_2O
500°C	Ru, RuO, RuOH*, Ru_2, Ru_2O

*Signals of low intensity

Table 2

Pyrolysis Temperatures	Ion cluster
Room temperature	Ir, IrC, IrC_2, IrCl, IrO•HCl
80°C	Ir, IrC
150°C	Absence of signals for all Ir clusters
250°C	Ir, IrO, IrOH, IrCO
350°C	Ir, IrO, IrOH, $IrOH_2$, IrCO
400°C	Ir, IrC, IrO, IrCO,IrSiO
450°C	Ir, IrO
500°C	Ir, IrO, IrSiO, $IrSiO_2$

In the case of $RuCl_3 \cdot xH_2O$ ion intensity curves for different clusters from samples heated at higher temperatures indicate that complex transformations occur. Above 350°C, for instance, peaks due to chlorinated and carbonyl compounds are absent. The signals due to hydroxylated species are still present in samples heated at 450°C and 500°C, although with smaller intensities, compared with samples heated at lower temperatures.

For $IrCl_3 \cdot xH_2O$ the pyrolytic process is definitely more complicated. As shown in table 2, signals due to hydroxylated species, IrC^+, IrC_2^+, $IrCO^+$ are present. From decomposition temperatures of 80°C the ionic yields of Iridium containing clusters are very low. For samples heated at 250°C the only significant signals are those due to hydroxylated, oxide and carbonylic species. The latter clusters are also present in spectra obtained for samples heated at 450°C and 500°C. The IrOH species are absent starting from the pyrolysis temperature of 400°C.

The anomalous behaviour of the pyrolytic process of $IrCl_3{\bullet}xH_2O$ can be explained admitting that at low temperatures chains based on Iridium-oxychloride, chemically inert also under ion bombardment (absence of peaks concerning chlorine-containing clusters) are formed. This mechanism is confirmed by isothermal curves that show a slow weight loss for the temperatures greater than 200°C [8].
In all the mass spectra are present signals due to residual organics and to substrate; signals due to alkaline metals and chlorine have also been found.

3.2. In-depth analysis

The figures 1, 2 and 3 show the depth-profiles of secondary ions with m/z=102, 116, 118 and 119 obtained from $RuCl_3{\bullet}xH_2O$ precursor at different pyrolysis temperatures;

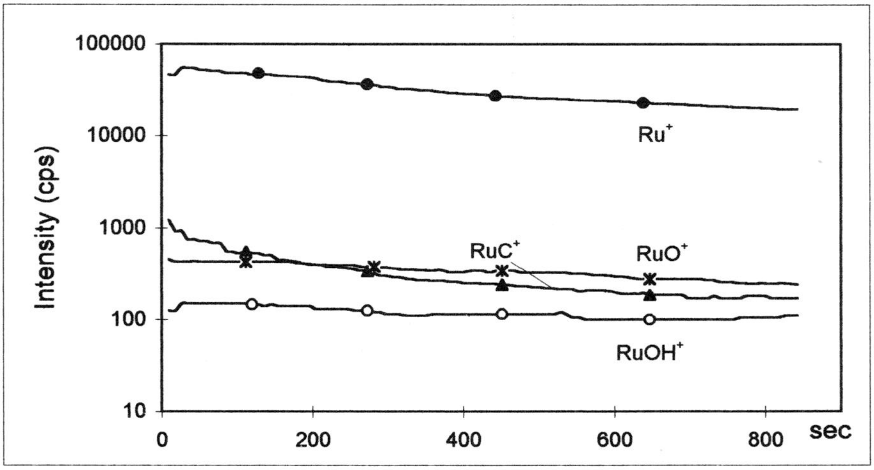

Figure 1: In-depth profiles for the sample prepared at 150°C.

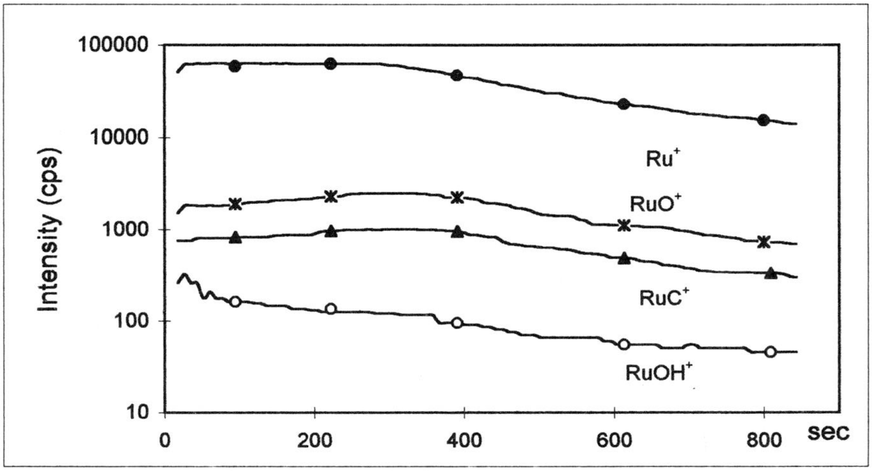

Figure 2: In-depth profiles for the sample prepared at 500°C.

as can be observed both at 150°C and 500°C temperatures, the profile of the Ru^+ ion at m/z=102 shows a segregation of Ruthenium at the surface and a continuous decreasing in the film-substrate interface. Analogous behaviour is exhibited by the species at m/z=116 ($^{104}RuC^+$ partially overlapped with $^{100}RuO^+$) while the fragment RuO^+ (m/z=118) and $RuOH^+$ (m/z=119) show significative modification in the surface region. At room temperatures (Figure 3) the very intense Ru^+ signal decreases slowly going in the inner layers, while the ionic yields of the oxide and hydroxylate species are correlated; on the contrary at higher temperatures (Figure 1,2) RuO^+ and $RuOH^+$ have different distributions and are characterised by different curve shapes and by $RuOH^+/RuO^+$ intensity ratio decreasing.

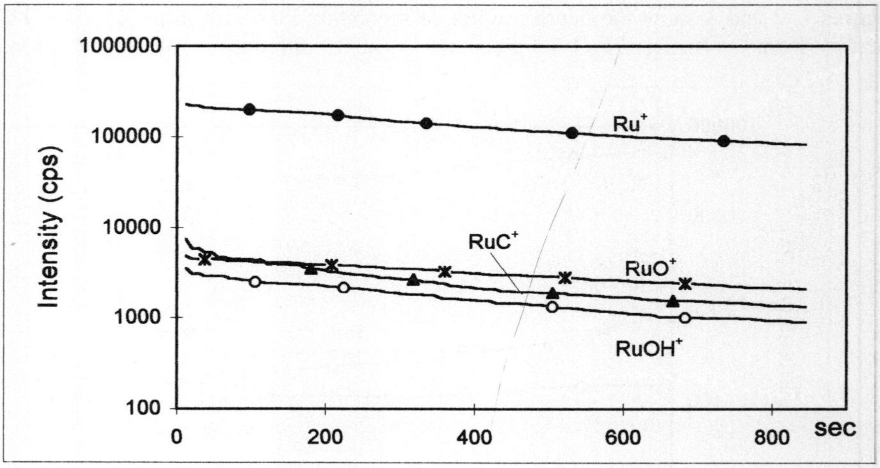

Figure 3: In-depth profiles for the sample prepared at room temperature.

In the case of $IrCl_3{\bullet}xH_2O$ precursor instead, the profile with m/z=193 show that Iridium is segregated near the substrate of amorphous silica, as illustrated in Figure 4;

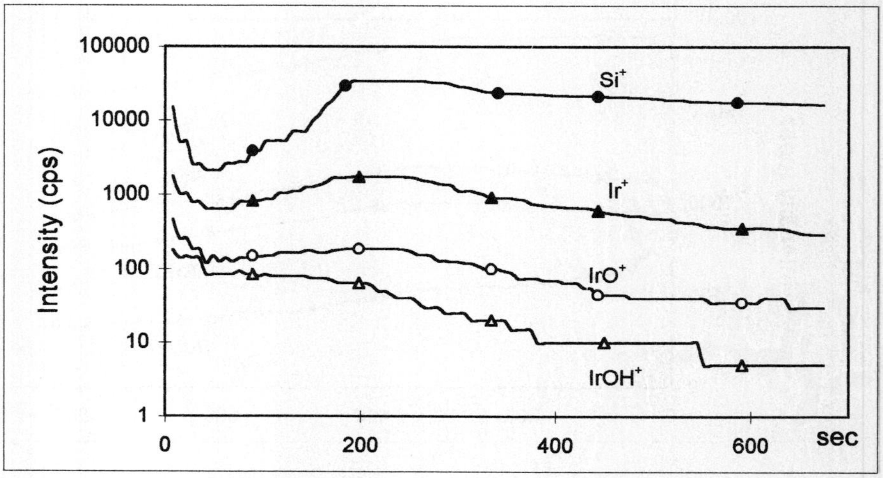

Figure 4: In-depth profiles for the sample prepared at 350°C.

in particular the shapes of Ir^+ and Si^+ profiles are similar. In these matrices an high reactivity between Iridium and Silicon is confirmed also by the presence of secondary ions $IrSiO^+$ and $IrSiO_2^+$ (see Table 2). Once again at room temperature (Figure 5) the species IrO^+ and $IrOH^+$ show the same profiles, while are different at 350°C.

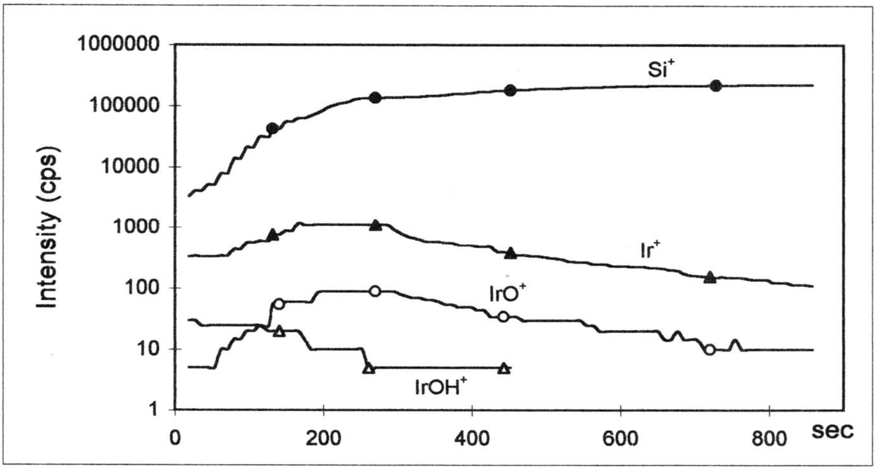

Figure 5: In-depth profiles for the sample prepared at room temperature.

The profile of C^+ at m/z=12 and not reported in figures show in all cases a larger concentrations in the near-surface region and a drastic decrease in the bulk of the films.

Conclusion

The studies of pyrolytic processes of the precursor salts $RuCl_3{\bullet}xH_2O$ and $IrCl_3{\bullet}xH_2O$ showed that the final result of RuO_2 and IrO_2 electrodes preparation is complicated; in fact the complex reactivity of the hydrated chloride films inevitably involves a progressive extension of the pyrolytic oxidation from the near-surface-region towards the film/metal substrate interface. This is likely to cause encapsulation of part of the volume of the films by the reaction products in the outermost part, inhibiting exchange of oxygen from the pyrolysis atmosphere and gaseous reaction products from the film. Incorporation of chloride, carbon, hydroxylated species, important for the electrochemical performance of the film, can be accounted for along these lines.
The pyrolytic oxidation of the Ruthenium (III) hydrated chloride is less hindered by the mentioned encapsulation phenomena, compared with that of the Iridium precursor.
The kinetics of the film pyrolytic oxidation seems to involve in any case surface segregation of components, either from the film volume (e.g. TiO_2 from Ti substrates, SiO_2 from amorphous silica substrates). These inhomogeneities in the distribution of components may also significantly influence the catalytic properties of the oxide electrodes. The results obtained in this work indicate that SIMS is a quite powerful technique in the compositional characterisation of oxide electrocatalysts[11-15] and a more complete study of the aspects involved in their synthetic processes is in progress.

References

[1] S. Trasatti and G. Lodi, in S. Trasatti (ed.), Electrodes of Conductive Metal Oxides, Part A, Elsevier, Amsterdam, 1980, p. 301

[2] S Trasatti and G. Lodi, in S. Trasatti (ed.), Electrodes of Conductive Metal Oxides, Part B, Elsevier, Amsterdam,, 1981, p. 301

[3] A. Nidola, in S. Trasatti (ed.), Electrodes of Conductive Metal Oxides, Part B,Elsevier, Amsterdam, 1981, p. 521

[4] D.M. Novak, B. V. Tilak and B.E. Conway, in B.E. Conway and J.O'M. Bockris (eds.), Modern Aspects of Electrochemistry, Vol. 14, Plenum, New York, 1982, p. 195

[5] A. De Battisti, G. Battaglin, R. Koetz, G. Lodi and M. Cappadonia, Ext: Abstr. of the 173rd Meeting of the Electrochem. Soc., Atlanta, May 15-20, 1988; A De Battisti, G. Battaglin, R. Koetz, G. Lodi and M. Cappadonia, J. Electroche. Soc., 136 91988) 2596

[6] J. Kristòf, J. Liszi, A. De Battisti, A. Barbieri, P. Szabò, Materials Chem. and Phys., 37 (1994) 23-28

[7] A. Banedetti, P. Riello, G. Battaglin, A. De Battisti and A. Barbieri, J. of Electroanal. Chem., 376 (1994) 195-202

[8] G. Lodi and A. De Battisti, J. Electroanal. Chem., 277 (1990) 139-150

[9]R. Amadelli, Thesis, Ferrara, 1975

[10] G. Lodi, C. Bighi and C. Asmundis, Mater. Chem., 1 (1976) 177

[11] L. D. Burke and M. McCarthy, Electrichim. Acta, 29 (1984) 211

[12] G. Battaglin, A. De Battisti, A. Barbieri, A. Giatti and A. Marchi, Surf. Sci., 251/252 (1991) 73

[13]Ch Comninellis and G. P. Vercesi, J. Appl. Electrochem., 21 (1991) 136

[14] A. De Battisti, R. Brina, G. Gavelli, A. Benedetti and G. Fagherazzi, J. Electroanal. Chem., 200 (1986) 93

[15] L. D. Burke, O. J. Murphy, J. Electroanal. Chem.., 112 (1980) 39

[16] J. Kristòf, J. Mink, A. De Battisti and J. Liszi, Electroch. Acta, Vol. 39, (1994), 1531-1535

SECTION II

Physical Properties

Materials Science Forum Vol. 195 (1995) pp. 141-148

Magnetic Properties of Nanostructured Systems

A. Hernando and M. Vázquez

Instituto de Magnetismo Aplicado y Instituto de Ciencia de Materiales,
Universidad Complutense y CSIC, P.O. Box 155, E-28230 Las Rozas, Madrid, Spain

Keywords: Heterogeneous System, Granular Solids, Nanocrystalline Systems, Exchange Interactions

Abstract

In this work we introduce some considerations on the influence of volume distribution of magnetic cluster on the macroscopic behavior of heterogeneous magnetic systems. Two experiments illustrates the relevance of the capacity as exchange transmitter of the intergranular region in order to determine the exchange correlation length. It is shown that the exchange interaction between grains as well as between grain and matrix give rise to different types of average anisotropy and magnetization processes in the same sample as a function of temperature.

INTRODUCTION

The study of heterogeneous magnetic systems with local properties fluctuating in the scale of nanometers -i.e. amorphous materials, granular systems, nanocrystalline materials, multilayers- represents one of the most attractive topics in the field of present basic magnetic research [1-5]. Due to their heterogeneous structure, new and striking magnetic phenomena are observed that cannot be satisfactorily explained within the framework of the previous theories developed for monophase magnetic materials. In these new materials, the coexistence of different magnetic phases and the strength of magnetic interactions between them, carried through the interphase, mainly determine their magnetic response.

Amorphous ferromagnets are examples of uniform heterogeneous magnetic systems. The local easy axis is assumed to fluctuate in orientation with uniform correlation length of a few angstrons. The exchange coupling between the adjacent magnetic moment as well as the strength of the local anisotropy are uniform. Nanocrystals of a single well defined size, embedded in an amorphous matrix at constant distance between them form another uniform heterogeneous system. Granular solids composed of magnetic particles, of the same size, which are uniformly diluted in a non- magnetic matrix are also uniform heterogeneous systems.

Nevertheless, because of the distribution of particle size, particle distance, local anisotropy strength and any other relevant local parameter most of the heterogeneous systems are non-uniform. It is the aim of this paper to show that non-uniform heterogeneous systems composed of different phases, magnetic and non-magnetic, can be approximately treated from the concept of magnetic cluster or agregate, once the crystallite size of the magnetic phases is known. It will be shown that magnetic properties as coercivity and its thermal dependence as well as magnetorresistance are windows on a nanometric scale, which allow fine details of the nanostructure to be observed.

For any magnetic particle, with well defined local easy axes and size d, should exist a volume $v(>d^3)$ defined as the volume of the particle plus the volume of all the adjacent magnetic particles which as a consequence of magnetic interactions magnetizes cooperatively. We call this set of particles magnetic cluster. The magnetization process of any cluster with volume v is governed by its effective anisotropy k and exchange correlation length L. The clusters can be more weakly coupled, for instance through magnetostatic energies, between them. Notice however that the cluster definition implies that the interaction between the crystallites which form the cluster itself are much stronger than the interaction between diferent clusters. By neglecting this weak inter-agregate interaction, the magnetization process of the non-uniform system is then reduced to the sum of the magnetization process of all the independent magnetic clusters. The particular v distribution which determines some important aspects of the magnetic behavior depends on the fabrication technique, on the degree of immiscibility of the elements and on the nominal composition. In the next section of this article some general considerations on the characteristics of v distribution are outlined.

The aim of the present work is to define the relevant estructural parameters which characterise the uniformity and non-uniformity of heterogeneous systems and to analyze the correlation between their nanostructure and macroscopic magnetic behavior.

Two recent experiments carried out on soft Fe rich nanocrystalline materials are discussed. These experiments deal with the thermal dependence of the coercivity and the measurement of the Curie temperature of the amorphous matrix. In particular two consequences derived from the experimental results are pointed out i) Coupling between grains, favouring magnetic softness, occurs at temperature above the Curie temperature of the matreix and ii) The Curie temperature of the matrix is higher than the Curie temperature of an amorphous sample with the same composition

UNIFORM AND NON-UNIFORM HETEROGENEOUS MAGNETIC SYSTEMS.

1) Parameter defining the volume distribution of magnetic clusters

Let us initially consider a mixture of randomly distributed, magnetic and non-magnetic, clusters of two types of nanocrystals. Each cluster consists in a different number of nanocrystals. In general the crystallite size would spread over a range of values. For the sake of clarity let us assume that the size distribution of magnetic crystallites is sufficiently narrow to allow us to consider only an uniform size d. The number of magnetic crystallites, n, strongly coupled by exchange interactions in a particular cluster or agregate, determines both the agregate size, l, and volume $v=\alpha l^3(=nd^3)$.(Herinafter, it will be considered $\alpha=1$). It is important to note that the distribution in volume of the magnetic clusters, which will be assumed to be a normal distribution, determines the magnetic behavior of the sample. Magnetoresistance should be related to the total number of magnetic clusters and the magnetization curve should be specially sensitive to the fluctuations of volume around its average value. Let us show by using quite general considerations and estimations that for a constant volume fraction of magnetic atoms, the average volume of the magnetic clusters determine the magnetic and magnetoresistance behavior of the system. The average volume should define the total number of clusters as well as the width of the volume distribution.

Let x and $\langle v \rangle$ be the total volume fraction of the magnetic phase and the average volume of the magnetic clusters, respectively. The corresponding number of clusters per unit volume is $N^3=x/\langle v \rangle$. For a given x and $\langle v \rangle$ is possible to define the homogeneous distribution as that for which all the N^3 clusters have the same volume $\langle v \rangle$ and are uniformly distributed. The distance between adjacent clusters, $\langle \Lambda \rangle$, at the homogeneous configuration becomes approximately :

$$\langle \Lambda \rangle=\langle v \rangle^{1/3}[(1-\lambda)/\lambda] \qquad (1)$$

where $\lambda=x^{1/3}$. The number of different distances Λ between adjacent clusters, in the homogeneous configuration is of the order of $3N^2(N-1)$.

It will be shown that, when N as well as the distance between the centers of the clusters remain constant, the width of the volume distribution has an upper limit. In fact, it is possible to keep N constant, after introducing flutuations of v around $\langle v \rangle$, if the maximum volume $v_{max}(N)$ verifies than

$$v_{max}(N)=(1+<\Lambda>)^3=(1/\lambda)^3=1/N^3= <v>/x \quad (2)$$

When v takes values higher than $v_{max}(N)$, some clusters would overlap, so producing a decrease of N and an increase of $<v>$. Therefore, for a constant N there are different possible v distribution but for all of them the maximum v fluctuation, $(\Delta v)_{max}= v - v_{max}$ should verify

$$(\Delta v)_{max} < <v>[(1-x)/x] =<v>x^* \quad (3)$$

It is also possible to show that, when N and the volume of the clusters is kept unchanged, the width of the distribution of inter-clusters distances has an upper limit. Suppose that the clusters positions undergo deviations from the homogeneous distribution . If the fluctuation of distances between clusters, measured respect to $<\Lambda>$, is of the order of Λ the clusters would overlap. Therefore, it is verified that

$$(\Delta\Lambda)_{max}=<\Lambda>=<v>^{1/3} [(1-\lambda)/\lambda] \quad (4)$$

Relations Eq. 3 and Eq. 4 should be used to find the average values and distribution width of the magnetic anisotropy.

2) Magnetic anisotropy of clusters. Random anisotropy. Magnetoresistance.

If the size of magnetic nanocrystals, d, which build up the cluster, is smaller than the corresponding domain wall thickness, the effective magnetoctystalline anisotropy of the cluster, k^*, can be calculated by using random walk considerations [6-7]. The random anisotropy model yields the following dependence of k on n or v.

$$k^*(v)=k / (n)^{1/2}=kd^3 / (v)^{1/2} \quad (5)$$

where k is the magnetocrystalline anisotropy constant of the nanocrystals which form the cluster.

When the exchange correlation length L is larger than the cluster size, l, the cluster can be considered as a single domain in which the magnetization is uniform or a smooth function of position. As L is given by $A^2/(k^2d^3)$, the single domain condition can be written as

$$v < A^6/k^6d^9 \quad (6)$$

where A is the strength of the exchange interaction between neighbor magnetic moments.

If we consider an uniform magnetization, M_s, at each cluster, the effective magnetocrystalline anistropy field of the single-domain cluster should decrease with n or v as

$$H_{k^*}= (2kd^3/\mu_0 M_s)v^{-1/2}. \quad (7)$$

According to the v distribution H_{k^*} can be rewritten as:

$$H_{k^*}=<H_{k^*}>+\Delta H_{k^*} \quad (8)$$

with

$$<H_{k^*}>= (2kd^3/\mu_0 M_s)<v>^{-1/2}$$

$$\Delta H_{k^*}= (2kd^3/\mu_0 M_s)\left(-(\Delta v/2<v>^{3/2}) +3\Delta v^2/8<v>^{5/2} +...\right) \quad (9)$$

through relation Eq. 3, the following expression of ΔH_{k^*} is found

$$\Delta H_{k^*}= \langle H_{k^*}\rangle(-x^*/2+3x^{*2}/8+...) \qquad (10)$$

The blocking temperature, T_B, of the single-domain clusters becomes, through Eq. 5,

$$T_B=k^*v / 25K=kd^3v^{1/2} / 25K \qquad (11)$$

where K is the Boltzmann constant.

Ther average $\langle T_b\rangle$ is $(kd^3\langle v\rangle^{1/2} / 25K)$ and the distribution width results

$$\Delta T_b=\langle T_b\rangle (x^*/2 -x^{*2}/8+...) \qquad (12)$$

On the other hand if d is larger than the exchange correlation length the agregates will behave as a mixture of almost independent multi-domains, i. e. like normal polycrystalline materials.

As concerns magnetoresistance the influence of the nanostructure is mainly determined by the total number of magnetic clusters which act as scattering centers. This number is related to the parameter $\langle v\rangle$ through the relation $N=x/\langle v\rangle$. The maximum N depends on the size of the smaller magnetic nanocrystals, d, which form the clusters. Therefore the maximum possible N is of the order of x/d^3.

3) Physical factors which determine v distribution.

There are different methods of producing uniform heterogeneous systems. Devitrification from amorphous state is the usual procedure used to obtain Fe rich nanocrystalline materials. Segregation of immiscible elements from their metastable phases (previously achieved by sputtering, melt spinning or mechanical alloying) is the commom method of obtaining uniform granular solids.

However, when the formation of metastable phases corresponding to two immiscible elements, *a* and *b*, is only partially reached the resultant heterogeneous system is non uniform. This character is conserved during the segregation processes induced by thermal treatments. The morphology of such system should consist of clusters of element *a*, with impurities of type *b* ; clusters of element *b*, with impurities of element *a*, and metastable phases *a-b*. By annealing, the segregation of impurities and the decomposition of metastable phases would produce changes in the volume distribution of clusters and therefore in magnetic properties. In general the effective volume fraction of the magnetic phase, x, would also change with annealing. Consider that clusters of element *a* are magnetic and that some metastable phases *a-b* are also magnetic. Lett a_yb_{1-y} be the nominal composition of the alloy. A complete segregation of *a* and *b* elements would lead to y=x. However, a fraction of y atoms of element *a* are impurities, of type *b* clusters, and consequently non magnetics, other fraction are forming metastable phases which can be either magnetic with different Curie temperatures or non magnetic and finally the rest are forming the magnetic type-*a* clusters. Therefore for any configuration the magnetic volume fraction, x, can be different to the volume fraction of *a*-atoms,y. Changes in configuration should produce changes in x. By keeping unchanged the configuration, x should be a function of temperature, if the different magnetic phases exhibit different Curie temperatures.

Consider now that by increasing the temperature a fraction of the ferromagnetic phase evolves to a paramagnetic state. The overall effect should be a decrease of x. Nevertheless, the change produced in the total number of magnetic clusters, $N=x/\langle v\rangle$,, should depend on the morphology of the system. If the phase which has evolved to the paramagnetic state were forming independent clusters, N would decrease during the transition. But if the clusters of this phase were connected to other magnetic clusters, its transition to the paramegnetic state would increase N. This last effect has been clearly observed in nanocrystalline materials in which the number of magnetic clusters evolves from 1 to the total number of crystallites when the amorphous matrix undergoes the ferromagnetic to paramagnetic transition. At this point x decreases but N increases dramatically.The role of the matrix was to provide a perfect exchange coupling between the different crystallites. Therefore, in this example is pointed out that the role of the intergranular system is of maximum relevance in the magnetic properties of heterogeneous systems.

In the next section we will describe two experiments recently performed and reported which illustrates the enormeous influence of the interphase on the magnetic properties of heterogeneous systems.

TWO EXPERIMENTS ILLUSTRATING THE INFLUENCE OF THE INTERPHASE ON THE MAGNETIC PROPERTIES OF SOFT NANOCRYSTALLINE MATERIALS

As observed by Yoshizawa et al. [8] for the first time the Fe rich nanocrystals are soft two-phase materials which can be obtained by crystallization of conventional FeSiB amorphous alloys with small addition of Cu and Nb. After crystallization the sample consists of small nanocrystals embedded in the residual amorphous matrix. The magnetic softness has been explained by Herzer [7] and is related to the ratio of the exchange correlation length (or domain wall thickness), L, to the orientation fluctuation length, d, of randomly distributed local easy axes, which in this case is the average crystallite size. For L>>d the macroscopic structural anisotropy averages out and the domain wall can move without hindrances. Moreover, for a critical crystallized volume fraction, x, the average magnetostriction vanishes; thus magnetoelastic contributions to the macroscopic anisotropy also become negligible. A phenomenological theory which takes into account the two-phase character of the sample has been recently developed [9] by considering that the exchange coupling between magnetic moments of adjacent crystallites takes place through the amorphous matrix.

The exchange A between crystallites has been substituted by γA, where γ is defined as

$$\gamma=\exp(-\Lambda/L_{am}) \qquad (13)$$

where L_{am} is the exchange correlation length of the amorphous matrix

Recently, Hernando and Kulik [10] and Slawska-Waniewska et al. [11] have measured the thermal dependence of coercivity for nanocrystalline samples obtained by annealing Fe-Si-B-Cu-Ta and Fe-Zr amorphous alloys, respectively. It was found that for any particular pair of x and d values, which depend on the annealing conditions, the coercive field shows an abrupt maximum, Fig. 1. The existence of this maximum evidence the presence of the factor γ in the expression of the exchange between crystallites.As T raises and approach to the Curie temperature of the matrix, L_{am} increases, but suddenly decreases when the matrix becomes paramagnetic. At this point γ decreases and the exchange between grains is so weak that can not overcome the crystalline anisotropy. The maximum occurs at a temperature, T_p, which is close to the Curie temperature of the amorphous, $T_c{}^a$, matrix, for those values of x which correspond to Λ>5 nm. Nevertheless, when x increases and the corresponding distance between the surfaces of adjacent crystallites verifies Λ<5 nm the maximum of coercivity takes place at temperatures well above $T_c{}^a$ i. e. $T_p=T_c{}^a+100^{\circ}$ C. This behavior, which can be well understood in the framework of the two-phase theory developed in [9], clearly shows that exchange interactions take place at temperatures well above $T_c{}^a$.

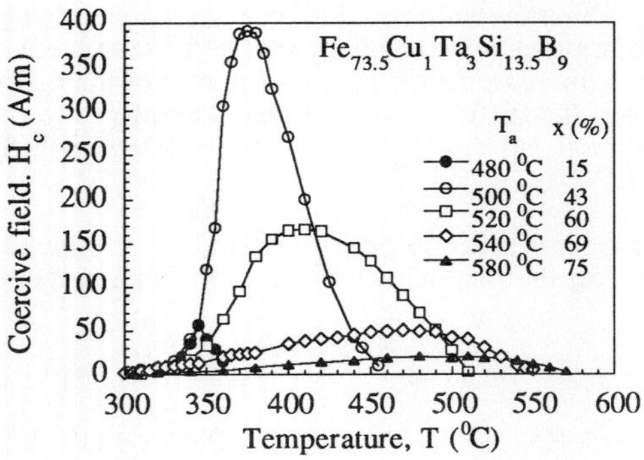

FIG. 1. Temperature dependence of the coercive field after one hour annealing at different temperatures and therefore at different crystallized volume graction, x.

Another interesting result showing the influence of the interaction between magnetic phases with different Curie temperature is pointed out with the following experiment. Amorphous ribbons, with the following compositions, were obtained by melt spinning technique: (A) $Fe_{96-x}B_xNb_3Cu_1$, with x=19, 20, 22.5, and 25; (B) $Fe_{77}B_{22-x}Nb_xCu_1$, with x=3, 4, 5, 6, and 7; (C) $Fe_{74.2}B_{20.2}Nb_{4.5}Cu_{1.1}$, $Fe_{70.7}B_{22.9}Nb_{5.1}Cu_{1.3}$ and $Fe_{67.1}B_{25.7}Nb_{5.7}Cu_{1.4}$.

In order to achieve different steps of the nanocrystallization process, samples of $Fe_{77}B_{18}Nb_4Cu_1$ were submitted to subsequent heat treatments performed under an Ar atmosphere, at temperatures ranging from 455 to 700°C.

The heat treatments give rise to an ultrafine structure of α-Fe crystallites, with approximately uniform grain size (10-12 nm) but increasing the volume fraction with the annealing temperature from 8 to 40%. The volume fraction of crystallized Fe, x, is accurately measured from the relative resonant area of the Mössbauer subspectra. The nominal composition of the intergranular amorphous region was determined by subtracting the relative number of crystallized Fe atoms, obtained from x, to the composition of the starting amorphous ribbon [12].

The thermal dependence of M_S for nanocrystalline samples exhibits the typical behavior of two-phases systems. The lower Curie temperature, which corresponds to the intergranular amorphous region T_C^a (<Curie temperature of crystallites=T_C^{cr}), can be obtained by determining the intersection point of the steepest tangent to $M_S(T)$ curve, with the T axis and with the magnetization curve of the crystallites extrapolated down to temperatures T< T_C^a, respectively [10]. It has been observed than T_C^a increases when the volume crystallized of Fe increases. The effective crystallized volume was determined by Mossbauer spectroscopy. Taking into account that the Nb and B content of the amorphous matrix grows with the heat treatments as well as considering the drastic influence of Nb atoms in decreasing the Curie temperature of the amorphous alloys, we are not able to explain this result.

A new experiment was performed in order to resolve this important question: Is the origin of the enhancement of the Curie temperature of the interphase any compositional changes of the amorphous matrix, or is the magnetic nature of the interphase changed by the presence of α-Fe particles?

Amorphous ribbons with the same nominal composition that the corresponding to the intergranular amorphous region were prepared (Series C) and their Curie temperature, $T^*{}_C^a$, were measured using the same procedure. It has been found differences between the thermal dependence of the saturation magnetization for nanocrystalline samples and that corresponding to an amorphous ribbon of the same composition that the interphase. It is remarkable that the Curie temperature of the intergranular region, T_c^a, is higher than the corresponding to an amorphous ribbon, $T^*{}_c^a$, of the same composition. The difference is large enough to discard errors in both composition estimation and measuring procedure.

Table 1 summarises the relevant experimental results corresponding to different nanocrystalline states of the $Fe_{77}B_{18}Nb_4Cu_1$ alloy, obtained after different annealing temperatures. Volume fraction of crystallized Fe (x), average grain size (d), average matrix thickness (L), and the estimated composition of the matrix are shown. The estimation of compositon was done by using the Mossbauer results which allow us to know x and thereby, through the initial compositon of the sample, the compositon of the residual matrix. For details see ref.[12] .

The corresponding T_c^a-$T^*{}_c^a$ for the states described in Table 1 are shown in Fig.2.

Tann (°C)	% at.Fe E. Moss	Estimated amorphous matrix composition	d (nm)	L (nm)
as cast	0	$Fe_{77}B_{18}Nb_4Cu_1$		
455	8	$Fe_{75.5}B_{19.2}Nb_{4.3}Cu_1$	9	12
475	14	$Fe_{74.2}B_{20.2}Nb_{4.5}Cu_{1.1}$	10	11
510	22	$Fe_{73.5}B_{20.7}Nb_{4.5}Cu_{1.2}$	10	7
530	28	$Fe_{70.7}B_{22.9}Nb_{5.1}Cu_{1.3}$	10	5
555	32	$Fe_{69.5}B_{23.9}Nb_{5.3}Cu_{1.3}$	10	4
580	32	$Fe_{69.5}B_{23.9}Nb_{5.3}Cu_{1.3}$	11	4
590	39	$Fe_{67.1}B_{25.7}Nb_{5.7}Cu_{1.4}$	12	4

TABLE I. Estimated amorphous interphase composition and relevant parameters of the $Fe_{77}B_{18}Nb_4Cu$ alloy, in the as-cast state and in nanocrystalline samples ennealed at different temperatures.

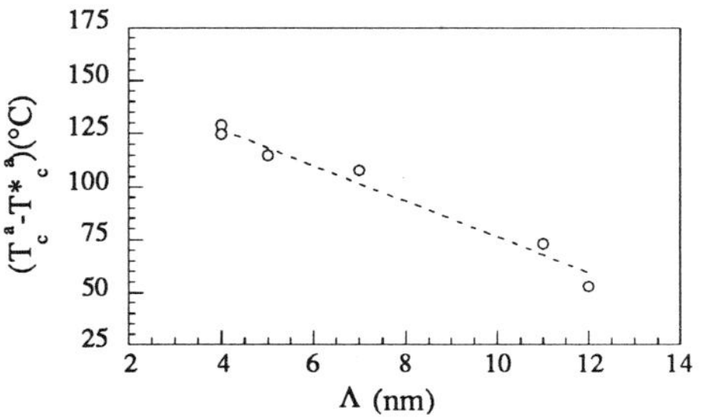

FIG. 2. Difference between the Curie temperatures of the intergranular region in nanocrystalline samples and the Curie temperatures of the amorphous sample with the same composition as the interphase, $(T^a{}_c - T^{a*}{}_c)$, versus the interphase thickness.

CONCLUSIONS

The influence of the exchange interactions between different magnetic phases determines the average volume of the magnetic clusters. According to the ratio between the exchange correlation length of the clusters and its average volume, different types of magnetism can be achieved in the same system. Temperature is the more important physical property which governs this ratio and therefore the macroscopic character of heterogeneous magnetic systems, in which the local properties fluctuates at the nanometric scale.

REFERENCES

[1] A.E. Berkowitz, A.P. Young, J.R. Mitchell, S. Zhang, M.J. Carey, F.E. Spada, F.T. Parker, A. Hutten ang G. Thomas, Phys. Rev. Lett. **68** (1992) 3745.

[2] J.Q. Xiao, J.S. Jiang and C.L. Chien, Phys. Rev. Lett. **68** (1992) 3749.

[3] J.R. Childress and C.L. Chien, J. Appl. Phys. **70** (1991) 5885.

[4] A. R. Yavari, P.J. Desré and T. Benameur, Phys. Rev. Lett. **68** (1992) 2235.

[5] M. El Ghannami, C. Gómez-Polo, G. Rivero and A. Hernando, Europhys. Lett., **26** (9) (1994) 701.

[6] R. Alben, J.J. Becker, and M.C. Chi, J. Appl.. Phys. **49**, 1653 (1978).

[7] G. Herzer, IEEE Trnas. Magn. **25**, 3327 (1989); Mater.Sci. Eng. **A133**, 1 (1991); Phys.Scr. **T39**, 307 (1993).

[8] Y. Yoshizwa, S. Oguma and K. Yamauchi, J. Appl.. Phys. **64**, 6044 (1988).

[9] A. Hernando, M. Vázquez, T. Kulik and C. Prados, Phys. Rev. B 51, 6, 3581 (1995)

[10] A. Hernando and T. Kilik, Phys. Rev. **49**, 7064 (1994).

[11] A. Slawska-Waniewska, P. Nowicki, H.K. Lachowicz, P. Gorría, J.M. Barandiarán, and A. Hernando, Phys. Rev.. B **50**, 6465 (1994).

[12] A. Hernando, I. Navarro, and P. Gorría, Phys. Rev. B **51**, 3281 (1995).

Materials Science Forum Vol. 195 (1995) pp. 149-154

Icosahedral Order in Transition Metal Clusters

G. D'Agostino

ENEA C.R.E. Casaccia, Dip. Innovazione sp 59, CP 2400 Roma, I-00100, Italy

Keywords: Cluster, Molecular Dynamics, Transition Metals, Icosahedron and Tight Binding

Abstract

The present paper is aimed at studying the transition from the molecular to the bulk behaviour of (fcc) transition metals. To this purpose, the cohesive energies of cluster of different size and shape have been computed by means of a many-body tight binding potential.

The resulting energies for the different structure and size have been reported and interpreted according to simple nuclear fits. It turned out that clusters consisting of some thousands atoms already exhibit the typical fcc behaviour.

First neighbours distance contractions have been reproduced as expected from many experimental measurements.

Introduction

One of the most interesting problems related to the cluster physics concerns the transition from the molecular to the bulk behaviour. It has been quite recently clarified that argon molecules turn from the icosahedral to the fcc structure once they reach a critical size Nc that is about 3500 [1]. The method employed in the above paper, conjugated with a simple tight binding approach to the cohesive energy evaluation, has allowed, the estimate of the critical size of both the Cu [2] and Au [3] clusters. It turned out that transition fcc metals reach their stable fcc configuration at an early stage. Pure and binary metal clusters have been also investigated by means of the embedded atom method leading to very interesting results [4].

The purpose of the present paper is to extend the previous calculations in the tight binding scheme to all the fcc (face centered cubic) transition elements.

A proper understanding of the cluster physics should, in principle, relate to both the electronic and the ionic degrees of freedom. Although several efforts have been devoted to develop suitable techniques to treat the problem, only few results have been obtained since. Actual 'ab initio' methods have been applied to small clusters (see for example self consistent Hartree-fock electronic calculation for Cu_{13} Al_{13} and Ni_{13} [5] or for dimers and trimers [6]). Unfortunately, the former methods are very time consuming and hence not suitable to be applied to systems of hundred particles. On the other hand, quite recently, the well established 'Car-Parrinello' method has been improved in order to deal with transition metal systems, leading to interesting results especially to what concerns the liquid copper [7]. Nevertheless no proper simulation of extended metallic clusters has been performed by means of that method. From a different point of view many authors developed approximate methods to obtain electronic binding energy of transition and alkalin metals (e.g. [8]). Since they deal with the electronic degrees of freedom only, the latter methods do not allow any prediction of the stable structure of the metallic clusters. Therefore, it has also been introduced a hybrid procedure, handling both the electronic and the ionic degrees of freedom [9]. The former procedure allows very accurate reproduction of both the electronic magic numbers and the structures of the aggregates and hence, in principle, should be the elective tool. Unfortunately, the method above is not based on an analytical potential and hence it does not fit the purposes or the present paper. Furthermore, the effects related to the closing of the electronic shell are expected to become less relevant when increasing the size.

In conclusion, a classical approach has been assumed and, hence, the atoms have been treated as point-like particles interacting by means of a (manybody) potential.

The potential

The interaction potential employed in the present paper exhibits the following analytic form:

$$U = \sum_{i=1}^{N} U_i = \sum_{i=1}^{N} (Ur_i - Ua_i)$$

$$Ur_i = \sum_{j=1}^{N} A\exp(-p(r_{ji}/r_0 - 1))$$

$$Ua_i = \xi\sqrt{\sum_{j=1}^{N} \exp(-2q(r_{ji}/r_0 - 1))}$$

where N is the number of atom in the aggregate, r_0 is the lattice parameter at zero Kelvin and a fifth neighbours cut-off has been intended throughout. A, ξ, p and q represent phenomenological parameters that characterise the element to be simulated and are currently derived from of its most relevant properties. The values applied to the former parameters result from a best fit based on the reproduction of the cohesive energy and many elastic modules [10].

The form of the potential, in principle, could be justified empirically by its capability to reproduce relevant physical features. Nevertheless, as it has been clarified by Finnis and Sinclair [11], there are two different ways to explain the dependence of the cohesive energy on the atomic positions. In the tight binding scheme the potential is simply a truncation to the second order of the moment expansion of the hoping integrals assumed to behave exponentially. On the other hand, following the embedded atom method approach, the terms under the square root represent the contribution to the local charge density over the i-th atom. Again, they have been assumed to decrease exponentially.

Low temperature stability

The cohesive energy of an aggregate of fixed geometry is an increasing function of its size. Nevertheless, also the structure and the surface termination strongly influence the stability of the small aggregates. The main aim of the present session is to discuss the behaviour of the cohesive energy for system of the same shape (that is the lattice structure and the termination) as increasing the size.

In principle, one is not allowed to assume the configuration of very small clusters to be similar to the bulk one. However, sufficiently extended system have to exhibit the lattice structure of the element into account (fcc in the present case). Therefore there must be a "critical size" where the cluster turns from the molecular to the bulk behaviour. That is the smallest size at which the bulk structure is settled down.

To find out the most cohesive structure of a cluster of given size one should in principle span all possible configurations of such system. Owing to the great number of degrees of freedom involved, one has no chance to perform an exhaustive scanning. Therefore, one is forced to search a local minimum in the neighbours of some 'a priori' selected structures. The high level of compactness and co-ordination makes the icosahedral package one of the most favourite initial structures. On the contrary, as representative of the fcc structures two different terminations have been selected, namely the troncoctahedron and cuboctahedron [1,12]. As Ino first demonstrated (by means of analytical calculation) the troncoctahedron is the most cohesive termination of the fcc structures. Nevertheless, in the present paper the cuboctahedral clusters have been also considered since they share the closed shell size with the icosahedral structure and, hence, a direct comparison is provided.

There exist many other attractive initial structures, such as the hcp and the decahedral ones (partly treated elsewhere [13]); they have not been presently considered and represent the subject of a future work.

To the purpose of achieving the classical equilibrium states, each configuration of the selected shapes and sizes has been constructed and quenched to its local minimum by means of the steepest descend algorithm.

The resulting configurations (of the different elements, structures and sizes), all exhibit a contraction in the first neighbours average distance as was expected from some experimental observation [15]. The values of the first neighbour average distance have been reported in table I whereas the resulting cohesive energies for all the transition fcc metals have been reported in table II.

STR	<N>	Ni	Cu	Ag	Rh	Ir	Pd	Pt	Au
Cub	13	2.4814	2.4544	2.7564	2.6505	2.6516	2.6061	2.6182	2.7063
Cub	55	2.4658	2.4997	2.8219	2.6660	2.6824	2.6778	2.6978	2.7952
Cub	147	2.4712	2.5157	2.8433	2.6725	2.6929	2.6981	2.7170	2.8161
Cub	309	2.4758	2.5237	2.8515	2.6759	2.6980	2.7079	2.7282	2.8285
Cub	561	2.4785	2.5284	2.8575	2.6785	2.7011	2.7150	2.7343	2.8363
Cub	923	2.4805	2.5320	2.8625	2.6799	2.7031	2.7184	2.7405	2.8436
Ico	13	2.4251	2.4185	2.7730	2.5878	2.5883	2.5682	2.6367	2.6702
Ico	55	2.4491	2.4924	2.8126	2.6334	2.6476	2.6692	2.6726	2.7680
Ico	147	2.4543	2.4946	2.8171	2.6389	2.6626	2.6742	2.6949	2.8042
Ico	309	2.4516	2.4980	2.8227	2.6434	2.679	2.6804	2.7020	2.8076
Ico	561	2.4564	2.5018	2.875	2.6463	2.6668	2.6801	2.7067	2.8149
Ico	923	2.4541	2.5025	2.8232	2.6486	2.6661	2.6865	2.7074	2.8132
Tro	38	2.4425	2.460	2.7813	2.6359	2.642	2.6265	2.6359	2.7644
Tro	201	2.4790	2.5227	2.8507	2.6784	2.6969	2.7025	2.7237	2.8274
Tro	586	2.4805	2.5289	2.8576	2.6800	2.7006	2.7154	2.7350	2.8435
Tro	1289	2.4847	2.5352	2.8665	2.6833	2.7051	2.7241	2.7454	2.8536
fcc	Bulk	2.4911	2.5562	2.8885	2.6891	2.7146	2.7485	2.7747	2.8843

Table I: First neighbours average distances (in Å units) for the different elements and structures. The label Cub, Ico and Tro refer to cuboctahedral, icosahedral and troncoctahedral clusters respectively.

In order to characterise the behaviour of the cohesive energy coherently and to extrapolate it to the missing sizes nuclear fits of each type of structure have been performed in the following form:

$$U / N = V_0 + aN^{1/3} + bN^{2/3}$$

where U/N represents the cohesive energy (i.e. the energy per particle) and V_0, a, and b are the volume, surface and curvature specific contributions respectively. In table III the resulting values of the preceding fits are reported for the different fcc transition metals.

It is worth stressing that Ino's prediction about the stability of the trococtahedral terminated structure turn out to be correct. The result is not trivial owing to the large number of hypotheses involved into Ino's argument that, in principle, applies only to very large systems of atoms interacting by means of a pair potential with a first neighbours cut-off.

As it has been already stressed above, a useful estimate of the critical size Nc is represented by the amount of atoms where the icosahedral and the troncoctahedral curves cross. The intersecting values (Nc) are presented in last two lines of table III together with the values one would have obtained considering the cuboctahedral termination instead of the troncoctahedral one (Ñc).

Shape	Ni	Cu	Ag	Rh	Ir	Pd	Pt	Au
Cub 13	-2.6033	-2.4530	-2.2872	-3.7682	-4.8856	-3.2090	-4.9291	-3.2400
Cub 55	-3.3007	-2.2930	-2.5517	-4.5297	-5.6795	-3.4960	-5.2939	-3.4630
Cub147	-3.6158	-3.1040	-2.6693	-4.8741	-6.0359	-3.6231	-5.4556	-3.5430
Cub309	-3.7950	-3.2029	-2.7347	-5.0685	-6.2355	-3.6935	-5.5450	-3.5990
Cub561	-3.9102	-3.2652	-2.7762	-5.1930	-6.3625	-3.7382	-5.6016	-3.6320
Cub923	-3.9903	-3.3038	-2.8048	-5.2792	-6.4503	-3.7690	-5.6407	-3.6550
Ico 13	-2.6725	-2.6100	-2.3432	-3.8939	-5.0553	-3.2665	-4.9936	-3.2730
Ico 55	-3.3366	-2.9710	-2.5822	-4.6069	-5.7731	-3.5243	-5.3235	-3.4680
Ico 147	-3.6361	-3.1300	-2.6887	-4.9212	-6.0933	-3.6398	-5.4719	-3.5550
Ico 309	-3.8076	-3.2200	-2.7478	-5.1001	-6.2727	-3.7037	-5.5538	-3.6030
Ico 561	-3.9153	-3.2760	-2.7853	-5.2141	-6.3865	-3.7443	-5.6057	-3.6550
Ico 923	-3.9908	-3.3150	-2.8111	-5.2928	-6.4651	-3.7722	-5.6416	-3.6710
Tro 38	-3.1855	-2.8747	-2.5208	-4.4199	-5.5805	-3.4670	-5.2616	-3.4320
Tro 201	-3.7128	-3.1611	-2.7130	-4.9899	-6.1653	-3.6731	-5.5215	-3.5860
Tro 586	-3.9274	-3.2750	-2.7887	-5.2196	-6.3978	-3.7538	-5.6229	-3.6450
Tr1289	-4.0437	-3.3336	-2.8288	-5.3432	-6.5220	-3.7964	-5.6768	-3.6760
Bulk	-4.4350	-3.5440	-2.9600	-5.7520	-6.9280	-3.9360	-5.8530	-3.7790

Table II: Cohesive energies (in eV units) for the various structures and sizes. The labels Ico, Cub and Tro refer, orderly, to Icosahedral, cuboctahedral and Troncoctahedral clusters.

A	Ni	Cu	Ag	Rh	Ir	Pd	Pt	Au
Tronc.								
V_0	-4.4349	-3.5436	-2.9600	-5.7725	-6.9483	-3.9597	-5.8558	-3.7789
a	4.2759	2.2453	1.4063	4.67883	4.61157	1.4908	2.0627	1.0868
b	-0.2552	0.0101	0.2368	-0.4486	-0.0506	0.2886	0.27367	0.2599
Icos.								
V_0	-4.4102	-3.5295	-2.9539	-5.7243	-6.9001	-3.9275	-5.8411	-3.7707
a	4.0786	2.0621	1.3736	4.17398	4.19673	1.4940	1.9202	1.1148
b	0.01738	0.2354	0.1476	0.30316	0.33257	0.14207	0.17172	0.1290
Cuboc.								
V_0	-4.4347	-3.5434	-2.9604	-5.7718	-6.9484	-3.9271	-5.8518	-3.7798
a	4.3323	2.3106	1.4982	4.77695	4.80635	1.6076	1.8687	1.2063
b	-0.0623	0.1183	0.2006	-0.1604	0.09906	0.24253	0.38818	0.1511
Nc	330	1555	450	650	450	80		
Ñc	674	1900	2750	1675	1830	1385	865	620

Table III: Volume, surface and curvature specific energies (in eV units) resulting from the nuclear fits. The labels Icos., Cuboc. and Tronc. refer to the icosahedral, cuboctahedral and troncoctahedral structures orderly. Last line exhibit the critical sizes resulting from the comparison of the icosahedral behaviour with the troncoctahedral (Nc) and the cuboctahedral (Ñc) one respectively.

So far only T=0K equilibrium condition has been considered. When dealing with non null temperatures the cohesive energy should be replaced by the free energy in any comparison. To the purpose of calculating the free energy of the cluster at very low temperature, the different phonon distributions have been evaluated. Since the effects concerning the freezing of the ionic degrees of freedom become relevant at low temperature, the vibrational degrees of freedom must be treated at the quantum level. As a consequence, the vibration contribution to the free energy becomes that of a gas of phonons [14]:

$$F = K_b \sum_{\omega i} \log(2\sinh(\hbar\omega_i / 4\pi K_b T))$$

where ω_i represent the proper frequencies of the system and the sum is carried out over all the non null frequencies. Therefore, in principle, the critical size predicted by means of the cohesive energy may change in consequence of a phonon dressing at non zero temperatures. Preliminary results concerning the Cu and Au clusters only, exhibit very week changes in the critical size tending to increase it of just some tenths of atoms.

A final comment is in order concerning the actual frequency of the above structures in the physical samples. Apart form the relevant point above, there is an other more important argument that causes the present results to be not directly transferable to the experimental samples. The values in table II (and the derived critical sizes) refer to the isolated clusters and, hence, they have no trace of the environment where they are nested. Therefore the present results, although providing interesting suggestions, do not directly apply to:

- nanophase materials;
- clusters enveloped in matrices, such as the ones in polydisperse EXAFS samples [15]

or to

- clusters with ligates, such as those obtained with the Schmidt process [16].

Moreover, the very week difference in the volume specific energies V_0 suggests the hypothesis of a co-existence of various structures in all the previous samples. In fact, the existence of very high potential barriers between the different structures does not allow to achieve the thermodynamic equilibrium under the kinetic conditions involved in the preparation processes.

Conclusions

Results of molecular dynamics simulation of different transition metal clusters have been presented. Structural and thermodynamic properties of the low energy configuration has been evaluated by means of a simple tight binding potential.

First neighbours distance contraction of some percent has been observed for all the considered elements and structures in agreement with many experimental measurements. As expected the former contraction decrease with increasing the cluster size.

The critical size of all the fcc transition elements have been estimated as the cross over between fcc and icosahedral energetic curves. An amount of about thousand atoms turned out to be sufficient for the transition metal clusters to exhibit the typical bulk behaviour.

Finally, owing to the very thin difference in the cohesive energy of the different structures, long range icosahedral order is to be expected in many physical samples.

References:

[1] L. Boyer and J. Q. Broughton Phys. Rev. B 42, 11461 (1990).
[2] G. D'Agostino Phil. Mag. B 68, 903 (1993).
[3] G. D' Agostino, A. Pinto and S. Mobilio Phys. Rev. 48, 14447 (1993).
[4] J.M. Montejano-Carrizales, M.P. Iñiquez and J. A. Alonzo Phys. Rev. B 49, 16649 (1994).
[5] C Elsässer, M Fähnle, E H Brandt and M. C. Bohm
J. Phys. F 17, L301 (1987).
[6] C. W. Bauschlicher, Jr. S. R Langhoff and H. Partidge J. Chem. Phys. 91, 2412 (1989).
[7] A. Pascarello,K. Lasonen, R. Car,R. Lee and D. Vanderbilt Phys. Rev. Lett. 69,1982 (1992).
[8] W. D. Knight, K. Clemenger, W. A. SAunders, M. Y. Chou and M. L. Cohen

Phys. Rev. Lett. 52, 2141 (1984).
[9] O. B. Christensen, K. W. Jacobsen, J.K. Norskov and M. Manninen Phys. Rev. Lett. 66, 2219 (1991).
[10] F. Cleri and V. Rosato Phys. Rev. B 48, 22 (1993).
[11] W. Finnis and J. Sinclair Phil. Mag. A 50, 45 (1984).
[12] B. W. van de Waal J. Chem. Phys. 98,4909 (1993).
[13] J. Uppenbrink and D. J. Wales J. Chem. Phys. 96, 8520 (1992).
[14] A. A. Maradudin, E. W. Monroll, G. H. Weiss and I. P. Ipatova *"Theory of lattice dinamics in the harmonic approximation"*, Solid State Phys. Suppl.3 (New York: academic).
[15] P. A. Montano, C. K. Shenoy, E. Alp, E. Shulze and J. Urban, Phys. Rev. Lett. 56, 2076 (1986).
[16] G. Schmid, R. Pfeil, R. Boese, F. Bandermann, S. Meyer, G.H. M. Calis and J.A.M. Velden Chem. Ber. 114, 3639 (1981).

Materials Science Forum Vol. 195 (1995) pp. 155-160

Near Melting Properties of Copper Clusters

G. D'Agostino and V. Rosato

ENEA C.R.E. Casaccia, Dip. Innovazione sp 59, CP 2400 Roma, I-00100, Italy

Keywords: Copper, Cluster, Melting, Molecular Dynamics and Tight Binding

Abstract

The present paper is aimed at studying the physics of small copper cluster at the atomic level in an extended range of temperature including the melting transition. The results have been obtained by means of molecular dynamics assuming an icosahedral structure for the classical ground states and a Finnis-Sinclair interactive potential. The celebrated phenomenon of phase coexistence has been widely discussed. It turned out that the most suitable order parameter to evidence the phase transition is not the density but the diffusivion coefficient of the system.

Introduction

During last decades a lot of efforts have been devoted to the physics of the small aggregates. The oldest interest in the subject arises from the crystal growth. At this respect, the attention was focussed on the structure of the grain at the very early stage of the nucleation process. The developing of new techniques of production also introduced a rather technological interest mainly related to the cluster design of new materials. Furthermore, even the fundamental physics of the phase transitions has been involved due to the discovering of the celebrated coexistence phenomenon in argon clusters [1].

The present paper is aimed at studying the thermodynamic and the structural properties of copper clusters in a wide range of temperature including the melting transition.

Several physical properties of the small aggregates are related to their electronic degrees freedom [2]. Therefore, in principle, one should include in the approach the electronic quantum behaviour. Due to essentially technical reasons concerning the number of particles in the aggregates (from 13 to 147), one is forced to simplify the treatment of the electronic degrees of freedom. In the present case the atoms have been considered classical point-like particles thus confining the electron influence into an interatomic effective many-body potential. A more accurate description of many phenomena directly related to the active presence of the electrons (such as the existence of magic numbers) has been obtained treating the electrons as a gas in suitably shaped ellipsoid [3].

Simulation Layout

As it was stated above, a classical molecular dynamics approach has been assumed. As a consequence, the interatomic potential becomes the only relevant issue that may influence the results. In the present paper a Finnis-Sinclair type [4] has been selected since it had provided good results for both bulk and defective systems [5]. The analytic form of the cohesive energy may be derived from both the embedded atom method [6] and the tight binding approach [7]. Disregarding its origin, the potential consists into a repulsive part of Born-Mayer type and an attractive many-body one. In the present model it depends only on four parameters A, ξ, p and q that have been fitted on bulk properties:

$$U = \sum_{i=1}^{N} U_i = \sum_{i=1}^{N} (Ur_i - Ua_i)$$

$$Ur_i = \sum_{j=1}^{N} A\exp(-p(r_{ji}/r_0 - 1))$$

$$Ua_i = \xi\sqrt{\sum_{j=1}^{N}\exp(-2q(r_{ji}/r_0 - 1))}$$

where N is the number of atom in the aggregate and r_0 is the lattice parameter at zero Kelvin. A fifth neighbours cut-off has been applied throughout.
The equation of motion have been integrated employing a time-increment of 10-15 s. No boundary conditions have been applied in order to reproduce the environment of the free clusters in the vacuum.

Classical Ground State

Prior to deal with the behaviour of clusters near their melting transition one has first to specify their structure and shape at low temperature. Following previous results [8], icosahedral clusters of 13, 55 and 147 atoms have been selected as suitable ground states. Table 1 reports the cohesive energy and the diameters of the configurations above calculated at T=0K. All the former structure resulted to be mechanically stable with 3N-6 positive vibration frequencies. In agreement with some EXAFS experiments, the interatomic distances clusters resulted to be contracted with respect to the bulk equivalent ones.

Size	diameter (Å)	U/N (eV)
13	4.81	-2.61
55	9.65	-2.97
147	14.53	-3.13

Table 1. Size diameter and cohesive energy of the cluster ground states.

Thermodynamic Properties

The initial clusters have been heated gradually up to about a thousand Kelvin Degrees. The mean values of the cohesive energy per particle (averaged over about a ns) have been reports in fig. 1

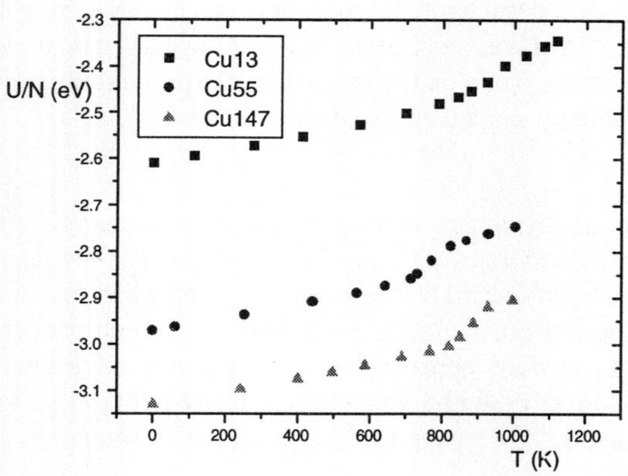

Fig. 1 Caloric curves of the three different aggregates. The cohesive energy per particle is reported as a function of the temperature.

As it can be seen, only changes in the slope are observed corresponding to the regions where the melting transition seems to take place. In order to enhance the effect due to the phase transition, the specific heat has been evaluated by direct derivation of the caloric curve. In fig. 2 the different specific heat-s have been compared. The specific heat has been also evaluated by means of the temperature fluctuation [9] leading to similar unless not identical results. Finally, as it was expected, the atomic heat of all the clusters tends to 3K at very low temperatures.

From the point of view of statistical mechanics, only systems having an infinite number of particle may exhibit a phase transition since thermodynamic potential of systems with a finite number of degrees of freedom are analytic. Never the less at the computational level systems with finite degrees of freedom exhibit neat jumps in the thermodynamic potentials thus reflecting the presence of the phase transition in the thermodynamic limit.

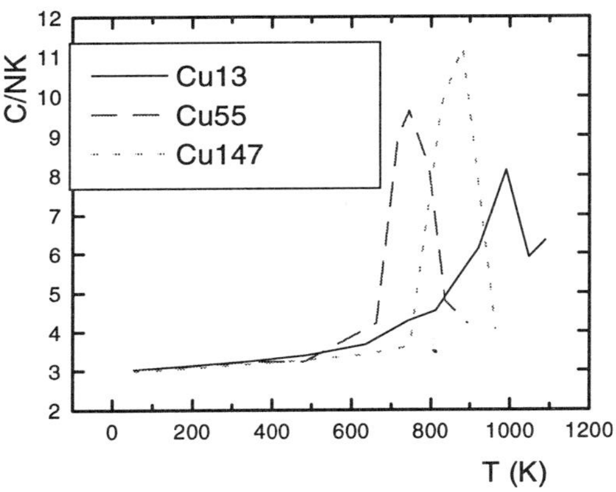

Fig. 2 The different specific heat-s as a function of the absolute temperature. The values are adimensional since they have been divided by the number of atoms N (N-2 actually due to traslation al and rotational zero modes) and the Boltzman constant K.

In order to clarify whether a melting like phenomenon occurs or not, it has been considered the specific volume (or conversely the density) of the aggregate. Since the volume of a non periodic structure is not well defined, an operative definition had been provided. In the present paper, the volume occupied by a system has been regarded as the space covered by the smallest convex polyhedron containing all the atoms inside. Alternative definitions can be obtained by means of the first neighbours average distance or analogous quantities. In fig. 3, the volume thermal profiles has been reported. The curves relative to the Cu_{55} and Cu_{147} show a neat change of the slope in correspondence of the interval where also the cohesive energy changes. This is consistent with the idea of a solid-liquid transition. On the contrary, the curve relative to the Cu_{13} behaves linearly. Moreover, that aggregate exhibits a reduction of the thermal dilatation with respect to the others. This effect seems to be related to the high level of compactness of that structure.

To the purpouse of clarifying the nature of the phase transition, it has been evaluated the Lindenman function δ that is the average fluctuation of the atomic relative distances. Such parameter represents a faithful indicator of the atomic mobility. As it can be seen from fig. 4 (that refers to Cu_{13}) the δ exhibits a neat change in its slope. Therefore, the observed phase transition seems to be essentially related to a change in the diffusion activity. As a consequence, the most sensitive order parameter of the melting transition in the small aggregates is not the density but the atomic mobility.

So far the global thermodynamics approach has been concerned. By that means, basically, three different thermodynamic regions have been identified:

- A solid-like state, at low temperatures (i.e. for the Cu13 below 650K) where the atoms occupy assigned positions in a defined structure
- A liquid-like state, at high temperatures where the atoms are free to move in the space occupied by the aggregate
- An intermediate region, characterised by very quick changes in the thermodynamic quantities.

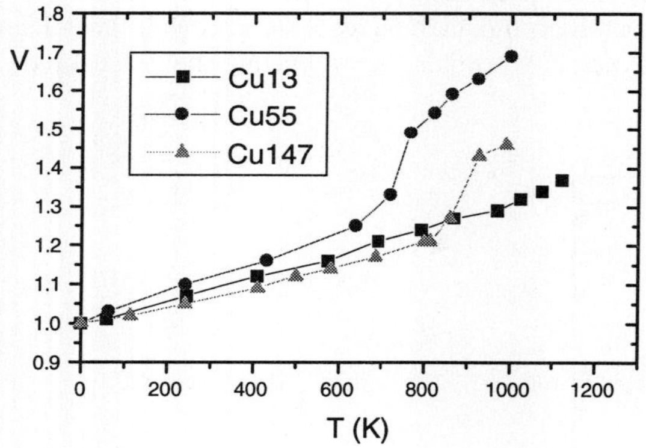

Fig. 3 Relative volume expansion.

Berry was the first to point out that the kinetic aspects of the phenomenon allow a simple interpretation of all the regions in terms of just two phases.

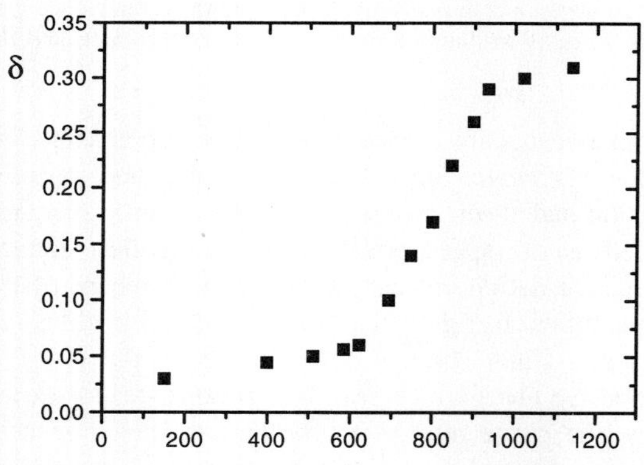

Fig. 4 The Lindenmann variable thermal profile.

From general statistical considerations one expects the various quantities to fluctuate around their average value with an amplitude that increases with decreasing the size. What is not expected (for large systems) is that during the time any thermodynamic quantity fluctuates around two o more modal values. That is exactly what has been observed for the temperature, the density and for all the related quantities in the intermediate region. More precisely, the small systems oscillate from solid-like configurations to liquid-like and vice-versa. Therefore, in the intermediate region, all the mean thermodynamic properties result from an average over the two phases weighted by the rates of time spent in each of them. To outline the presence of both the solid and the liquid phase, the intermediate region is often referred to as coexistence region.

Fig 5 and 6 report the evolution of the temperature along two sample trajectories representative of the typical behaviour. As it may be seen, the duration of the persistence in a phase is an increasing function of the cluster size. Therefore, one expects the life time of the temporary phase to tend to infinite as achieving the thermodynamic limit.

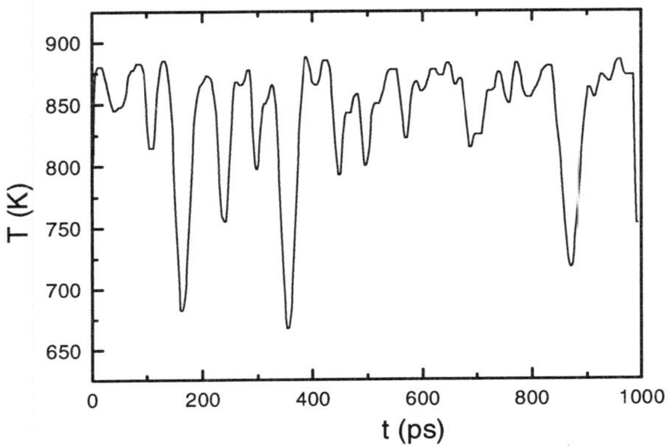

Fig. 5 Temperature fluctuation for the Cu13 simulated cluster. Each point represent a average over 100 steps (i.e. a ps).

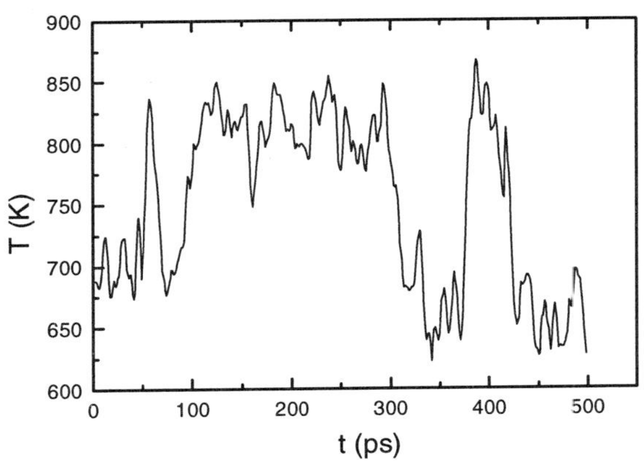

Fig. 6 The Temperature fluctuation for a Cu55 cluster in the coexistence region.

It has been already emphasised that the most relevant order parameter characterising the phase transition is the diffusivity. The molecular dynamics provides a tool to explore the nature of the diffusivion at the atomic level. It has been observed that the elemental event contributing to the diffusivion in coexistence region is a solid-like mechanism that allows the atoms to jump from a stable position of a lattice into an equivalent one. As the temperature is driven toward its highest values in the coexistence region, the number of jumps per time increases thus leading to the typical continuos diffusion mechanism which is proper of the liquids. A further analysis, that has been performed for the Cu_{13} only, demonstrate that any configuration in the coexistence region leads, when quenched, to a perfect icosahedron, thus implying that the mechanism of diffusion does not allow to exit from the attractive basin of that structure. On the contrary when quenching configurations from the liquid phase, a large number of isomers is detected.

As it was already stressed, in the coexistence region the cluster temperature swaps between two modal reference values. In order estimate the most likely modal values, the distribution of instantaneous temperature has been estimated performing simulations of about a ns. The resulting histograms have been then fitted by means of a bivariate distribution. Finally the modal values have been reported on the caloric curve. As it can be seen form fig. 8 the modal averages represent the continuation of the solid and the liquid curve respectively. Similar behaviours resulted for the other clusters.

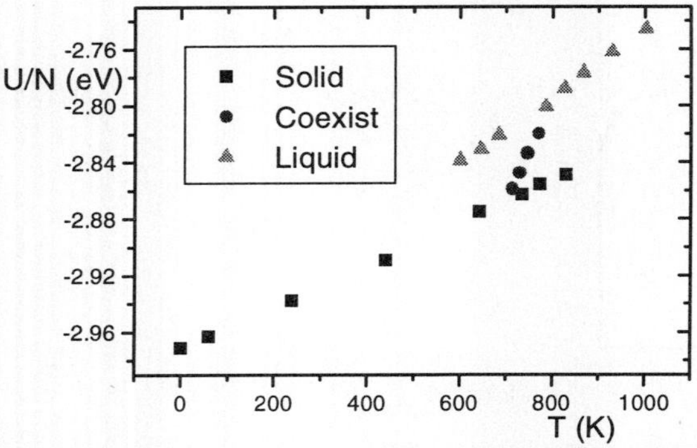

Fig. 7 The cohesive energy of the Cu_{55}. The modal average in the coexistence region have been reported. As it can be seen they prosecute the liquid and the solid curves.

Conclusions

The present paper extends former results relative to argon cluster to copper. The most important variations are represented by the reproduction of the fist neighbours contraction and the absence of the evaporation phenomena which were observed in the previous simulations. The former result is essentially a consequence of the different analytic form of the potential that exhibits a many-body attractive term. Whereas the absence (within the duration of the simulations i.e. a ns) of the evaporation effects is probably related to the increase in the cut-off that, in the present work, extends up to the fifth neighbours.

Finally, since the coexistence phenomenon takes place also employing the present (manybody) potential it seems to be strictly related to the size of the system rather than to the interaction details.

References:

[1] A. Berry, J. Jllinek e G. Natanson Phys. Rev. A 30, 919 (1990).
[2] E.D. Knight, K. Clemenger, W. A. Heer, W.A. Saunders,M.Y.Chou, M. Cohen Phys. Rev. Lett. 52, 2141 (1984).
[3] O.B. Christensen, K. W. Jacobsen, J.K. Norskov and M. Manninen Phys. Rev. Lett. 66, 2219 (1991).
[4] W. Finnis and J. Sinclair Phil. Mag. A 50, 45 (1984).
[5] F. Cleri and V. Rosato Phys. Rev. B 48, 22 (1993).
[6] M.S. Daw and M.I. Baskes Phys. Rev. Lett. 50, 1285 (1983).
[7] D. Tomanek, A.A. Aligia and C.A. Balseiro Phys. Rev. B 32, 3051 (1985).
[8] G.D'Agostino Phil. Mag. B 68, 903 (1993).
[9] J.L. Lebowitw, S.K.Percuz and L.Verlet Phys. Rev. 153, 150 (1967).

Materials Science Forum Vol. 195 (1995) pp. 161-166

Thermodynamic Properties and Optical Characterization of Metal Nanoparticles in Dielectric Matrix

P. Cheyssac[1], M. Geddo[2], R. Kofman[1], P.G. Merli[3], A. Migliori[3], A. Stella[1] and P. Tognini[2]

[1] Laboratoire de Physique de la Matière Condensée, URA 190, Université de Nice Sophia Antipolis, F-06108 Nice Cedex, France

[2] Dipartimento di Fisica „A. Volta", Università degli Studi di Pavia, Via A. Bassi 6, I-27100 Pavia, Italy

[3] Istituto di Chimica e Tecnologie dei Materiali e dei Componenti dell'Elettronica del Consiglio Nazionale delle Ricerche, Via de'Castagnoli 1, I-40126 Bologna, Italy

Keywords: Nanoparticles, Surface Melting, Effective Medium

Abstract

In this paper we intend to give new results on thermodynamic properties of Pb and Sn nanoparticles. We intend also to present a characterization method based on the effective medium theories, that allows a careful knowledge of the structural properties of the samples, depending just on the composition and not on the quantum dimensions of the particles.

Introduction

The fundamental properties of metal and semiconductor nanoparticles (i.e. particles with dimensions of a few nanometers) have been extensively studied in the last ten years, showing the different behavior of these systems with respect to the bulk. In particular, one observes: in the case of semiconductors, a blue-shift of the absorption and photoluminescence peaks, the strong non-linear properties, and, in the case of metal nanoparticles, a decrease of the melting temperature with size, the appearance of the so-called surface melting and of phase fluctuations [1]. Here we concentrate our attention particularly on these thermodynamic properties, showing the results obtained by TEM measurements at various temperatures.

However, it has been previously shown [2] that it is important to carry out such a study also through an optical technique, but, in this case, one immediately has to deal with the following situation: depending on the growth technique utilized, the particles can be enclosed in an amorphous or crystalline matrix, or they can lean against a substrate. When the optical properties are investigated, it is not trivial to separate what is related to the quantum dot dimensions from what just depends on the changes of the sample structure.

On the other hand, it is well known that using the effective medium theories it is possible to have some useful information about the sample structural characteristics. The interesting feature is that any effective medium theory prediction depends on the dimensions of the single particles: comparing these predictions with the experimental results it is possible to conclude that any difference is due to dimension reduction effects.

Experimental data were collected using the well-known Brewster angle technique [3], i.e. measuring the reflectivity of the samples versus incidence angle. Since the condition that allows the application of an effective medium theory is $\lambda >> d$ (i.e. the wavelength of the radiation must be much greater than the dimensions of the particles), and we studied nanoparticles with average radius between 35 and 200 Å, we utilized infrared radiation, with $\lambda = 9.4$ μm.

It should be noted that these theories are usually applied to experiments carried out at near-normal incidence in the visible or near-infrared spectral region, so our approach allows not just a quite accurate characterization of the samples, but also a check of the validity of the various effective medium theories [4].

Sample preparation and thermodynamic properties

The samples were grown in ultra-high vacuum by an evaporation-condensation technique [1], using a sapphire substrate for the optical analysis and a copper grid covered with carbon for the TEM measurements. A SiO layer is evaporated on the substrate, followed by the metal (PB or Sn) evaporation: an aggregation process allows the formation of the nanoparticles instead of a film. A second SiO layer is deposited to cover, and then protect, the particles.

The spatial distribution of the particle sizes obtained in this way is $\sigma \approx \langle d \rangle /4$.

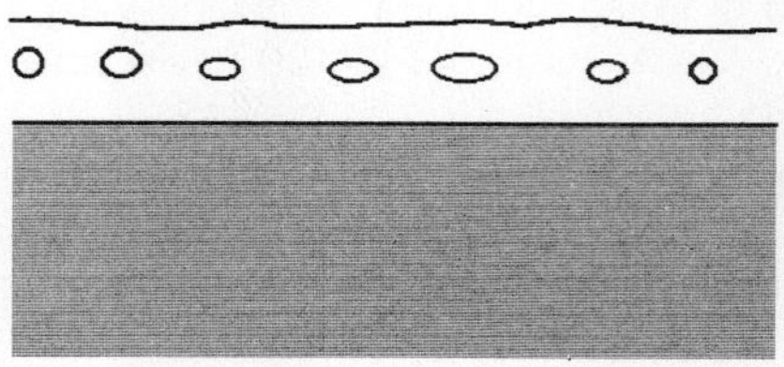

Fig. 1-Dispersion of the particles in the SiO matrix

HREM, DFEM, BFEM, SAD measurements in the range 300-600K were performed on these samples, in order to study nanoparticle thermodynamic properties.

A decrease of the melting temperature Tm with the size of the particles has been highlighted (e. g. in the case of Pb, when the radius is less than 2-3 nm, Tm is about 200K less than the bulk value).

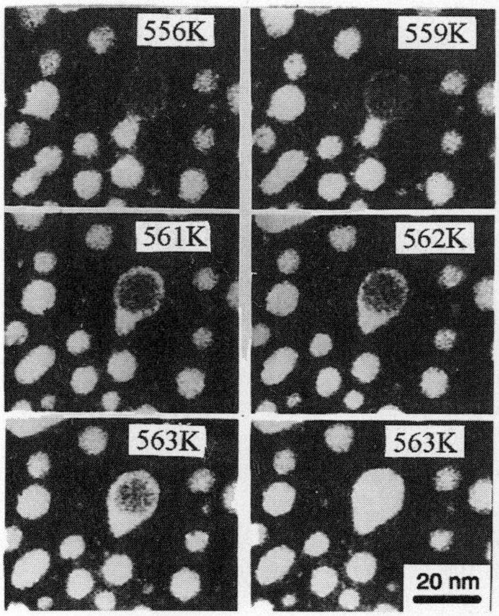

Fig. 2-DFEM pictures showing the progressive thickness increasing of the liquid layer as a function of temperature. The grey contrast is created by the liquid layer while the black one reflects the solid part.

One has also observed that the melting process starts at the cluster surface [1], giving evidence of the appearance of a liquid layer well below the melting temperature (Fig. 2).The thickness of this layer depends on the size, shape and crystalline orientation of the clusters; it increases when approaching Tm, and when Tm is reached the particles melt abruptly.

Phenomenological models, no matter what the starting hypothesis are, predict a change of Tm following the relation $1\text{-Tm}/T_0 \approx r^{-1}$, where r is the particle radius and T_0 the bulk melting temperature. A theory due to Semenchenko takes into account also the presence of the liquid layer in thermodynamic equilibrium conditions, giving:

$$T_m = T_0 \exp\left\{-\frac{2\gamma_{sl}}{\rho L r}\right\} \qquad (1)$$

where ρ is the volume mass, L the latent heat of fusion and γ_{sl} the interfacial free energy between solid and liquid. All these theories are in good agreement with experimental data when $r \geq 10$nm, but they fail when r is under this value. Pseudopotential [5] and tight-binding calculations for clusters containing up to 1000 atoms are been carried out: they reproduce quite correctly the decrease of Tm and also the surface melting.

When the radius is less than 5 nm, one observes spontaneous phase fluctuations of the nanoparticles at room temperature (Fig. 3), with a frequency that increases by decreasing the size [2]. By means of the so-called many body glue model, the latent heat of fusion is shown to decrease with size, vanishing and yielding a glass-liquid transition under a certain critical size.

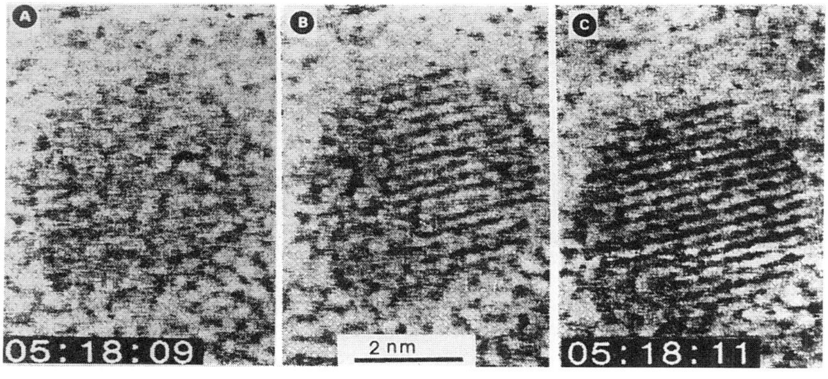

Fig. 3-HREM images from videotape; the interval between the frames is 0.04 s. The appearance of a crystalline structure is showed.

Optical characterization

The analysis was performed through the Brewster-angle technique: p-polarized radiation strikes the samples with a variable incidence angle, the reflected component is collected and plotted versus incidence angle. A description of the experimental apparatus (Fig. 4) is given elsewhere [4]; we just recall that the radiation source is a CO_2 laser emitting p-polarized radiation (i.e. with E in the incidence plane) at a wavelength of 9.4 μm.

From the Fresnel formulas one has for the p component of the reflectivity, if the medium is non absorbing:

$$R_p = \left|\frac{\tan(\vartheta_i - \vartheta_t)}{\tan(\vartheta_i + \vartheta_t)}\right|^2 \qquad (2)$$

If $\theta_i+\theta_t=\pi/2$, $R_p=0$ and this value of the incidence angle is known as Brewster angle. If the medium is absorbing, R_p still has a minimum different from zero; the incidence angle corresponding to this minimum is the pseudo-Brewster (pB) angle.

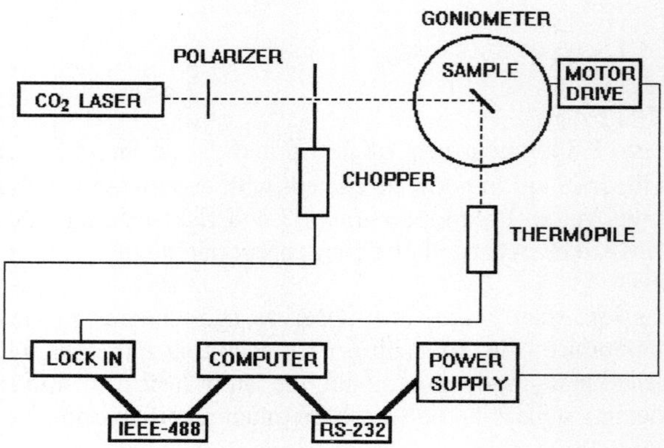

Fig. 4-Experimental apparatus

The pB angle position is very sensitive to the structural composition of the sample investigated. From the knowledge of this value and of the overall lineshape collected R_p vs. θ_i, it is possible, through a fit process, to obtain the optical constants n and k of the sample.

The samples are composed of two layer: the sapphire substrate and the effective layer (metal particles+SiO); such multi-layered systems are treated within the Rouard method [6], which gives the optical constants of a system from that of each single layer. In that way it has been possible to apply an effective medium analysis on the composite medium made of metal particles embedded in the SiO matrix.

There are many effective medium theories available in the literature [7]; the most frequently used are the Maxwell-Garnett and the Bruggeman theories. The first one uses the Lorentz local-field corrections to the effective field, whereas the second one is a self-consistent theory. Both of them are based on the assumption that the inclusions are randomly distributed in the matrix, i.e. that the particle distribution is 3-dimensional.

We tried to reproduce the experimental data using models based on these theories, taking into account also the shape of the particles, their orientation and the polarization of the incident radiation, but we did not obtain a good agreement (Fig. 5), since, due to the growth conditions, the particles distribution is planar and not a randomly one.

A theory based on a 2-dimensional distribution of the inclusions in the matrix is that of Yamaguchi [8]: the optical properties of the effective medium depend on:
-n and k of the particles and of the matrix;
-thickness of the film;
-filling factor;
-shape and orientation of the particles;
-arrangement of the particles in the matrix.
Using this model, one has a better agreement with the experimental reflectivity data, as one can see in Fig. 5.

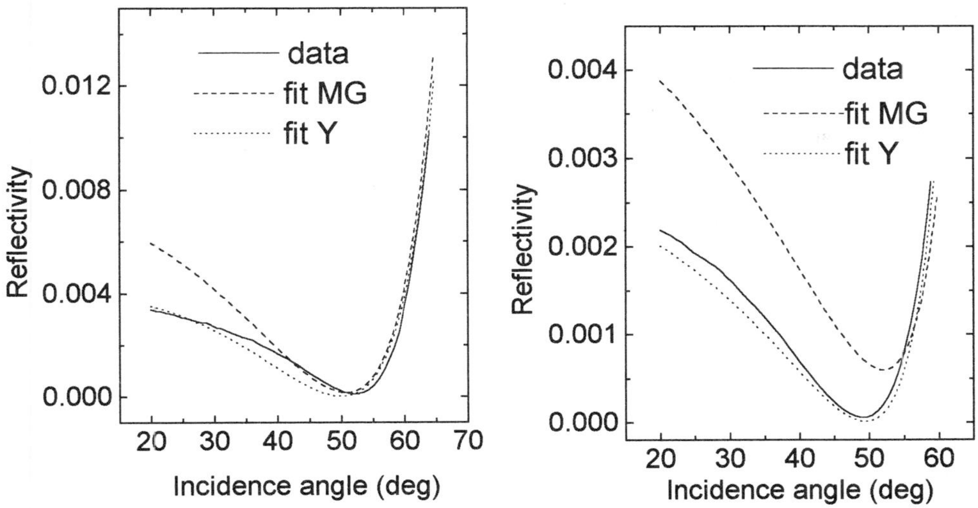

Fig. 5-Comparison among experimental data (data), prediction of the Maxwell-Garnett theory (M-G) and prediction of the Yamaguchi theory (Y). Samples containing: a) Pb nanoparticles with $\langle r\rangle \approx 36$ *Å; b) Sn nanoparticles with* $\langle r\rangle \approx 60$ *Å*

One obtains a tensorial dielectric function ε; the anisotropy is due to the 2-dimensional distribution and to the orientation of the particles:

$$\varepsilon_{para} = \varepsilon_m \frac{\varepsilon_p (F_{para} + f) + \varepsilon_m (1 - F_{para} - f)}{\varepsilon_m F_{para} + \varepsilon_p (1 - F_{para})} \quad (3)$$

$$\varepsilon_{perp} = \frac{1}{\varepsilon_m} \frac{\varepsilon_p F_{perp} + \varepsilon_m (1 - F_{perp})}{\varepsilon_p (F_{perp} - f) + \varepsilon_m (1 - F_{perp} - f)} \quad (4)$$

where parallel and perpendicular are referred to the film plane, ε_m and ε_p are the matrix and metal dielectric functions, f is the filling factor and F are the effective depolarization factors:

$$F_{para} = L_1 - \frac{\varepsilon_{sub} - \varepsilon_m}{\varepsilon_{sub} + \varepsilon_m} \frac{1}{\eta^3} \left(\frac{a_1}{a_3}\right)^2 \frac{1}{24} - 0.716 \frac{\varepsilon_m}{\varepsilon_p + \varepsilon_m} \frac{d_W}{d} \quad (5)$$

$$F_{perp} = L_3 - 2 \frac{\varepsilon_{sub} - \varepsilon_m}{\varepsilon_{sub} + \varepsilon_m} \frac{1}{\eta^3} \left(\frac{a_1}{a_3}\right)^2 \frac{1}{24} + 0.716 \frac{2\varepsilon_p}{\varepsilon_p + \varepsilon_m} \frac{d_W}{d} \quad (6)$$

L_1 and L_3 are the geometrical depolarization factors of the ellipsoids used to represent the particles, ε_{sub} is the dielectric function of the sapphire substrate, η is the ratio between the distance dipole-image dipole and the rotational axis, a_1 and a_3 are the principal axis of the ellipsoids, d_W is the mass thickness and d is the distance among the particles. In the above expression the second term accounts for the interaction of every dipole with its substrate image, the third term accounts for the interaction of every dipole with all the other dipoles.

Conclusions

The aim of this paper consists essentially in the presentation of a method which allows to solve the problem caused by the presence of an embedding matrix and of a substrate when studying the optical properties of a nanoparticle system. An effective medium theory has been employed in order to perform a careful characterization of the samples, pointing out the crucial role played by the quasi-planar distribution of the Pb and Sn inclusions.

The interest of metal clusters is due to their thermodynamic properties; a review of them is given, particularly of the so-called surface-melting and of the evidence of phase fluctuations.

References

[1] R.Kofman, P.Cheyssac, A.Aouaj, Y.Lereah, G.Deutscher, T.Ben-David, J.M.Pennison and A.Bourret, Surf.Sci. **303**, 231 (1994).
[2] A.Stella, A.Migliori, P.Cheyssac and R.Kofman, Europhys. Lett. **26**, 265 (1994).
[3] M.Geddo, D.Maghini, A.Stella and M.Cottini, J. Appl. Phys. **58**, 4733 (1985).
[4] P.Tognini, M.Geddo, A.Stella, P.Cheyssac and R.Kofman, to be published.
[5] F.Ercolessi, W.Andreoni and E.Tosatti, Phys.Rev.Lett. **66**, 911 (1991).
[6] A.Vasicek, Journ. de Phys. **11**, 342 (1950).
[7] C.G.Granqvist and O.Hunderi, Phys. Rev. B **18**, 2897 (1978).
[8] S.Berthier, J.Phys. **4** 303 (1994).

Materials Science Forum Vol. 195 (1995) pp. 167-172

Thermomagnetic Analysis of the Phase Formation in Fe-Ge Compounds Obtained by Mechanical Alloying

F. Albertini[1], A. Paoluzi[1], L. Pareti[1], L. Nasi[2], G. Salviati[1] and G. Calestani[3]

[1] Istituto Maspec del C.N.R., Via Chiavari 18/A, Parma, Italy

[2] Dipartimento di Fisica, Università di Parma, Viale delle Scienze, Parma, Italy

[3] Dipartimento di Chimica Fisica e Inorganica, Università di Bologna, Viale Risorgimento 4, Bologna, Italy

Keywords: Mechanical Alloying, Thermomagnetic Analysis, Fe-Ge Compounds

Abstract. The hexagonal phases $Fe_{2.9}Ge_2$ ($B8_2$, P6/mmc) and Fe_3Ge (DO_{19}, P6/mmc) have been obtained in single phase status by mechanical alloying followed by heat treatments. The alloying and the phase formation processes have been monitored by using thermomagnetic analysis.
In both phases the easy magnetization direction lies in the basal plane of the hexagonal structures at RT. The anisotropy field values are 11 kOe for $Fe_{2.9}Ge_2$ and 5.8 kOe for Fe_3Ge. A spin reorientation transition plane to axis takes place in Fe_3Ge at T=150 °C.
The combination of magnetometric measurements and thermomagnetic analysis has proved to be a powerful and direct method to follow the preparation process of ferromagnetic systems by means of mechanical alloying.

Introduction. Magnetic and structural characteristics of Fe-Ge compounds have been widely studied in the past [1, 2, 3, 4]. In particular $Fe_{3+x}Ge$ (DO_{19} hexagonal) and $Fe_{3+x}Ge_2$ ($B8_2$ hexagonal) phases have uniaxial crystal symmetry, high Curie temperature and saturation magnetization which make them interesting for possible application in the field of permanent magnet as low cost materials, non containing rare-earth. Both phases exist in a wide stoichiometric range. This make it possible to vary the magnetic properties by changing the stoichiometry within the phase existence range. However in the FeGe phase diagram only a congruent point exists, corresponding to the composition $Fe_{3.34}Ge_2$ and it has not been possible to obtain, by standard melting techniques, single-phase samples having different composition from the congruent stoichiometry.
The aim of this work is to prepare single-phase specimens of composition $Fe_{2.9}Ge_2$ e Fe_3Ge by mechanical alloying and to study their magnetic properties. Thermomagnetic analysis and magnetic moment measurements have been used to monitor step by step the solid state solution and the phase formation processes.

Experimental. The milling process was performed in two steps: a preliminary one (mixing) up to 6 hours in a vibrating ball mill, using a ball-to-powder weight ratio of 2:1 and a more energetic second one using a planetary ball mill Pulverisette-7 with a ball-to-powder weight ratio of (6-10):1. Starting powder elements had the following average size: Fe 10μm e Ge 100μm. See tab. I for more details.
Powder XRD patterns were collected with an automatized Philips PW1050 diffrattometer using Co K_a radiation.
SEM microscopy was carried out in a 360 Stereoscan Cambridge Instrument in the secondary electron image mode.
Thermomagnetic analysis (TMA) was performed by means of AC susceptometer (RT-1300 °C).
TMA consists in the measure of the temperature dependence of the initial susceptibility. It is a very fast, reliable and sensitive method to measure the Curie temperature of a sample, to control the stoichiometric homogeneity (defined by the shape of the transition) and to make a phase analysis, provided the different phases in the specimen are ferromagnetic. Amounts of secondary phases smaller than 2% can be detected.
Magnetic moment measurements were performed in a VSM (LNT-1000 °C).

The singular point detection technique (SPD) [5] was used to measure the temperature dependence of the anisotropy field. In order to enhance the SPD peak, that for easy-plane samples is found in d^3M/dH^3, specimens have been prepared orienting powders in a magnetic field by using a rotating system and fixing powders with wax.

Table I a. First step (mixing) - Vibratory mill

Stoichiometry	Mixing hours	WR=ball weight / powder w.	N.of balls,material	Working vol. (cm^3) / Jar vol. (cm^3)
$Fe_{2.9}Ge_2$	5	2	4 steel	44.5/50
Fe_3Ge	6	1.5	3 steel	45/50

Table I b. Second step (milling) - Planetary mill Pulverisette-7

Stoichiometry	Milling hours	WR=ball weight / powder w.	N.of balls, material	Working vol. (cm^3) / Jar vol. (cm^3)
$Fe_{2.9}Ge_2$	1	5.6	4 steel	5.2/10
Fe_3Ge	1h 45'	9.3	3 tungsten carbide	5.7/10

Results

The solid state solution. TMA has been performed at different mixing and milling time (fig. 1). The susceptibility signal of the starting powders (Fe+Ge) is exclusively due to Fe and consists in a sharp transition at $T=T_c=770$ °C.
A change in the susceptibility curve shape has been observed by increasing mixing time which indeed produces a broadening of the TMA curves, corresponding to a continuous distribution of T_c values, lower than that of Fe. This is due to the occurrence of a progressive dissolution process of Ge in α-Fe which takes place even in this preliminary mixing process.
The alloying process was found to start within 1 hour mixing. With increasing working time the knee of the curve moves to lower temperature and the amount of free iron decreases. The knee represents the lowest T_c values that is the maximum concentration of Ge which is dissolved in Fe while ΔT_c between the knee and zero signal is proportional to the concentration interval of the solution.
Free iron disappears after the milling process has been completed that is after 5 hours mixing plus 1 hour milling for $Fe_{2.9}Ge_2$ and 6 hours mixing plus 1h 45' milling for Fe_3Ge (fig. 3).
The slope of the $\chi(T)$ curve gives information on the profile of the concentration of Ge atoms inside Fe particles. A large $\chi(T)$ slope implies a Ge concentration which varies slowly from the outer to the inner particle zone, while a small slope is caused by a rapidly decreasing of Ge concentration.
Magnetic moment measurements performed at RT in a maximum field of 19 kOe were done at different stages of the process. A slight decrease of the magnetic moment can be observed for increasing milling time (6% in $Fe_{2.9}$+Ge_2 and 9% in Fe_3+Ge). This decrease in the magnetic moment can be ascribed to the concourse of different causes; first of all to a possible oxidation of Fe because the mixing process is done in air, in addition to the increased specific surface area and number of defects, furthermore a very limited entrance of Fe atoms in the Ge lattice cannot be excluded.

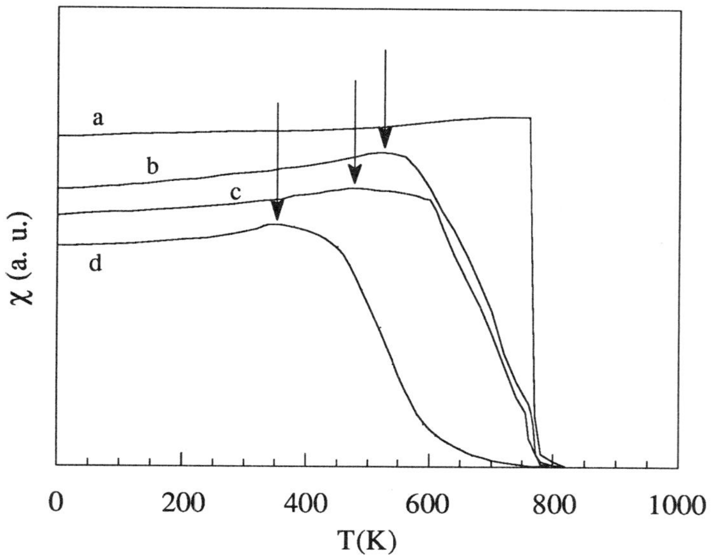

Figure 1. Temperature dependence of the a.c. susceptibility χ (TMA patterns) for $Fe_{2.9}$+Ge_2 system at different MA time. a: t = 0; b: t = 2h mixing; c: t = 5h mixing; d: t = 5h mixing + 1h milling. Fo every stage an arrow indicates the lowest T_c of the solution which corresponds to the higher Ge concentration.

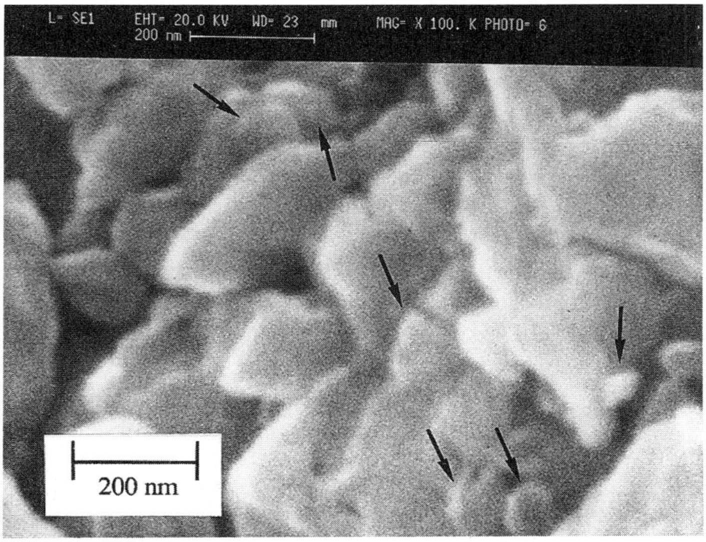

Figure 2. SEM image of $Fe_{2.9}$+Ge_2 solid solution after 5h mixing and 1h milling.

A comparison of susceptibility and magnetic moment measurements indicates that the main dissolution process consists in a progressive entrance of Ge in the α-Fe lattice, the reverse process being of negligible entity.
X-ray diffraction analysis confirms the occurrence of the Fe-Ge solution, evidenced by a decrease of α-Fe lattice parameters and indicates that some amount of free Ge is still present at the end of the milling process.
SEM observation indicates that after milling grain dimensions are in the range 50-100 nm (fig. 2).
Larger agglomerates could be the result of magnetic aggregations. In fact from the values of anisotropy an M_s (see below) it can be deduced that actual particles are single-domain, being 3-4 μm the critical domain size.

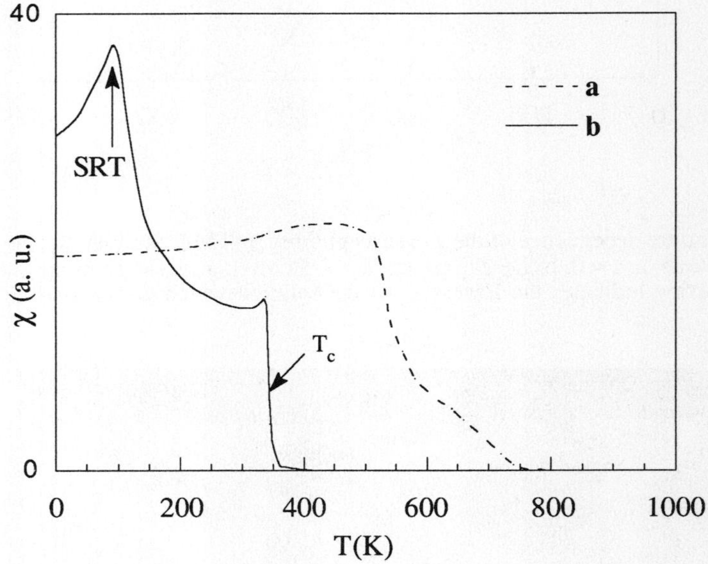

Figure 3. Temperature dependence of the a.c. susceptibility (TMA pattern) for the system Fe_3+Ge. a: after 6h mixing + 1h 45' milling (solid solution). b: after 36 h at 950°C + 24 h at 870°C annealing. The spin reorientation transition (SRT) temperature and the Curie temperature T_c are indicated by arrows.

Table II. Annealing parameters and XRD results.

Sample	Heat treatment time, temperature	Structure	Lattice parameters (Å)
$Fe_{2.9}Ge_2$	1h, 800°C	$B8_2$ (Ni_2In)	a=3.9919(3) c=5.0020(4)
Fe_3Ge	36h, 950°C+ 24h, 870 °C	DO_{19} (Ni_3Sn)	a=5.174(2) c=4.222(1)

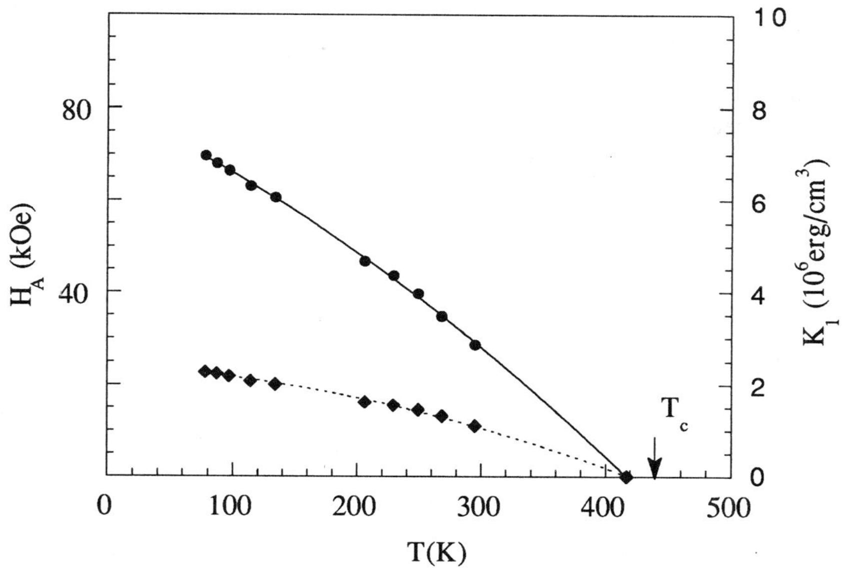

Figure 4. Temperature dependence of anisotropy field H_A and anisotropy constant $K_1 = H_A M_s/2$ for $Fe_{2.9}Ge_2$.

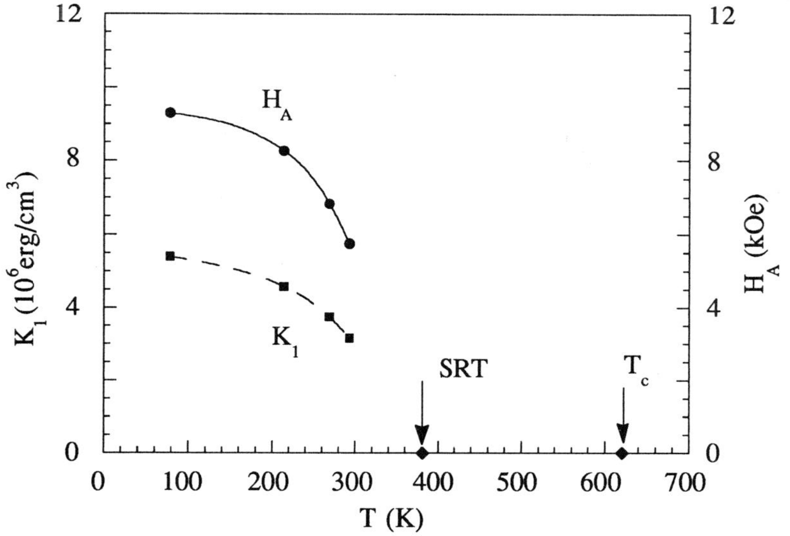

Figure 5. Temperature dependence of anisotropy field H_A and anisotropy constant $K_1 = H_A M_s/2$ for Fe_3Ge.

Fe_3Ge and $Fe_{2.9}Ge_2$ phases. The phases Fe_3Ge and $Fe_{2.9}Ge_2$ have been obtained after heat treatments of the solid solutions (see table II). By TMA analysis it has been possible to follow the Fe-Ge phase formation for different values of the annealing temperature and time.The conditions to get single phase specimens have been defined. The specimens have been characterized by X-ray diffraction analysis. The results confirm the absence of impurity phases; the refined lattice parameters are reported in table II.
The Curie temperatures values are 144°C for $Fe_{2.9}Ge_2$ and 348°C for Fe_3Ge while the saturation magnetization values at RT are respectively 63.2 and 131.1 emu/g.
The easy magnetization direction at RT lies in the basal plane for both samples. For $Fe_{2.9}Ge_2$ the anisotropy field ($H_A=2K_1/M_s$) decreases almost linearly with increasing temperature and extrapolates to zero at $T=T_c$ (fig. 4).
For Fe_3Ge the anisotropy extrapolates to zero at $T<<T_c$ (fig. 5). This behaviour indicates that a spin reorientation transition (SRT) from plane to axis takes place around $T/T_c=0.4$. This occurrence is confirmed by the peak observed at T=150 °C in the $\chi(T)$ curve (fig.3).

Conclusions. The hexagonal phases $Fe_{2.9}Ge_2$ ($B8_2$, P6/mmc) and Fe_3Ge (DO_{19}, P6/mmc) have been obtained in single phase status by mechanical alloying followed by heat treatments.
After the milling process (1h for $Fe_{2.9}Ge_2$ and 1h and 45' for Fe_3Ge) the samples consisted of a solid solution of Ge in α-Fe with the presence of small amount of free Ge. The necessary heat treatments have been performed in the range 800-970 °C.
By using thermomagnetic analysis the alloying and the phase formation processes have been monitored at the different stages.
In both phases the easy magnetization direction lies in the basal plane of the hexagonal structures at RT. The anisotropy field values are 11 kOe for $Fe_{2.9}Ge_2$ and 5.8 kOe for Fe_3Ge.
A spin reorientation transition plane to axis takes place in Fe_3Ge at T=150 °C.
The combination of magnetometric measurements and thermomagnetic analysis has proved to be a powerful and direct method to follow the preparation process of ferromagnetic systems by means of mechanical alloying.

References
[1] K.Kanematsu, K. Yasukochi, J. Phys Soc. Japan **18**, 920 (1963).
[2] K.Kanematsu, J. Phys Soc. Japan **20**, 36 (1965).
[3] E.Adelson and E.Austin, J. Phys. Sol. Chem. **26**, 1795 (1965).
[4] F.Albertini, A.Paoluzi, L.Pareti, G.Turilli, A.Y.Ermakov, N.Mushnikov, O.Moze, G.Calestani, J. Magn.Magn.Mat. **140-144**, 141 (1995).
[5] G. Asti, S. Rinaldi, J. Appl. Phys. **45**, 3600 (1974)

Materials Science Forum Vol. 195 (1995) pp. 173-178

Magneto-Optical Effects on γ-Fe_2O_3 Nanometric Particles in Polyvinylic Alcohol

L. D'Amico[1], F. D'Orazio[1], J.L. Dormann[2], D. Fiorani[3]
F. Lucari[1] and E. Tronc[2]
[1] Dipartimento di Fisica, Università di L'Aquila, Italy
[2] ICMAT-CNR Area della Ricerca di Roma, Italy
[3] Lab. de Magnétisme et Mat. Magnétique, CNRS, Bellevue, Meudon, France
Keywords: Magneto-Optics, Magnetic Nanoparticles

Abstract: We have examined the magneto-optical properties of a series of samples made of γ-Fe_2O_3 nanometric particles dispersed in polyvinylic alcohol at different concentrations. Faraday rotation spectra are measured. They show the presence of characteristic iron ion transitions whose amplitude are correlated with the γ-Fe_2O_3 content. At low wavelength we observe a magneto-optical transition connected to the polymer matrix. Faraday rotation as a function of temperature shows that the magnetization tends to saturate at a sufficiently low temperature which depends on concentration.

Introduction

Recently, the behavior of magnetic nanoparticles and the effect of interparticle interactions on the superparamagnetic relaxation has received a great interest in literature [1]. These studies have been carried on with different experimental techniques and theoretical approaches [2 - 6]. In this paper we have addressed our attention on the system of γ-Fe_2O_3 particles dispersed in a non magnetic matrix (polyvinylic alcohol) [6 - 8]. This system shows a superparamagnetic behavior with blocking temperatures of the order of tens of Kelvin and activation energies around thousands of K depending on the particle size and on the concentration. Magneto-optical measurements can give a useful contribution to the study of their magnetic properties because of the peculiarity of the probe utilized, because the photon energy is of the same order of those mentioned above, and because its interaction time is different from that of other techniques.

Materials

The samples under study belong to a large family of specimens, all prepared by a sol-gel method [7]. They consist of γ-Fe_2O_3 particles dispersed in polyvinylic alcohol with specific values of particle diameter, concentration, and state of aggregation. All these parameters are controlled varying the chemical conditions (bath composition, precipitation conditions and pH) during the preparation process.

The four samples, measured by magneto-optics, are characterized by the name 4D and are choosen in such a way that they possess similar characteristics except the concentration of the particles. For all of them the particle diameter is 7 nm and the pH during the preparation was 2.5.

Table 1. Sample characteristics.
The blocking temperature is obtained by a.c. susceptibility measurements with a.c. field of 1 Oe and 20 Hz.

Sample	Concentration R	Thickness d (μm)	Blocking Temp. T_B (K)
4D C/1	1.2	25.4 ± 2.5	113.0
4D C/5	0.24	48.1 ± 3.8	81.3
4D C/25	0.048	85.6 ± 1.9	59.5
4D C/50	0.024	64.4 ± 2.5	50.5

This ensures that the particles are well dispersed in the alcohol with average distances depending

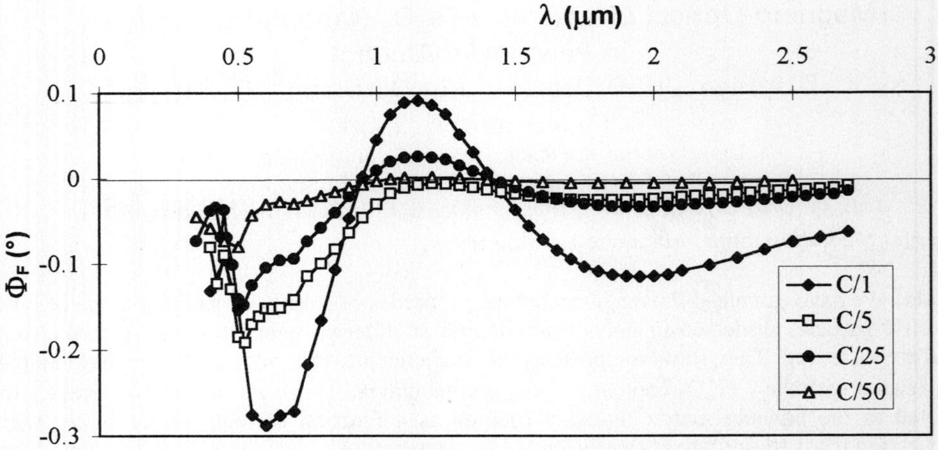

Figure 1. Faraday rotation as a function of wavelength, measured at 290 K on the four samples as indicated in the legend.

only on the total concentration of the oxide. The concentration R (indicated as the ratio between the mass of the oxide and the mass of the polyvinylic alcohol) and the thickness d of the samples are reported, together to their denomination, in Table 1 These samples were also characterized by a.c. susceptibility measurements and in the last column the blocking temperature T_B, obtained at 1 Oe and 20 Hz, is also shown [8].

Experiments

We performed very high sensitivity Faraday effect measurements. The experimental apparatus permits to vary the wavelength λ between 0.3 and 2.7 μm, the magnetic field H, applied parallel to the propagation incident direction of the light, up to 3.5 kOe, and the temperature of the sample from room temperature down to 4 K. The measurements were performed averaging for 20 seconds the signal proportional to the light polarization under magnetic field and comparing it with that without field. The Faraday rotation (FR) is proportional to the local magnetization and it shows a spectrum which depends on the optical transitions occurring principally on the magnetic ions. Therefore, the results show maxima (positive or negative, depending on the choice of the positive versus of the angle of rotation) which can be attributed essentially to the presence of Fe^{3+} ions in the particles. The intensities of the spectra are connected to the quantity of γ-Fe_2O_3 in the sample. The spectra were recorded at 290 and 4 K and compared between them. Measurements were also performed as a function of temperature at specific wavelengths. From their results it was possible to follow the temperature dependence of the sample magnetization.

Results and interpretation

The FR spectra obtained on the four samples at 290 K are shown in Fig. 1. All spectra have a similar behavior characterized by a principal structure between 0.5 and 1 μm where the FR reaches the highest values, a maximum with opposite sign localized at 1.2 μm and finally a wider peak centered around 2 μm. The values obtained in the first structure and in the final region appear to be correlated to the concentration of γ-Fe_2O_3 in the four samples. This correlation is less evident at 1.2 μm.

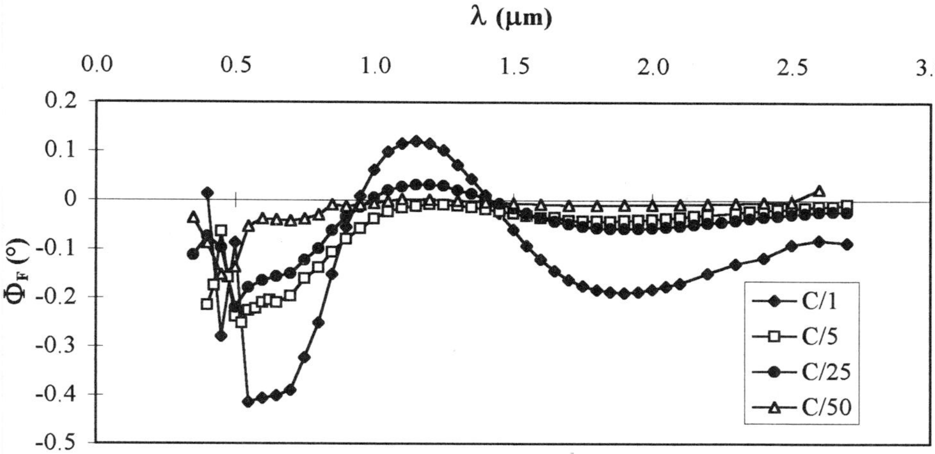

Figure 2. Faraday rotation as a function of wavelength measured at 4 K on the four samples as indicated in the legend.

In Fig. 2 the corresponding results obtained at 4 K are reported. Here, the shape of the spectra are very similar to those obtained at room temperature and the correlation with the γ-Fe_2O_3 concentration is maintained.

As an example, Fig. 3 shows the comparison between the results at the two temperatures on sample 4D C/5. It is evident that only the structure at lower wavelengths and the region around 2 μm vary significatively with temperature. Therefore, only these correspond to transitions connected with the energy levels of the ions in the γ-Fe_2O_3 particles. In fact, following a microscopic description of FR, the population of the lower energy levels increases as the temperature decreases and thus both the FR amplitude and the magnetization increase. Consequently, the maximum at 1.2 μm can be attributed to the superposition of the tails of the contiguous transitions. Similar comparisons

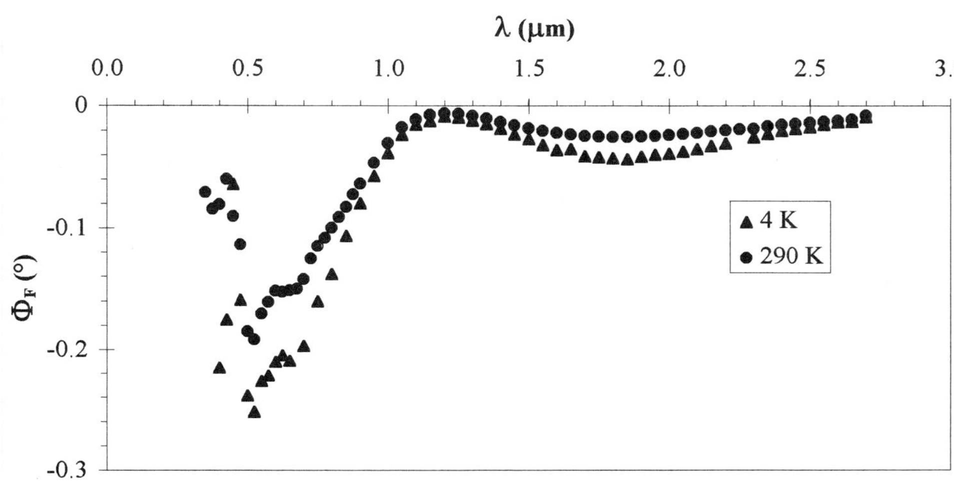

Figure 3. Comparison of the Faraday rotation on sample 4D C/5 at 4 and 290 K.

operated on the results obtained on the other samples confirm this interpretation.

To better understand the results relative to the structure at low wavelengths, we calculated the specific FR defined as:

$$\Phi_{F\ spec} = \frac{\Phi_F}{d \cdot \frac{R}{1+R}}, \tag{1}$$

which normalizes the results for all samples with respect to their thickness and amount of γ-Fe_2O_3. The resulting spectra, from the measurements at 4 K, are reported in Fig. 4. The differences between amplitudes of different samples are strongly reduced (compare with Fig. 2) at all wavelengths, except below 0.6 μm, confirming that the above renormalization is consistent with the known values of particle concentration in the samples. Therefore, we can attribute the FR signal below 0.6 μm to the presence of the polymer matrix, since the relative amplitude in this region increases for the less concentrated samples, where the relative concentration of polymer increases. Moreover, a detailled inspection of this structure reveals the presence of two peaks, which can be well localized in energy. The second one is always centered around 0.75 μm (1.8 eV) while the position of the first peak depends on the concentration. The values of wavelengths and energies of this transition are reported in Table 2 for each sample.

Table 2. Energy of the first transition observed in Fig. 4 for the different samples.

Sample	λ (μm)	Energy (eV)
4D C/1	0.57	2.18
4D C/5	0.52	2.38
4D C/25	0.50	2.48
4D C/50	0.48	2.58

The differences among the values are significant: the energy of this transition decreases as the concentration, and then the interaction energy among the magnetic particles, increases. These findings are confirmed also when applying Eq. (1) to the results at 290 K.

To explain this result, we speculate, as a possible hypothesis, that this peak originates from an inter-ion transition between electronic states one in the alcohol molecules and the other in the iron ion at the particle surface. Therefore, the increasing amplitude can be explained by the relative

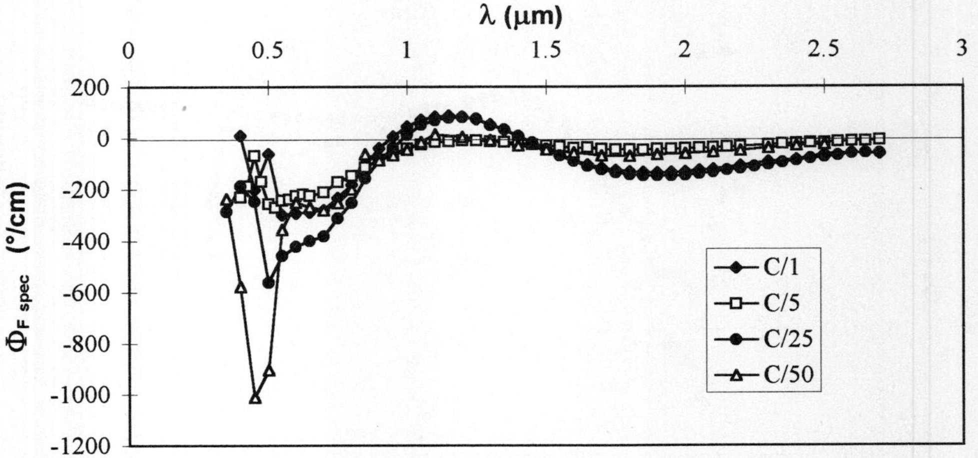

Figure 4. Specific Faraday rotation spectrum (from Eq. 1) at 4 K on the four samples as indicated in the legend.

increase of the amount of polymer surrounding the particles. The energy variation can be explained by assuming a corresponding variation of the energy level in the molecule when more than one particle is in its proximity. The involvement of iron ions is necessary in order to explain the observed strong magneto-optical signal.

To follow the temperature dependence of the magnetization, we performed FR measurements at fixed wavelength and varying the temperature. The wavelengths were chosen in correspondence of the maxima of the spectrum. Figure 5 represents an example of such a measurement. Since the FR absolute value is proportional to the magnetization, Fig. 5 shows its increase as the temperature decreases till it tends to a saturating behavior under a specific temperature T*.

To interpret our results we attempted a best fit of the experimental data relative to the high temperature interval, utilizing the Langevin expression:

$$\Phi_F(T) \propto M(T) \propto \coth\left(\frac{M_p \cdot H}{K \cdot T}\right) - \frac{K \cdot T}{M_p \cdot H}, \tag{2}$$

where M_p is the average magnetic moment of the particles and K is the Boltzmann constant.

The solid line in Fig. 5 indicates the result of the fit in the temperature region where the data were considered, while the dotted line represents its extrapolation at lower temperatures. The difference between the latter and the experimental points in this region can be attributed to the increasing influence of magnetic anisotropy which becomes predominant with respect to the thermal energy as the temperature is lowered, determining a slowing down of the particle moment fluctuations. As a consequence, the magnetization, and then the FR, is limited by the blocking process which prevents the free reorientation of the magnetic moments. In Fig. 5 the broken line indicates a linear approximation obtained from the data at the lowest temperatures. The intercept between broken and dotted lines is indicative of the characteristic temperature T*.

The estimated values of T* from the measurements on the different samples and wavelengths range between 120 and 70 K, increasing with oxide concentration. This trend is consistent with that of the blocking temperatures obtained from a.c. susceptibility measurements performed without static magnetic field on the same samples (see Table 1), and it is attributed to the effect of the

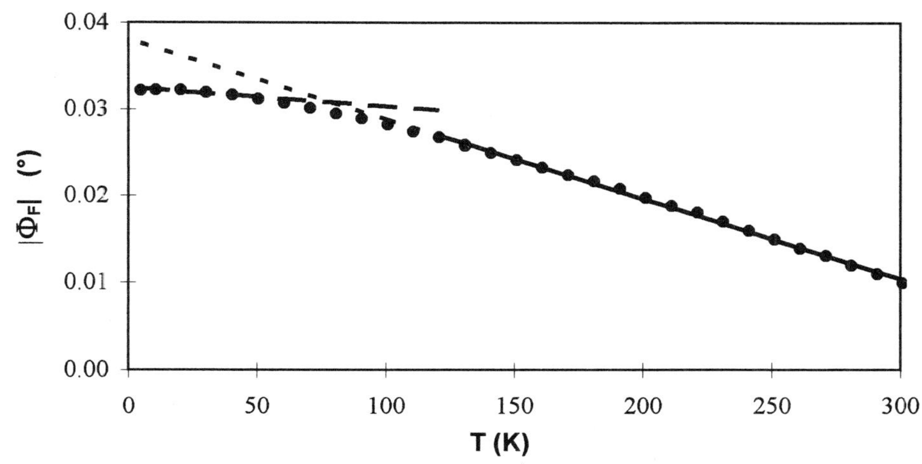

Figure 5. Absolute value of Faraday rotation vs. temperature measured on sample 4D C/25 at λ = 2.0 μm. The full circles are the experimental points and the lines are obtained from the model illustrated in the text.

interactions between the magnetic moments of the particles. However, a detailled interpretation of the values of T* is prevented by the large experimental errors caused by misalignment of the optical line with respect to the sample originated by the thermal expansion of the cryostat during the sweep in temperature of most of the experimental runs.

The fitting with Eq. (2) furnishes also a rough estimate of M_p, although with equally large experimental errors, which indicate that the average particle moments are smaller, by a factor of 5-20, than those that can be estimated considering a spherical fully magnetized γ-Fe_2O_3 particle with a diameter of 7 nm. However, we have to consider that spin misalignment on the surface should lead to a decrease of the particle moment.

In conclusion, the results on γ-Fe_2O_3 nanoparticles indicate that Faraday Rotation spectra are very sensitive to the thermal fluctuations of particle moments and to the effect of inter-particle interactions.

---

[1] J. L. Dormann and D. Fiorani eds., *Magnetic Properties of Fine Particles* (North-Holland, Amsterdam, 1992).

[2] S. Mørup and E. Tronc, Phys. Rev. Lett. **72**, 3278 (1994).

[3] R. Ferré, B. Barbara, D. Fruchart and P. Wolfers, J. Magn. Magn. Mat. **140-144**, 385 (1995).

[4] M. El-Hilo, K. O'Grady and R. W. Chantrell, J. Magn. Magn. Mat. **140-144**, 359 (1995).

[5] M. Henmion, G. Belluard, I. Mirabeau, J. L. Dormann and M. Nogues, Europhys. Lett. **25**, 43 (1994).

[6] E. Tronc, P. Prené, J. P. Jolivet, F. D'Orazio, F. Lucari, D. Fiorani, M. Godinho, R. Cherkaoui, M. Nogues, and J. L. Dormann, Hyperfine Interact. **95**, 129 (1995).

[7] J. P. Jolivet, and E. Tronc, J. Colloid Interf. Sci. **125**, 688 (1988).

[8] L. D'Amico, F. D'Orazio, J. L. Dormann, D. Fiorani, F. Lucari and E. Tronc, Mater. Sci. Forum, this issue.

Materials Science Forum Vol. 195 (1995) pp. 179-184

A.C. Susceptibility Measurements on Nanometric γ-Fe_2O_3 Particles

L. D'Amico[1], F. D'Orazio[1], J.L. Dormann[2], D. Fiorani[3]
F. Lucari[1] and E. Tronc[2]

[1] INFM, Dipartimento di Fisica, Università di L'Aquila, Italy

[2] Lab. de Magnétisme et Mat. Magnétique, CNRS, Bellevue, Meudon, France

[3] ICMAT-CNR Area della Ricerca di Roma, Italy

Keywords: Magnetic Nanoparticles, Superparamagnetism, A.C. Susceptibility

Abstract: We have studied a number of samples made of γ-Fe_2O_3 nanometric particles dispersed in polyvinylic alcohol. The temperature dependence of low field a.c. magnetic susceptibility has a maximum corresponding to a superparamagnetic blocking temperature, which changes with frequency (5 Hz < ν < 10 kHz) following an Arrhenius-like law for all samples. The fitting parameters of this dependence result to be correlated to the compositional characteristics: state of aggregation, concentration, and size of the particles. Physically meaning values of the pre-exponential factor are obtained only in diluted samples consisting of non interacting particles. The behavior of concentrated samples is satisfactorily described by the model of Dormann *et al.* [J. Phys. C **21**, 2015 (1988)] which accounts for interparticle interactions.

Introduction

Magnetic nanoparticles have recently attracted a large interest from both experimental and theoretical points of view [1]. In particular, their properties are affected by the reduced size and their mutual interactions. The dimensions are such that each particle can be considered single domain. This implies that two energy minima, corresponding to opposite directions of the magnetization, exist. These minima are separated by an energy barrier which, being proportional to the particle volume, may become comparable to the thermal energy even at relatively low temperatures. This mechanism introduces a transition at a so called blocking temperature T_B above which the magnetization is free to fluctuate between different easy directions (superparamagnetism). Magnetic interparticle interactions strongly affect the fluctuations of the magnetic moments modifying T_B and its frequency dependence [2 , 3]. In this work, we describe the results of a.c. magnetic susceptibility measurements at different frequencies on γ-Fe_2O_3 particles dispersed in polyvinylic alcohol. We have examined different samples characterized by different preparation conditions leading to different mean particle size and to different aggregation states.

Experimental details

The samples were prepared by a sol-gel method reported elsewhere [4 , 5]. In Table 1 we list the characteristics of the specimens which were obtained by varying the preparation parameters. In particular, precipitation conditions of γ-Fe_2O_3 affected the average particle size. The particles were successively dispersed in an aqueous solution whose pH determined the degree of aggregation, which is known to produce isolated particles at pH $\approx$2, chains at pH $\approx$4, up to flocculation, at pH >5. Finally, the proportion between particles and polyvinylic alcohol added to the solution determined the concentration R defined as the oxide/polymer mass ratio. This parameter is related to the average interparticle distance. The samples shown in Table 1 are grouped accordingly to emphasize the dependence of the experimental observation on the above mentioned preparation parameters. Sample series 35A is characterized by a common particle size and concentration, but different aggregation state; sample series 4D is characterized by different concentration and same size and aggregation conditions. The last two samples in the list, together with sample 4D C/50, represent a comparison at

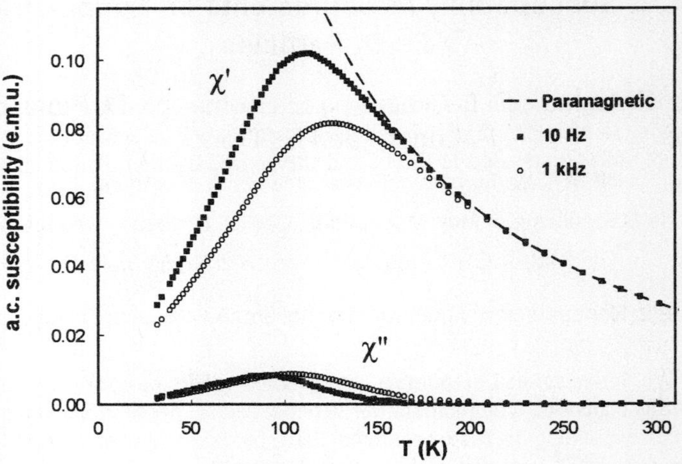

Figure 1. Typical temperature dependence of a.c. magnetic susceptibility. The sample is 4D C/1. The plot shows that the temperature of the maximum of the real part, representing the blocking temperature, depends on frequency.

different average particle size with constant concentration and aggregation state.

A.c. magnetic susceptibility was measured using a LakeShore 7000 Susceptometer. This apparatus allowed us to perform experiments as a function of temperature (between 15 and 300 K), at low fields (1 Oe), and frequencies ranging from 5 Hz to 10 kHz.

Results

Figure 1 shows the typical temperature dependence of the a.c. susceptibility of the materials under study. The effect of changing the frequency is also evidenced in the graph. From now on, we

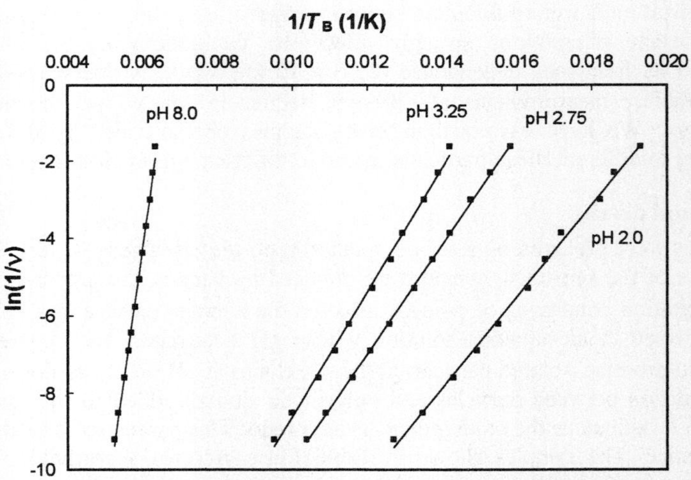

Figure 2. Blocking temperature as a function of frequency for sample series 35A C/50 prepared in solutions with different pH. The straight lines are obtained by fitting the experimental data points with Eq. (1).

will focus only on the real part of the susceptibility. The temperature T_M at which the maximum occurs is proportional to the average blocking temperature $< T_B >$, the exact relationship depending on the particle size distribution function. For simplicity, we will assume $T_M = T_B$ in the following.

For each sample and for each frequency we performed a polynomial fit of the region around the maximum in order to establish precisely the temperature and the susceptibility of the maximum. The exact number of data points used in the fit and the degree of the polynomial was determined in each case using conventional statistical criteria.

In Figs. 2-4 we report the frequency dependence of T_B for all samples grouped according to the criteria discussed in the previous section. In all cases, it is clear that the behavior may be described by an Arrhenius-like law [6]:

$$\tau = \frac{1}{\nu} = C \exp\left(\frac{E_b}{k_B T_B}\right), \qquad (1)$$

where k_B is the Boltzmann constant and E_b is the energy barrier for magnetization reorientation.

In particular, from Fig. 2 we notice that the blocking temperatures and the slope of the lines increase with the degree of aggregation. Figure 3 indicates that a similar trend is obtained as the oxide concentration is raised. Finally, the plots in Fig. 4 suggest also a similar correlation with the particle size, showing steeper lines for larger particles. This gives a clear indication that both the increase of particle size and interparticle interactions lead to an increase of the effective anisotropy energy barrier.

In order to interpret all these results, we need to work out the dependence of the parameters in Eq. (1). For isolated, non interacting particles, the coefficient C is given by [6]:

$$C = \tau_0 = \frac{\pi^{1/2}\mu}{2\gamma_0 E_a}\left(\frac{k_B T}{E_a}\right)^{1/2}, \qquad (2)$$

where μ is the particle magnetic moment, γ_0 the gyromagnetic ratio and E_a the anisotropy energy.

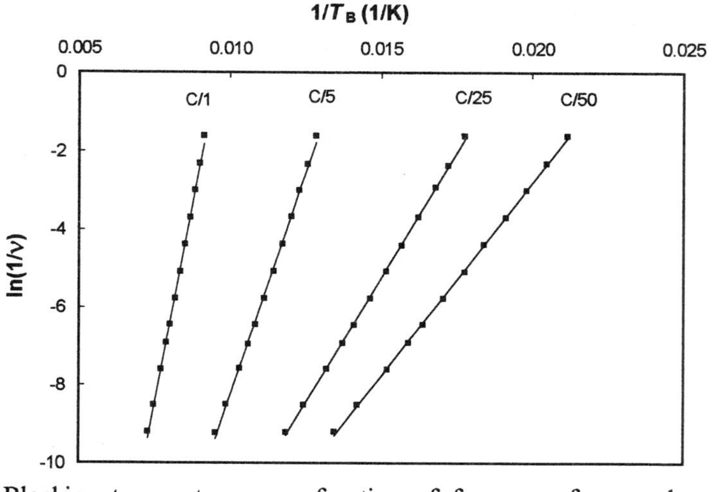

Figure 3. Blocking temperature as a function of frequency for sample series 4D characterized by different concentration (see Table 1). The straight lines are obtained by fitting the experimental data points with Eq. (1).

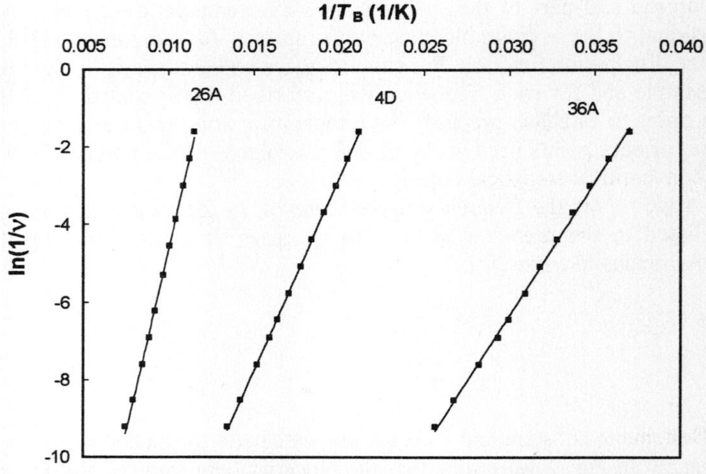

Figure 4. Blocking temperature as a function of frequency for samples with similar pH (2.5) and concentration ($R = 0.024$) but different particle size (see Table 1). The straight lines are obtained by fitting the experimental data points with Eq. (1).

The latter, which coincides with the energy barrier in this case, is given by

$$E_a = E_b = K_t V \ , \tag{3}$$

where K_t is the total anisotropy energy constant for uniaxial anisotropy. Although also τ_o is a function of temperature, it can be taken as a constant since such dependence is negligible with respect to that in the exponential of Eq. (1). For ferromagnetic and ferrimagnetic particles τ_o is of the order of 10^{-9} - 10^{-11} s.

The values of C and E_b/k_B obtained by fitting the experimental data with Eq. (1) are reported in Table 1. For most of the samples, the value of C is consistent with the expected values of τ_o. However, the most concentrated or aggregated samples have C constant values which cannot physically be identified as τ_o values and, therefore, the Arrhenius law does not hold. This result is interpreted as an increased effect of interparticle interactions whose contribution to the total anisotropy energy cannot be neglected [2, 7].

Various models have been used to account for the additional interaction energy term E_{int}. Dormann *et al.* [2] assumed an assembly of particles with volume distribution, disordered arrangement and easy axes in random directions and proposed a model to account for interparticle interactions.

For dipolar interactions and considering only the effect of first neighbors particles, the anisotropy energy is given by:

$$E_a = K_U V + E_{int} \ ,$$

$$E_{int} = M_{nr}^2 n_1 b_1 V \, L\left(\frac{M_{nr}^2 V a_1}{k_B T}\right), \tag{4}$$

where K_U is the anisotropy constant relative to magnetostatic and magnetocrystalline energies, L is the Langevin function, n_1 is the number of first neighboring particles, M_{nr} is the non relaxing

Table 1. Characteristics of the samples. In the first four columns the denomination and preparation parameters are indicated. The concentration R is defined as the oxide/dry polymer mass ratio. The particle diameter $<D>$ is defined from the average particle volume, assuming spherical shape. In the last four columns we report the results obtained from the interpretation of a.c. susceptibility measurements (see text); the last two are related to samples which manifest interparticle interactions.

Sample	R	pH	$<D>$ (nm)	E_b/k_B (K)	C (s)	n_1b_1/a_1	VM^2a_1/k_B (K)
35A C/50	0.024	8.0		6760	$3.5\cdot10^{-20}$	20	275
35A C/50	0.024	3.25		1610	$1.9\cdot10^{-11}$	-	-
35A C/50	0.024	2.75		1410	$4.0\cdot10^{-11}$	-	-
35A C/50	0.024	2.0		1190	$2.4\cdot10^{-11}$	-	-
4D C/1	1.2	2.5	7	4120	$7.5\cdot10^{-18}$	17	186
4D C/5	0.24	2.5	7	2300	$2.5\cdot10^{-14}$	9	151
4D C/25	0.048	2.5	7	1290	$2.1\cdot10^{-11}$	-	-
4D C/50	0.024	2.5	7	990	$1.5\cdot10^{-10}$	-	-
36A C/50	0.024	2.5	4	690	$1.8\cdot10^{-12}$	-	-
26A C/50	0.024	2.5	11	1890	$5.9\cdot10^{-11}$	-	-

magnetization, $a_1 = V/d^3$, d the average interparticle distance (taken from the centers of spherical particles) and b_1 a corrective factor, which tends to compensate the simplifications of the model, but is of the order of a_1. The first order expansion of the Langevin function gives:

$$E_{\text{int}} = n_1 b_1 \left(M_{\text{nr}}^2 V - \frac{k_B T}{a_1} \right), \tag{5}$$

valid for relatively *strong* interactions, i.e. if $M^2 V a_1 / k_B T \geq 2$. Evaluating Eq. (5) for $T = T_B$, Eq. (1) can be rewritten:

$$\tau = \tau_o \exp\left(-\frac{n_1 b_1}{a_1} \right) \exp\left(\frac{E_b}{k_B T_B} \right) \approx \tau_o \exp(-n_1) \exp\left(\frac{E_b}{k_B T_B} \right), \tag{6}$$

$$E_b = V\left(K_U + M_{\text{nr}}^2 n_1 b_1 \right).$$

An Arrhenius-like law is still maintained but with a decrease of the pre–exponential factor $C \approx \tau_o \exp(-n_1)$, depending on the number of first neighbor particles, and an increase of the slope (i.e. of the energy barrier E_b) in the log τ vs. $1/T$ plot. This is consistent with the plots shown in Figs. 2-4 and the values reported in Table 1 obtained by fitting the curves. These experiments are in agreement with previous results on similar samples, whose properties were investigated in a more extended frequency interval, including Mössbauer spectroscopy measurements [7].

Examining the first group of samples (series 35A) we notice that the constant C is essentially equal for the three less aggregated samples with an abrupt decrease for the flocculated one, consistently with an expected increase of neighbors. Moreover, the energy barrier E_b increases progressively with the pH, showing a large increase for pH = 8, as expected for increasing interactions between particles. For the sample with pH 8, containing particles with the largest mutual interaction, we use Eqs. (6) to calculate, from the fitting values of C and E_b, the parameters in the last two columns of Table 1. We assume for VK_t and τ_o the values of E_b and C relative to sample 35A pH 2, representing the less interacting particle system. The value of n_1b_1/a_1 is consistent with the fact that $b_1 \approx a_1$ with a number of first neighbors close, or slightly larger (when a distribution of particle size is present), than the closed packed structure for identical spheres (i.e. 12). Notice that a similar calculation cannot be done for samples with intermediate pH values. In fact, the model leading to Eq. (6) does not apply since the arrangement is far from being regular and the interactions are probably occasional and do not involve all particles.

In the second group of samples (series 4D) there is a progressive increase of E_b and decrease of C as the concentration increases and, as a consequence, the interaction is raised. The values reported in the last two columns of Table 1 for the most concentrated samples are referred to sample 4D C/50 taken as the non interacting case. They indicate, for this series, the presence of smaller and less dense aggregates with respect to the 35A series. This is evidenced by the lower value of the n_1b_1/a_1 term, and by the value in the last column. In fact, the latter does not scale with the concentration, as it would be suggested by the dependence on a_1.

Finally, the variations of E_b corresponding to the plots of Fig. 4 are consistent with an increasing volume of the magnetic (non interacting) particles of the three samples, determining an increase of the energy barrier. Using Eq. (3), a different total anisotropy constant is obtained for the three samples. This reveals that other terms, in addition to magnetocrystalline anisotropy, play a role in determining the total anisotropy energy (e.g. surface anisotropy).

Conclusions

A.c. susceptibility measurements at different frequencies allowed us to investigate the dynamical behavior of γ-Fe_2O_3 particles as a function of their size and of their mutual interaction energy. The results indicate that for samples consisting of isolated particles the frequency dependence of the blocking temperature follows an Arrhenius law, the energy barrier increasing with increasing the mean particle size. For samples consisting of particles close to each other or of aggregates of particles, the data are qualitatively described by a model, which accounts for interparticle interactions.

[1] J. L. Dormann and D. Fiorani eds., *Magnetic Properties of Fine Particles* (North-Holland, Amsterdam, 1992).

[2] J. L. Dormann, L. Bessais, D. Fiorani, J. Phys. C **21**, 2015 (1988).

[3] S. Mørup and E. Tronc, Phys. Rev. Lett. **72**, 3278 (1994).

[4] J. P. Jolivet, R. Massart, and J. M. Fruchart, Nouv. J. Chim. **7**, 325 (1983).

[5] J. P. Jolivet and E. Tronc, J. Colloid Interf. Sci. **125**, 688 (1988).

[6] W. F. Brown, Phys. Rev. **130**, 1677 (1963).

[7] E. Tronc, P. Prené, J. P. Jolivet, F. D'Orazio, F. Lucari, D. Fiorani, M. Godinho, R. Cherkaoui, M. Nogues, and J. L. Dormann, Hyperfine Interact. **95**, 129 (1995).

Materials Science Forum Vol. 195 (1995) pp. 185-190

Magnetoelastic Coupling in the Amorphous and Nanocrystalline $Fe_{73.5}Cu_1Nb_3Si_{13.5}B_9$ Alloy

E. Bonetti[1], L. Del Bianco[1] and P. Tiberto[2]

[1] Dipartimento di Fisica, Università di Bologna, Istituto Nazionale per la Fisica della Materia, GNSM-CNR, V. le Berti Pichat 6/2, I-40100 Bologna (Italia)

[2] Istituto Elettrotecnico Nazionale G. Ferraris, Istituto Nazional per la Fisica della Materia, GNSM-CNR, C. so Massimo d' Azeglio 42, I-10125 Torino (Italia)

Keywords: Ferromagnetic Alloys, ΔE-Effect, Magnetomechanical Damping, Nanocrystallization, Mechanical Spectroscopy

Abstract

The magnetoelastic behaviour of melt spun amorphous alloy $Fe_{73.5}Cu_1Nb_3Si_{13.5}B_9$ was studied by performing room temperature Young's modulus and internal friction measurements in an applied varying magnetic field. Similar measurements were performed on samples of the same composition after being submitted to Joule heating treatments. These data have been correlated with internal friction and modulus measurements vs. temperature. The experimental results provide information on the microstructural evolution of the annealed samples from the amorphous phase to the nanocrystalline phase and allow one to establish a strict relationship between structural configuration and magnetoelastic behaviour. The maximum ΔE effect is observed in samples where the Joule heating treatment induced the onset of the nanocrystallization process.

Introduction

The $Fe_{73.5}Cu_1Nb_3Si_{13.5}B_9$ nanocrystalline alloy has been widely studied in recent years by virtue of its excellent soft magnetic properties and in prospective of the possible technological applications [1].

The nanocrystalline phase is obtained by annealing an amorphous precursor generally produced by ultrarapid quenching techniques [2].

The crystallization process is mainly controlled by the minority elements which, slowing down the transformation kinetics, play the decisive role for the achievement of nanometer sized grains [3]. The magnetic softness of the alloy is reached when the microstucture consists of α-FeSi nanocrystals of typically 10 nm in size embedded in a residual amorphous matrix [4]. Actually, as described by the random anisotropy model advanced by Herzer [3], the magnetic properties are strongly determined by the structural features of the nanophase which, in turn, depend directly on the thermal history experienced by the sample [5,6].

Particularly efficient for obtaining samples with high magnetic permeability, due to a small and homogeneous grain size, is the Joule heating technique [7].

The aim of this paper is to provide information about the modifications in the magnetic configuration accompanying the structural evolution of the sample from the amorphous to the nanocrystalline state, as induced by the Joule heating treatment.

In this sense, correlated measurements of the anelastic and magnetoelastic properties have proved to be very suitable tools of investigation thanks to their high sensibility to the details of the structural- magnetic configuration [8,9].

Experimental

The ribbons of composition $Fe_{73.5}Cu_1Nb_3Si_{13.5}B_9$ were produced by melt spinning. The peripheral velocity of the rotating wheel was 32 m/s. Strips cut from the same ribbon were submitted to DC Joule heating [10] in vacuum for fixed time (t = 4 s) under electrical currents I = 2.8 A (samples **A**), I= 3 A (samples **B**) and I = 3.2 A (samples **C**)

Samples for mechanical spectroscopy were cut in shape of parallelepiped [cross section ~ (0.02 x 2) mm^2, length ~ 6 mm], from as-cast ribbons and from strips previously submitted to Joule heating treatments. Internal friction (Q^{-1}) and Young's modulus (E) measurements were performed through a vibrating reed technique apparatus, operating in diffusion pumped vacuum in the frequency range $10^2 \div 10^4$ Hz, as a function of temperature, during furnace annealing (heating rate 4 K/min), and as a function of the strength of a magnetic field (applied along the longitudinal axis of the samples) at 300 K.

Results and discussion

It's well known that the variation of the elastic modulus with magnetization (ΔE effect) reflects the addition of an extra magnetostrictive strain component to the ordinary elastic strain as a result of stress induced changes in the domain configuration of the sample [11]. In amorphous ferromagnetic alloys, this contribution to the strain is of prominent importance, due to the enhanced domains mobility with respect to coarse grained materials. In the specific case of a sample submitted to an oscillatory uniaxial stress, the strain ε_S (as would be observed at saturation and hence of non-magnetic origin) is usually written as a complex quantity $\varepsilon_S = \varepsilon_{S1} - i\varepsilon_{S2}$. ε_{S1} is the component in-phase with the stress (elastic) whereas ε_{S2} is the out-of-phase component (anelastic).

In association with ε_S, the magnetostrictive strain ε_m must be considered. It also can be expressed in complex formalism as $\varepsilon_m = \varepsilon_{m1} - i\varepsilon_{m2}$. As in the above case of non-magnetic strain, ε_{m1} and ε_{m2} are respectively the elastic and anelastic components.

It's easily demonstrated [12] that the magnetoelastic contribution to the internal friction can be expressed as

$$\tan \phi = \varepsilon_{m2} / (\varepsilon_{m1} + \varepsilon_{S1}) \qquad (1)$$

(where ϕ is the loss angle) and that the dimensionless ratio usually used to represent the ΔE effect $\Delta E / E = (E_S - E) / E$ (where E_S and E are the elastic modulus values as measured at saturation and at a certain magnetization level respectively) can also be written as

$$\Delta E / E = \varepsilon_{m1} / \varepsilon_{S1} \qquad (2)$$

The results of elastic modulus and Q^{-1} measurements vs. applied magnetic field are reported in Fig. 1a-d. The as-quenched samples (Fig. 1a) show only a small ΔE effect reaching values of ΔE/E in the 0.05-0.08 range ($E/E_S \approx 0.93$).

Fig. 1b is relative to sample **A**: the lowest reached value of $E/E_S \approx 0.88$ corresponds to $\Delta E/E \approx 0.14$. The strongest magnetoelastic coupling was found in samples **B** (Fig. 1c). For these samples $\Delta E/E \approx 0.5 \div 0.6$: a remarkable result considering that $\Delta E/E = 0.57$, the highest value found for an alloy of very similar composition, was obtained after thermal annealing in transverse magnetic field at 420 ÷ 480 °C for 1 hour [13]. The magnetomechanical damping gives information on the extent of the losses, which accompany the magnetization process, due to the displacements of the non-180° Block walls and domain rotations [12]. The Q^{-1} trends are opposite to the modulus ones, reaching the maximum when E is minimum: this behaviour reveals the existence of a proportionality between the in-phase and the out-of-phase component of the magnetic strain.

The highest $Q^{-1} \approx 11 \times 10^{-3}$ was measured in samples **B** (Fig. 1c) In as-cast ribbons (Fig. 1a) Q^{-1} is generally in the $(4 \div 6)\times 10^{-3}$ range whereas, in samples **A,** it reaches the value of 8×10^{-3} (Fig. 1b). The internal friction progressively decreases with the increase of the magnetic field strength and, at $30 \div 40$ Oe, it drops to values included in the $(3 \div 4)\times10^{-3}$ range, both in the as-cast alloy and in samples **A** and **B**. Concurrently, the Young's modulus reaches the saturation value. In short, with approaching the magnetic saturation, the magnetomechanical losses gradually vanish and, at saturation, Q^{-1} represents just the non-magnetic background damping strictly related to ε_{S2}.

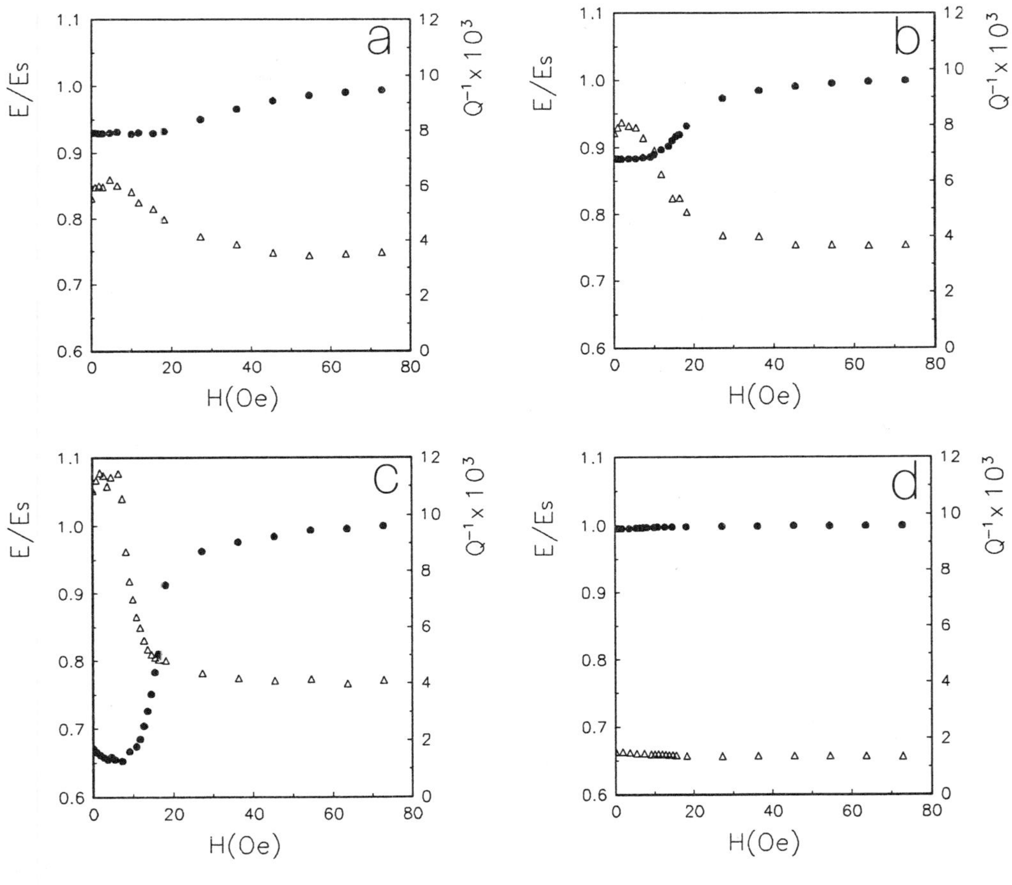

Fig. 1. Magnetic field dependence of the normalized Young's modulus E (circles) and internal friction Q^{-1} (triangles) for (a) as-cast sample, (b) sample heated at I = 2.8 A, (c) sample heated at I = 3 A, (d) sample heated at I = 3.2 A. E_S is the modulus value at saturation magnetization.

By Eqs. 1 and 2, it's easily found that the out-of-phase component of ε_m is only a small fraction of the in-phase component. The ratio $\varepsilon_{m2}/\varepsilon_{m1}$ is 0.037, 0.03 and 0.02 for as-cast and **A** and **B** samples respectively, indicating that the stress induced domains movements occurs with a good degree of reversibility.

Finally, sample **C** (Fig. 1d) shows neither appreciable ΔE effect nor obviously magnetomechanical damping. Noteworthy is that in this last case the Q^{-1} is lowered to 1.4×10^{-3}.

In amorphous ferromagnets, since the magnetocrystalline anisotropy averages out, the ΔE effect may be expressed as

$$\Delta E/E = (E_s - E)/E = (3E_s\lambda_s \sin^2\theta \cos^2\theta)/\sigma_i \qquad (3)$$

where λ_S is the saturation magnetostriction, σ_i is the internal stress and θ is the angle between the applied stress direction and the magnetization direction [14].

The $Fe_{73.5}Cu_1Nb_3Si_{13.5}B_9$ as-cast alloy is reported to have a saturation magnetostriction not unusually large ($\lambda_S \sim 20 \times 10^{-6}$) decreasing to near-zero value ($\lambda_S \sim 2 \times 10^{-6}$) by a thermal annealing at 540 °C for 1 hour, which induces the formation of a well developed nanocrystalline phase [2].

Accordingly, a strong magnetoelastic coupling should be measured in samples where the internal stresses are significantly reduced and λ_S is quite high.

The low value of the ΔE effect for the as-cast sample is easily explained in terms of the high internal stresses, usually present in melt spun ribbons. On the other hand, the disappearance of the magnetoelastic effect in sample **C** indicates that the thermal annealing was sufficient to accomplish the nanocrystallization process leading, as already pointed out, to the drop in the saturation magnetostriction.

The nanocrystalline alloy produced, through proper Joule-heating treatments, with maximum crystalline fraction of about 60-70 %, is seen to be thermally stable, at least up to about 900 K [8]. Moreover, due to the homogeneous grain size distribution, the interface configuration is expected to be structurally well equilibrated and a low defects content is easily predictable, as a consequence of the fast thermal annealing at high temperatures (of order of 10^3 K). Since the above considerations, the low Q^{-1} in sample **C**, with respect to the background damping of the as-cast amorphous alloy, is well interpreted in the hypothesis of fully nanocrystalline state of the sample, and provides further evidence of it.

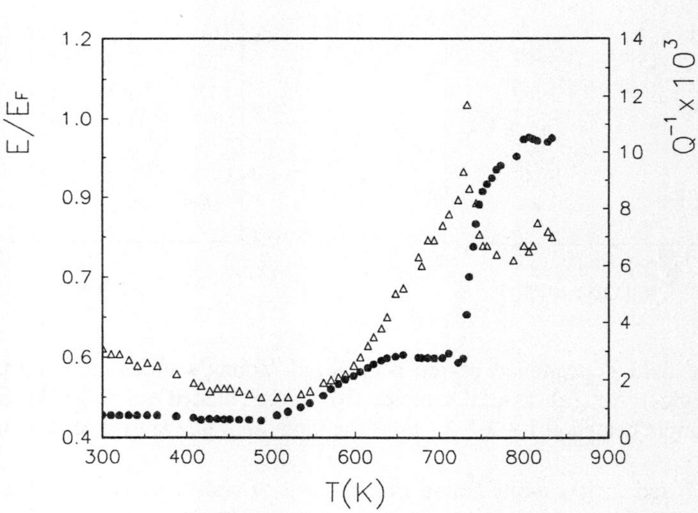

Fig. 2. Young's modulus (circles) and Q^{-1} (triangles) vs. temperature in the $Fe_{73.5}Cu_1Nb_3Si_{13.5}B_9$ as-cast alloy. E_F is the modulus at the end of measurement; vibration frequency at 300 K ~ 600 Hz

The possibility to understand the enhancement of the ΔE effect in samples **A** and **B,** with respect to the as-prepared alloy, implies the study of the structural modifications induced in the alloy by the Joule heating treatments. In this sense, particularly useful are the analysis of the elastic modulus vs. temperature measurements and the comparison with the anelastic data relative to the as-quenched ribbon.

The elastic modulus and the internal friction vs. temperature for the as-cast alloy are shown in Fig. 2.

Some features are noteworthy: from 300 K up to about the Curie temperature (~ 600 K) the spectrum is affected by magnetoelastic effects; in the intermediate temperature region (from 550 ÷ 600 K up to about 680 K) a modulus increase is experienced and finally, in the 680 ÷ 800 K range, a strong modulus variation occurs corresponding to the high peak in the Q^{-1} trend (softening stage).

As previously reported [15], the modulus increase at intermediate temperatures is linked to the progressive formation of iron rich zones, favoured by copper clusters formation, which act as nucleation sites for the formation of the nanocrystalline phase. It was also well established that the softening stage at elevated temperatures is connected to the nucleation and growth of the α-FeSi nanocrystals [15].

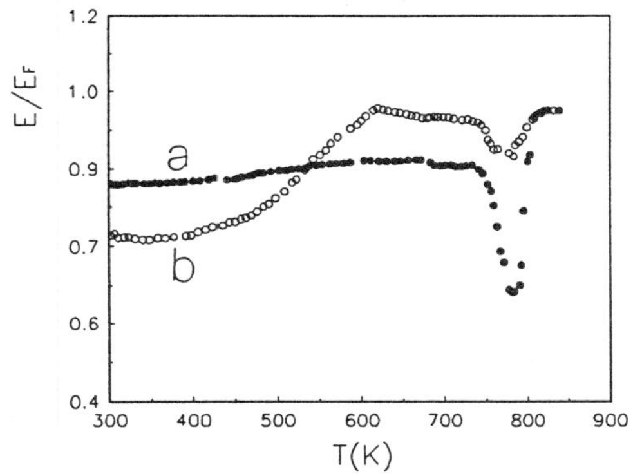

Fig. 3. Young's modulus vs. temperature in samples previously Joule heated at I = 2.8 A (full circles) and I = 3 A (open circles). E_F is identified as in Fig. 2.

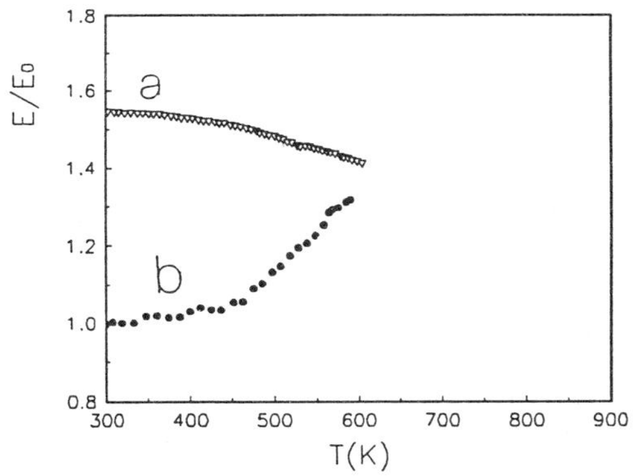

Fig. 4. Modulus vs. temperature, in sample **B** (heated at I=3 A), measured in saturating magnetic field (triangles) and in zero field (circles). E_0 is the modulus at 300 K.

The modulus vs. temperature behaviour of sample **A** (Fig. 3, curve a) shows evident differences with respect to the as-cast ribbon. The modulus rise at intermediate temperatures is almost completely disappeared. Faint traces of it are partially convoluted with a slight increasing trend of E with temperature, starting from 300 K, related to the non-negligible magnetoelastic effect experienced at room temperature (Fig. 1b). On the other hand a well defined softening stage is still clearly observable at 700 K.

So, the treatment at I = 2.8 A was sufficient to induce the decomposition stage in the amorphous phase but not to crystallize the ribbon. The enhancement of ΔE/E reveals that the process was accompanied by a relaxation of the quenched-in internal stresses.

The modulus vs. temperature spectrum relative to sample **B** (Fig. 3, curve b), is dominated in the low temperature region, by a strong increase of E, easily predictable since the low value of $E/E_S \approx 0.67$ found at 300 K (Fig. 1c). The magnetoelastic origin of this behaviour was confirmed by measuring the modulus vs. temperature up to 600 K under saturating magnetic field (H = 120 Oe). The result is plotted in Fig. 4 (curve a) together with the modulus measurement performed in zero field on the same specimen (Fig. 4, curve b). Clearly, the presence of the magnetic field eliminates any trace of magnetoelastic coupling

Again with reference to curve (b) of Fig. 3, a strong reduction of the softening stage at high temperature is evident.

All these elements indicate that, in sample **B,** the Joule heating treatment induced not only the decomposition stage but even the nucleation of nanocrystals.

Herzer stated that the magnetostriction of the nanocrystalline alloy may be approximately expressed by the sum of λ_S of the nanocrystalline phase (negative in sign) and λ_S of the residual amorphous matrix, weighted by their respective volume fractions [3].

Therefore, to account for the observed strong magnetoelastic coupling, the crystalline phase in sample **B** must constitute just a small volume fraction of the material: in this way λ_S is not appreciably reduced with respect to its as-quenched value.

As a matter of fact, as above pointed out, the background damping level is as high as in the as-cast ribbon revealing that sample **B** is, for majority, still in the amorphous phase.

In short, the treatment at I = 3 A allowed the transformation process from the amorphous phase to the nanocrystalline phase to be stopped just at the onset of the crystallization. As a consequence, while the quenched-in internal stresses were significantly annealed out, new stresses, connected with the formation of the nanocrystalline phase and typically due to the mismatch between adjacent crystalline grains, were not introduced if not at a very low extent.

References

[1] G. Herzer, H. Warlimont, Nanostruct. Mater., 1 (1992) 263.
[2] Y.Yoshizawa, S. Oguma, K. Yamauchi, J. Appl. Phys. 64, 6044 (1988).
[3] P. Allia, M. Baricco, P. Tiberto, F. Vinai, J. Appl. Phys. 74, 3137 (1993).
[4] G. Herzer , Mater. Sci. Eng. A133, 1 (1991).
[5] F. van Bouwelen, J. Sietsma, A. van den Beukel, J. Non-Cryst. Solids, 156-158, 567 (1993).
[6] M. Baricco, C.Antonione, P. Allia, P.Tiberto, F. Vinai, Mater. Sci. Eng. 179-180, 572 (1994).
[7] P. Allia, M. Baricco, M. Knobel, P. Tiberto, F. Vinai, Phil. Mag. B 68, 853 (1993).
[8] E. Bonetti, L. Del Bianco, P. Allia, P. Tiberto, F. Vinai, J. Appl. Phys., submitted.
[9] E. Bonetti, L. Del Bianco, G. Valdrè, P. Allia, M. Baricco, P. Tiberto, F. Vinai, J. Magn. Magn. Mater. 133, 357 (1994).
[10] P. Allia, M. Baricco, P. Tiberto, F. Vinai, Rev. Sci. Instrum., 64 (1993) 1053.
[11] R.M. Bozorth, in Ferromagnetism, Van Nostrand (New York, 1951), p. 684
[12] B.S. Berry, W.C. Pritchet, J. Appl. Phys., 47 (1976) 3295.
[13] Z. Kaczkowski, L. Malkinski, M. Muller, Key Eng. Mater. 81-83, 389 (1993).
[14] H.U. Künzi, in *Glassy Metals II*, edited by H. Beck and H.J. Guntherodt (Springer-Verlag, Berlin, 1983), p. 185.
[15] P. Allia, M. Baricco, E. Bonetti, L. Del Bianco, P. Tiberto, G. Valdrè, F. Vinai, Nanostruct. Mater. 3, 433 (1993).

Materials Science Forum Vol. 195 (1995) pp. 191-196

Mechanical Spectroscopy of the Nanostructured $Fe_{50}Al_{50}$-Alloy

E. Bonetti and G. Scipione

Istituto Nazionale per la Fisica della Materia (INFM), GNSM-CNR and Dipartimento di Fisica, Università di Bologna, Via Irnerio 46, I-40126 Bologna, Italy

Keywords: Nanocrystalline Phases, Mechanical Alloying, FeAl Intermetallic, Anelasticity

Abstract

Mechanical spectroscopy measurements have been performed on $Fe_{50}Al_{50}$ samples obtained by mechanical alloying (MA) pure iron and aluminium elemental powders. Two mechanical spectroscopy apparatuses including a vibrating reed technique and a torsion pendulum apparatus have been used to cover a wide range of frequencies: 10^{-1}-10^{4} Hz. Measurements of the dynamic Young's modulus M and elastic energy dissipation coefficient reveal the occurrence of two transient anelastic effects on samples obtained from cold-consolidated milled powders. The first at 570 K is related to structural transformations depending on the milling time. At short milling times the Al_5Fe_2 phase forms, while a nanostructured FeAl intermetallic compound develops for longer times. The second transient effect is related to magnetic order-disorder transition. In samples where a well developed ordered FeAl nanophase occurs, an anelastic relaxation peak attributed to grain boundary sliding mechanisms in the intermetallic compound has been observed.

Introduction

The mechanical properties of metals and compounds depend to a large extent on the nature and structure of the interfaces [1]. It was reported that the reduction of the average grain size down to nanometers has interesting consequences on mechanical properties both in the elastic and plastic regime [2]. Of particular interest is the strong enhancement of diffusivity in nanophase materials attributed to grain boundary mechanisms, which should have relevant consequences on the ductility at relatively low temperatures [3,4]. So the synthesis throw mechanical alloying (MA) of the iron aluminides such as $Fe_{50}Al_{50}$ should overcome some drawbacks typical of this system: poor room-temperature ductility and low high-temperature strength [5].

Among the several techniques that can be used to study the mechanical behaviour, mechanical spectroscopy is particularly sensitive to the evolution of the microstructure and suitable to acquire informations concerning defect mobility and atomic scale diffusion [6]. It was recently observed on nanostructured pure metals that the grain boundary-type anelastic relaxations represent a dominant mechanisms for the damping at relatively low temperatures in comparison with coarse-grained materials [7]. Moreover in the iron aluminides, the Zener and the grain boundary type anelastic relaxations, constitute an indirect and efficient probe of the Al diffusivity in the lattice and of the interfaces mobility respectively [8-10].

In the present work we have studied MA iron aluminium in the equiatomic composition. In particular we have focused our attention on the anelastic behaviour and structural transformation from Fe(Al) solid solution to FeAl intermetallic.

Experimental

Pure Fe and Al powders (325 mesh, ALFA products, 99.9 % purity) were mixed in the 1Fe:1Al atomic ratio and milled for selected times (2÷32 h) at room temperature in a hardened tool steel vial with a SPEX mixer mill, model 8000.

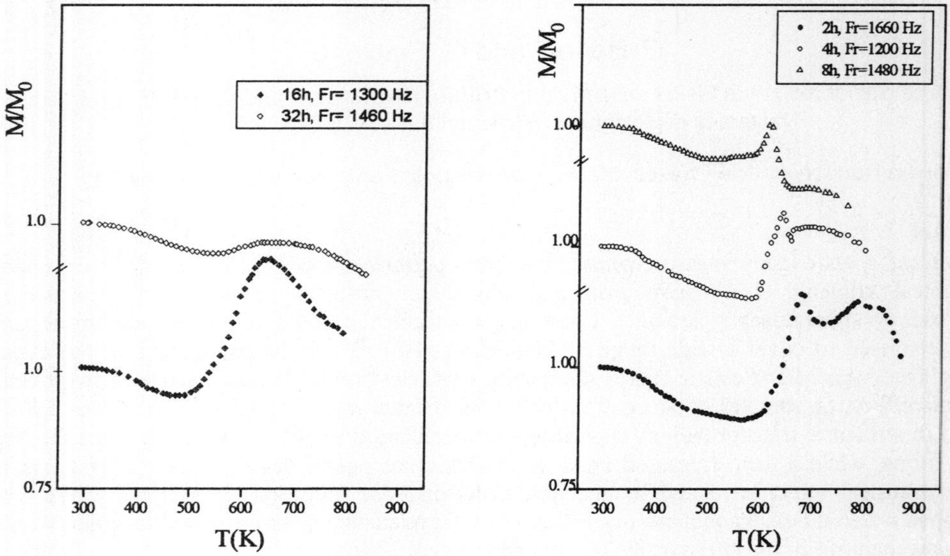

Fig.1. Dynamic Modulus M vs. temperature of the $Fe_{50}Al_{50}$ cold consolidated powders MA 2, 4 and 8 h (right) and 16 and 32 h (left) by using vibrating reed. M_0 is the room temperature dynamic modulus value.

The milling treatments were performed under argon atmosphere (oxygen < 5 ppm). Ethanol was used as a lubricant in a concentration of 0.05 ml/g of powder to avoid the sticking of aluminium particles on the walls of the vial [11].

The consolidation procedure consisted in cold pressing 0.2 g of milled powder in a specifically constructed die with a rectangular section 15x5 mm^2 by means of a piston under an axial pressure of about 1.5-2 GPa. In this way thin bar specimens with a thickness of 0.8 mm have been obtained.

The elastic energy dissipation coefficient Q^{-1} (internal friction) and the dynamic Young's modulus M were measured on cold-consolidated reed specimens by a vibrating reed technique, in the 10^2-10^4 Hz range and at a strain amplitude of 10^{-5}, and by an inverted torsion pendulum, in the 10^{-1}-30 Hz range and at a strain amplitude of 10^{-4}. All measurements were performed at a heating rate of 3 K/min.

Results and discussion

Figure 1 reports the modulus vs. temperature trends for specimens milled at different times. All the curves show an anomalous modulus effect with respect to the usual decreasing trend, starting at approximately 550 K. For temperatures higher than 790 K the modulus recovers the usual isoconfigurational decreasing trend.

The M trend appears quite different at shorter milling times (fig. 1 right). The modulus step occurring at 700 K in the samples MA 2 h shifts at lower temperatures by increasing milling times and simultaneously decreases in strength.

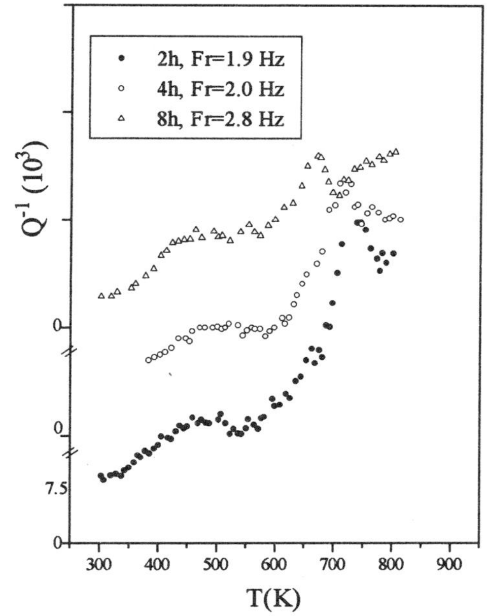

Fig. 2. Elastic energy dissipation coefficient Q^{-1} of the samples MA 2, 4 and 8 h performed by using an inverted torsion pendulum.

It is no more evident in samples MA in excess of 16 h (fig. 1 left). As observed previously [11-13] a progressive dissolution of aluminium in the bcc iron lattice occurs by increasing MA time.In the Fe(Al) solid solution, progressively aluminium enriched, the onset of the ferromagnetic-paramagnetic transition is shifted at lower temperature by increasing the aluminium content. So a shift of the Curie point at lower temperatures together with the modulus step is expected by increasing MA time, reaching room temperature at 34 % at. Al [14].
The elastic energy dissipation coefficient Q^{-1} measurements corresponding to the samples MA up to 8 h are reported in figure 2. For each curve a first broad peak at about 500 K for all samples and a second sharp peak (700 K for the sample MA 2 h) occur [12]. The temperature of the last one shifts downwards about 70 K from the samples MA 2 h to that MA 8 h. This behaviour is in complete agreement with that of the anomalous effect of the modulus (figure 1 right) examined above.
The Q^{-1} peak, together with the modulus step, disappears in the sample MA 32 h (fig. 1 left) [13].
On account of the observed trends of figure 2 it is reasonable to attribute the second Q^{-1} peak and modulus step at 700 K (sample MA 2 h) to the ferromagnetic-paramagnetic transition [14].
The anelastic behaviour of samples thermally stabilised can be observed in the second run of anelasticity measurements. The trend of dynamic Young's modulus and internal friction is reported in fig.3 for the sample MA 32 h. The modulus reveals a monotonic decreasing trend, suggesting that the anomalous trend during the first run is a transient effects and that the structural transformations involved are irreversible.

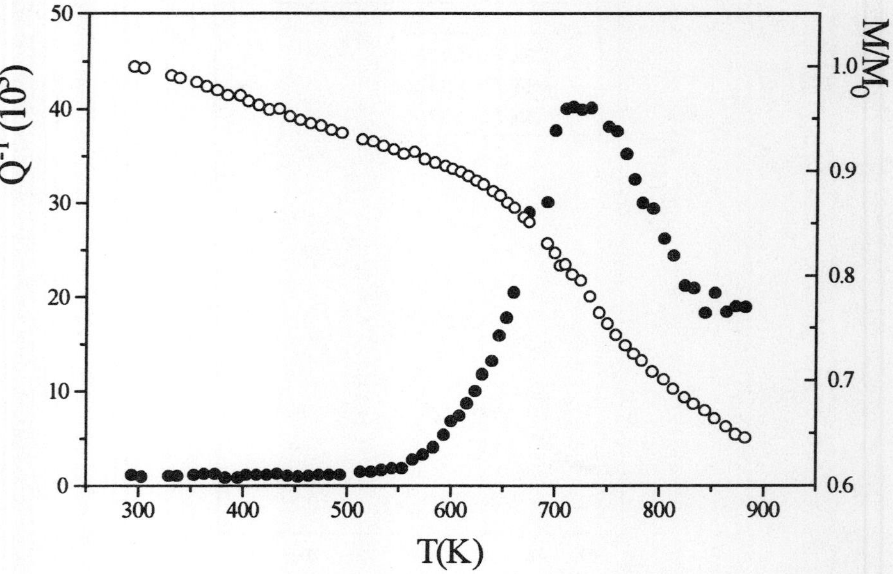

Fig.3. Elastic energy dissipation Q^{-1} (full circle) and Dynamic Modulus M (open circle) for the sample MA 32 h during a second run with raising temperature. Frequency at room temperature: Fr=4.6 Hz.

The Q^{-1} curve shows a strong reduction of the exponential background damping at low frequencies and superimposed on it a well defined relaxation peak.

This result suggests that the MA in the $Fe_{50}Al_{50}$ composition favours the formation of a metastable phase. A moderate thermal annealing gives rise to the formation of several stable phases depending on the previous milling time [12,15]. In fact in samples milled for times <16 h the Al_5Fe_2 phase forms, together with the residual unreacted Al and bcc iron. For samples MA for longer times, the nanostructured FeAl intermetallic compound as a majority phase and the carbide $AlFe_3C_X$ cubic phase are present [11].

The Q^{-1} peak of figure 3 is already detectable in the sample MA 16 h and its relaxation strength strongly increases in the specimen MA 32 h. At the same time the peak temperature is shifted towards lower values as a function of the grain sizes. The activation energy of the peak has been evaluated from the peak temperature shift with measuring frequency ω_i , giving the value H=(1.8±0.2) eV and a pre-exponential factor $\tau_0=10^{-14}$ s [12]. Moreover after background removal, a careful control of the peak shape by fitting the data with a lognormal distribution of relaxation times reveals a peak asymmetry and a width of the peak larger by a factor 2.5 than that of a Debye peak with comparable activation energy. In order to understand the origin of this anelastic peak, we have compared our measurements with those performed by Hren on iron-aluminium coarse-grained [10]. The author found a relaxation peak at 1000 K (1000 Hz), which was attributed to grain boundary sliding mechanisms. So taking into account that this mechanism seems dependent on grain sizes, because the peak temperature shifts towards higher values by increasing the grain size, we can conclude that the anelastic peak of fig. 3 corresponds to that of Hren and is associated to grain boundary relaxation in the nanostructured B2 FeAl intermetallic compound.

Conclusions

The measurements of the elastic energy dissipation coefficient on bulk $Fe_{50}Al_{50}$ obtained by cold consolidation of mechanical alloyed powders, indicate that on the as prepared samples the anelasticity spectra are completely determined by a) the occurrence of the ferromagnetic-paramagnetic transition in the specimens MA up to 8 h and b) structural transformations of the metastable phases induced by the previous MA.

When after moderate thermal ageing of the long time milled samples a well developed nanophase $Fe_{50}Al_{50}$ intermetallic compound forms, the anelasticity spectra show a well resolved relaxational damping peak probably associated to grain boundary sliding mechanisms in the nanophase.
Worth nothing the peak activation energy is substantially lower than that corresponding to grain boundary damping in similar coarse grained materials, as recently observed also on nanocrystalline metals.

Acknowledgements

The authors wish to thank Prof. G. Cocco for providing the facilities of the MA synthesis and Prof. S. Enzo for the useful discussions.

References

[1] R. W. Siegel and G. E. Fougere, *Nanophase Materials*, edited by G.C. Hadjipanais and R. W. Siegel, Kluwer Academic Publisher, 1994, p. 233.
[2] G.W.Nieman, J.R.Weertman, and R.W.Siegel, J.Mater. Res **6,** 1012 (1991).
[3] H.J.Holfler, R.S.Averback, H.Hahn, and H.Gleiter, J.Appl.Phys. **74** (6), 3832 (1993).
[4] S.Schumacher, R.Birringer, R.Strauss, and H.Gleiter, Acta Met. **37**, 2485 (1989).
[5] C.G.McKamey, J.H.Devan, P.F.Tortorelli, and V.K.Sikka, J.Mat.Res. **6**, 1779 (1991).
[6] A. Nowick and B. Berry, *Anelastic Relaxation in Crystalline Solids* (Academic Press, New York, 1972), p.61.
[7]
[8] C. Shyne and M. J. Sinnott, Trans. Metal. Soc. AIME **218**, 861 (1960).
[9] D. B. Fischbach, Acta Metall. **10**, 319 (1962).
[10] J. A. Hren, Phys. Stat. Sol. **3**, 1603 (1963).
[11] R. Frattini, L. Schiffini, G. Scipione, E. Bonetti and S. Enzo, *Processing, Properties, and Application of Iron Aluminides*, edited by J. H. Schneibel and M. A. Crimp, (TMS Pub., 1994), p. 79.
[12] E. Bonetti, G. Scipione, R. Frattini, S.Enzo, L. Schiffini, submitted to J. Appl. Phys.
[13] E. Bonetti, G. Scipione, R. Frattini, S.Enzo,G. Cocco, L. Schiffini, submitted to The 4th European Conference on Advanced Materials and Processes: EUROMAT 95, 25-28 September 1995, Venice/Padua.
[14] M. Yamamoto and S. Taniguchi, Sci. Rep. RITU A **8**, 112 (1956).
[15] R. Gupta, P.P. Macrì, R.Frattini, G.Principi, G.Scipione, L.Schiffini, submitted to Acta Met. Mater.

Materials Science Forum Vol. 195 (1995) pp. 197-202

Influence of Growth Condition and Heat Treatments on Transport and Magnetotransport Properties in Granular Co-Cu Metallic Systems

L. Pareti and G. Turilli

Istituto MASPEC C.N.R., Via Chiavari 18/A, I-43100 Parma, Italy

Keywords: Giant Magnetoresistence, Granular Metallic Films, CoCu, RF Sputtering

Abstract. Thin metallic films of Co, Cu and heterogeneous CoCu films have been prepared by RF sputtering. The influence of growth conditions and annealing temperatures on resistivity and giant magnetoresitence (for CoCu) has been studied. Resistivity in all the systems has been found to be very sensitive to the preparation conditions; however heat treatments at about 300°C eliminate every difference.

The largest spin dependent scattering ($\Delta\rho$) is observed in as deposited films which show a superparamagnetic behaviour of Co particles. However due to the relative behaviour of ρ and $\Delta\rho$, as a function of annealing temperature, the maximum $\Delta\rho/\rho$ value at 19k0e, is observed for Co particles in the ferromagnetic state.

Introduction. Granular metallic systems, which are composed by magnetic particles of nanometric size dispersed in a nonmagnetic conducting matrix, have been found to present effects of giant magneto-resistence (GMR) [1-4]. The GMR, firstly discovered in multilayers, composed by magnetic/nonmagnetic elements [5], consists in a large negative variation of the electrical resistivity when an external magnetic field is applied and it is generally ascribed to the interplay between spin dependent scattering in successive magnetic layers or nano-dispersed particles [4, 6,7]. Necessary conditions for the occurrence of GMR are a suitable distance among magnetic scattering centres, which must be smaller than electron mean free path, and the possibility to change the relative orientation of the magnetization in adjacent layers or particles.

Sputtering is a flexible technique widely used to produce both multilayer and heterogeneous granular films. In the present work a study has been done on the influence of working conditions (pressure, targets power) and growth procedure (see below) on the resistance and magnetoresitance of metallic (Co, Cu) and heterogeneous CoCu thin films. The effects of heat treatments at different temperatures has also been studied.

Sample preparation. Films as been deposited by RF sputtering in which targets are mounted vertically . A rotating carousel allows the substrate to rotate sequentially in front of the different targets. The substrate can either stand or oscillate in front of the target. The residual pressure in the sputtering chamber was 2 10^{-7} mbar.

Monoatomic metallic films of Co and Cu ave been prepared oscillating in front of the single target with different working pressure ranging from 1.9 10-3 to 2 10^{-4} mbar and different target power: 50, 80, 150, 300 w.The films thickness was in the range 100-200 nm.

Heterogeneous CoCu films of 150-200 nm, as been prepared by depositing successive layers of Co and Cu. Two series of samples denominated A and B, having the same composition $Co_{20}Cu_{80}$, were deposited sequentially at increased Co and Cu thickness: [Co 0.75Å/Cu 3Å], [Co 1.5Å/Cu 6Å], [Co 2.25Å/Cu 9Å].Samples A were prepared with the working pressure p=1.9 10^{-3} mbar and target power [Co 70w/Cu 105w], for B samples p=2 10-4 and power [Co 105w/Cu 160w]. A samples of composition $Co_{33}Cu_{67}$ was prepared with the sequence [Co 45Å/Cu 9Å] in the power and pressure conditions of samples A. Heat treatments has been performed under vacuum (10-6 mbar) for 20 minutes at T= 150, 200, 250, 375, 400, 425, 450 °C.

Experimental. The resistivity at RT and LNT ave been measured by van der Pauw technique in a magnetic field up to 19 kOe.

Magnetization curves M(H) as been performed at RT by using an high sensitivity gradient force magnetometer in a maximum field of 20 kOe.

The morphological characterization of Cu films as been performed by SEM.

The condition of high pressure and low power gives rise to a low deposition rate and results in high ρ values. Lower values of resistivity are instead obtained for deposition rate larger than 100 A°/minute were low Ar pressure and high R.F. power are used. Different resistivity values correspond to different grain size as it is shown in Fig. 3. In particular decreasing ρ values are associated to increasing grain size.

The effects, induced by pressure and power respectively are evidenced in Fig. 2 were the influence of heat treatments is also shown. The differences due to various growth conditions can be smeared out and same value of ρ can be obtained for all samples by annealing samples for 10 minutes above 300 °C. On the basis of previous observations it can be deduced that heat treatments result in a reduction of the defects number through the lattice relaxation and grains growth.

A similar evolution of ρ upon annealing is observed in granular bimetallic CoCu. The behaviour of ρ, as a function of the annealing temperature, for samples A and B is shown in Fig. 4. The different ρ values measured in the as deposited samples becomes substantially the same for annealing temperature larger than 250 °C. This means that the effects induced by different growth conditions are completely recovered. A further decrease of ρ is obtained by treating at higher temperature, due to Co grain growth.

Samples of composition $Co_{20}Cu_{80}$ which were grown sequentially with different layer thickness (Co=.75, 1.5, 2.25 °A) present slight difference in ρ and Δρ for the as deposited films, however such a difference disappear after annealing above 375 °C. Moreover the magnetization curve of these as deposited samples show a superparamagnetic behaviour. For a layer thickness of Co >= 4.5 °A a full ferromagnetic behaviour is observed in the magnetization curve.

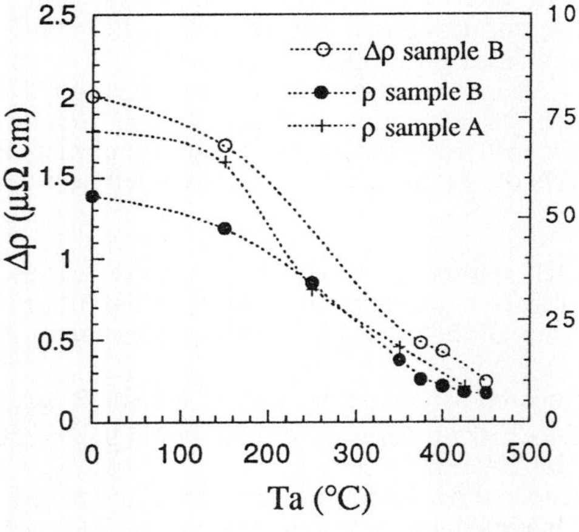

Fig. 4. Dependence of ρ and Δρ on annealing temperature Ta.

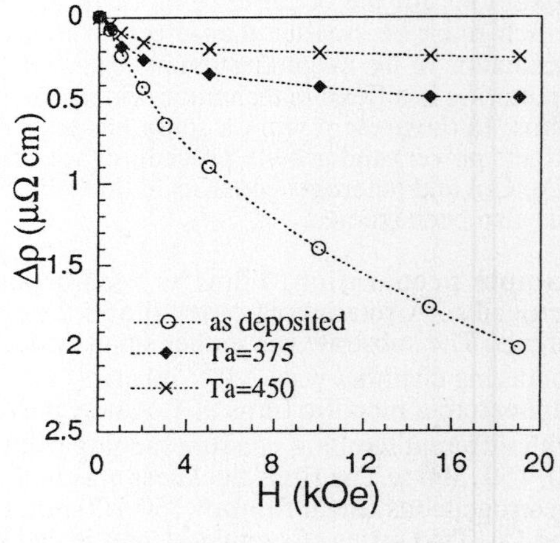

Fig. 5. Dependence of Δρ on applied magnetic field H for samples treated at different temperature

Both ρ and Δρ, were $\Delta\rho=\rho_0-\rho_H$, ρ_0 is the value at H=0 and ρ_H is the value at H=19kOe, which are reported in Fig 4, decrease monotonically increasing the annealing temperature. The behaviour of Δρ which represents the magnetic contribution to the resistivity, indicates that it should reach the zero value for an annealing temperature of 500 °C. This implies that the Co particles size is so large

Results and discussion. The study of the influence of growth conditions and heat treatments on the resistivity has been firstly done on single element Co and Cu films . Both working pressure and target power strongly influence the resistivity (ρ) of as deposited films. In Fig. 1, the resistivity of as deposited films is reported as a function of rate (Å/m), which depends on both pressure and power .

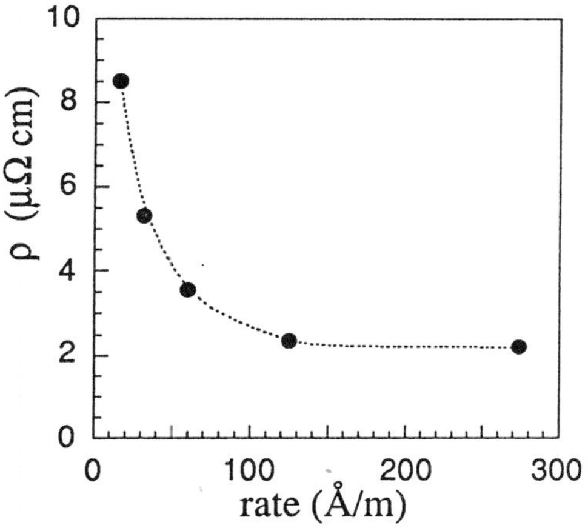

Fig. 1. Dependence of resistivity on deposition rate in Cu films.

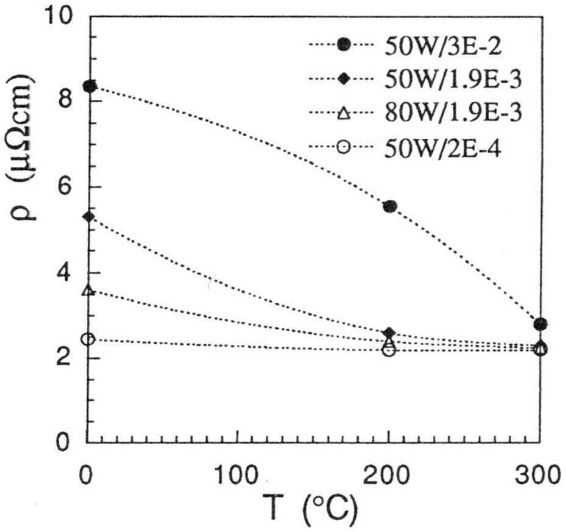

Fig. 2. Dependence of resistivity on annealing temperature for various values of power and working pressure.T=0 means as deposited samples.

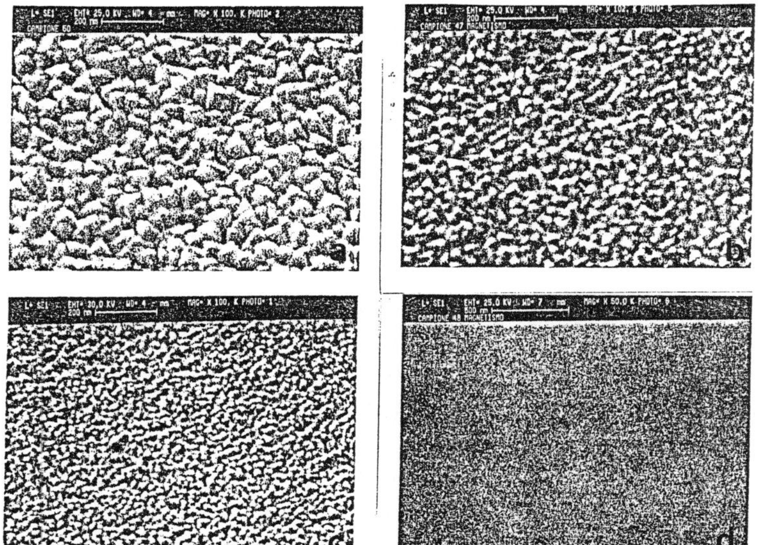

Fig.3. Dependence of grains size on working power and pressure conditions. a: 300w, 1.9 10^{-3}mbar b: 150w, 1.9 10^{-3}mbar; c: 50w, 1.9 10^{-3}mbar; d: 150w, 2 10^{-4}mbar

to make them ineffective for spin dependent scattering contribution. Moreover ρ approaches a saturation value which is close to the resistivity of the Cu matrix. The Δρ values measured at 293 K, as a function of the applied magnetic field H for sample B annealed at different temperature, are reported in Fig.5. It is evident that Δρ for the as deposited sample is far from saturation, as expected for superparamagnetic particles. The treated sample show lower Δρ value which are practically saturated for H<10kOe, even if e slight Δρ increase is observed at higher field in the sample annealed at 375 °C. This withnesses that a small fraction of superparamagnetic particles is still present in the specimen. The magnetization curves performed on the same samples at 293 K are reported in Fig. 6. The M(H) curve for the as deposited sample is far from saturation at H=19kOe in addition no hysteresis is observed, indicating a superparamagnetic behaviour. On the contrary the magnetization curves of annealed samples have a clear ferromagnetic behaviour which is characterised by e larger saturation magnetization a low saturation field and the onset of a coercive field.

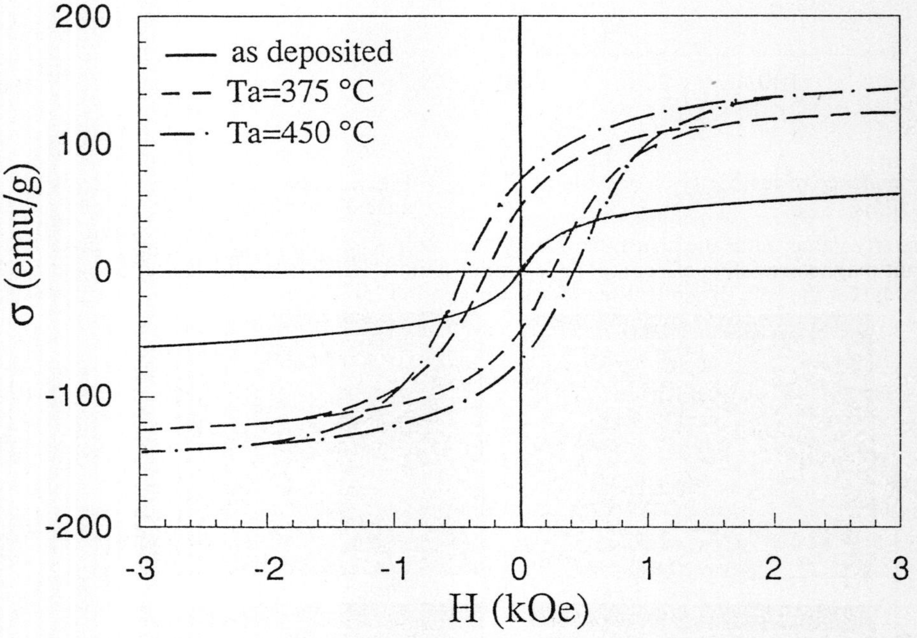

Fig.6. Magnetization curves at 293 k for samples treated at different temperature values.

The higher values of $\Delta\rho/\rho = \Delta\rho/\rho_{(H)}$ at H=19kOe (5.9% and 11% at RT and LNT respectively) were obtained for the sample annealed at 375 °C. The curves of Δρ as a function of applied field H, for the sample treated at 375 °C, are shown in Fig.7. The shape of the two curve performed at RT and LNT are similar even if a smaller high field contribution to Δρ, due to superparamagnetic residual particles, should be expected for the measurement performed at LNT.

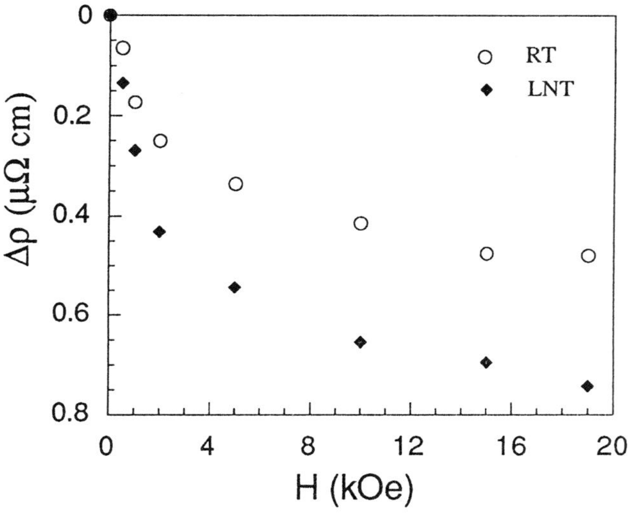

Fig. 7. Δρ at RT and LNT as a function of the applied field H for a sample annealed at 375°C

Large differences up to 50% in the Δρ(H) amplitude are observed, this implies a temperature dependence of the spin dependent scattering effect. This could be attributed to an increase of the mean free path of the conduction electrons with decreasing temperature , as withnessed by the reduction of the resistivity value. Thus the distribution in size and number of the Co particles, that gives rise to a significant magnetoresistive effect at RT, determine a more efficient magnetic scattering at LNT. In other words, for the spin dependent scattering, the relationship between electron mean free path and mean particles distance seems to be more favourable at LNT than RT.

Conclusions. Growth condition strongly influence the resistivity of as deposited Co, Cu, and Co-Cu films. High deposition rate result in a low resistivity and large grain size. However all effects on resistivity of the different growth condition can be cancelled by a 20 minutes annealing at 250 °C Samples of composition $Co_{20}Cu_{80}$ which were grown sequentially with different layer thickness, present slight difference in Δρ/ρ in the as deposited films, however such a difference disappear after annealing above 375 °C.

Increasing annealing temperature booth ρ and Δρ decrease. Their ratio Δρ/ρ which represent GMR increases. The best Δρ/ρ values (5.9% and 11% at RT and LNT respectively) ave been obtained by annealing the samples for 20 minutes at T=375 °C. The magnetic scattering contribution reaches zero for annealing above 500 °C. The same sample shows a marked increase up to 50% of the magnetic contribution to the resistance (Δρ) decreasing the measuring temperature from RT to LNT.

The largest spin dependent scattering (Δρ) is observed in as deposited films which show a superparamagnetic behaviour of Co particles. However due to the relative behaviour of ρ and Δρ , as a function of the annealing temperature, the maximum Δρ/ρ value at 19k0e, is observed for Co particles in the ferromagnetic state.

References

[1] A. E. Berkowitz, J. R. Mitchell, M. J. Carey, A. P. Young S. Zhang, F. E. Spada,F. T. Parker, A. Hutten and G. Thomas, Phys. Rev. Lett. 68, 3745 (1992)
[2] J. Q. Xiao, J. S. Jiang and C. L. Chien, Phys. Rev. Lett. 68, 3749 (1992)
[3] C. L. Chien, J. Q. Xiao and J. S. Jiang, J. Appl. Phys. 73, 5309 (1993)
[4] M. R. Parker, J. A. Barnard, D. Seale and A. Waknis, J. Appl. Phys. 73, 5512 (1993)
[5] M. N. Baibich, J. M . Broto, A. Fert, F. Nguyen van Dau, F. Petroff, P. Etienne, G. Kreuzet, A. Friedericich and J. Chazeles, Phys.Rev. Lett. 61, 2472 (1988)
[6] S. Zhang, Appl. Phys. Lett. 61, 1855 (1992)
[7] A. Fert, P. Grunberg, A. Barthelemy, F. Petroff, W. Zinn, J. Magn. Magn. Mater. 140, 1 (1995)

Materials Science Forum Vol. 195 (1995) pp. 203-208

Room-Temperature Giant Magnetoresistance of Melt-Spun Ribbons and Thin Films of Granular $Cu_{100-x}Co_x$

P. Tiberto[1], F. Vinai[1], P. Allia[2] and P. Rava[3]
[1] IEN Galileo Ferraris and INFM, C.so M. d'Azeglio 42, I-10125 Torino, Italy
[2] Politecnico di Torino, Dip. di Fisica, and INFM,
C.so Duca degli Abruzzi 24, I-10129 Torino, Italy
[3] Elettrorava S.p.A., Savonera, Torino

Keywords: Granular Magnetic Materials, Giant Magnetoresistance

Abstract

Bulk granular $Cu_{90}Co_{10}$ alloys displaying a negative giant magnetoresistance (GMR) have been produced by submitting melt-spun ribbons to conventional annealing and to dc Joule-Heating in vacuo to induce the precipitation of Co-cluster with size in the nanometer range. Room-temperature GMR values up to 7% al 20 kOe are observed in samples submitted to suitable dc Joule-heating treatments. Shorthand information on the distribution of particle radii is obtained by analysing the room-temperature anhysteretic magnetisation curves.

The magnetoresistance has been evaluated on as-prepared and annealed thin films of the same starting composition obtained by sputtering. The measured effect is always lower in sputtered films than in melt-spun ribbons. The effect of the substrate temperature during the film deposition on the GMR is studied.

Introduction

Granular magnetic systems, consisting of nanometer-sized particles of a magnetic metal (e.g., Fe, Co) dispersed in a non-magnetic metallic matrix (e.g., Ag, Cu), have been extensively studied in the last few years, owing to the occurrence of an effect of giant magnetoresistance (GMR) [1,2]. The GMR effect was first discovered in magnetic multilayers [3], and consists of a very large change in the electrical resistance under an external magnetic field. The observation, in granular solids, of a GMR comparable to the one found in multilayer systems stimulated the study of transport properties in magnetically heterogeneous media. Usually, heterogeneous thin films are produced by sputtering or coevaporation of immiscible magnetic and non-magnetic metals. Only recently, granular solids have been also produced by means of other non-equilibrium fabrication processes, such as mechanical alloying [4] or melt-spinning [5].

In both multilayered systems and granular solids, the field dependence of resistivity is generally interpreted by the so-called two-current model, based on the spin-dependent scattering of conduction electrons essentially at the magnetic/non-magnetic interfaces, and to a lesser extent within the magnetic regions [6]. In particular, the resistance of heterogeneous alloys is higher when the magnetic moments of neighboring magnetic entities are not aligned (balanced scattering of spin-up and spin-down conduction electrons), and the effect of the magnetic field is to align the magnetic moments, increasing the macroscopic magnetic ordering (preferred scattering of one spin direction, giving rise to a short-circuit effect), and therefore decreasing the resistance of the system.

In order to obtain the best GMR response, the as-produced materials are normally submitted to thermal treatments, to precipitate nanometer-sized particles of the magnetic element. Several theories have been proposed to quantitatively describe the GMR phenomenon and its dependence upon temperature, magnetic particle size and alloy composition [7,8]. In all existing models, one of the most important parameters is the magnetic particle size, which directly determines the spin-dependent scattering through the surface-to-volume ratio and interparticle distance.

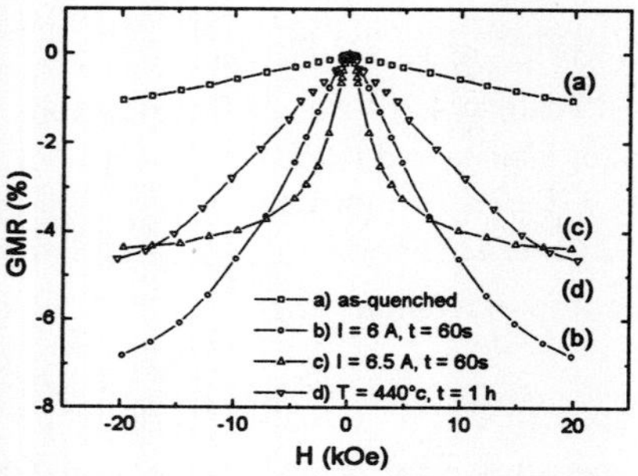

Fig. 1 Room Temperature GMR in selected melt-spun $Cu_{90}Co_{10}$ strips

The hysteretic magnetisation curve of granular magnetic materials may be viewed as the superposition of a suitable Langevin function and an ad-hoc analytical function describing a hysteretic contribution of given coercivity and squareness [10]. As known, the measured coercivity is often assumed to provide a good estimate of the size of fine particles above the so-called critical diameter for the appearance of superparamagnetism [11]. However, in the present case, different evidences point to a substantial inadequacy of this procedure to determine the particle size:

1) the critical diameter for superparamagnetism is not exactly known, because it depends on the value of the dominant magnetic anisotropy energy in Co particles. The latter could originate either from the particle's crystalline structure, or from their shape. The magnetic anisotropy energy could range in this case from about 4 10^5 erg/cm^3 for spherical particles of fcc Co, to 2 10^6 erg/cm^3 for slightly elongated particles of fcc Co, to 4 10^6 erg/cm^3 for spherical particles of hcp Co [12], leading to a room-temperature critical diameter for superparamagnetism of 17, 10, 7.6 nm, respectively;

2) the low coercivity of the assembly of Co particles may be explained: a) by assuming the presence of clusters of size larger than the critical size; b) by invoking the presence of interactions between particles whose size may be smaller than the critical size [13];

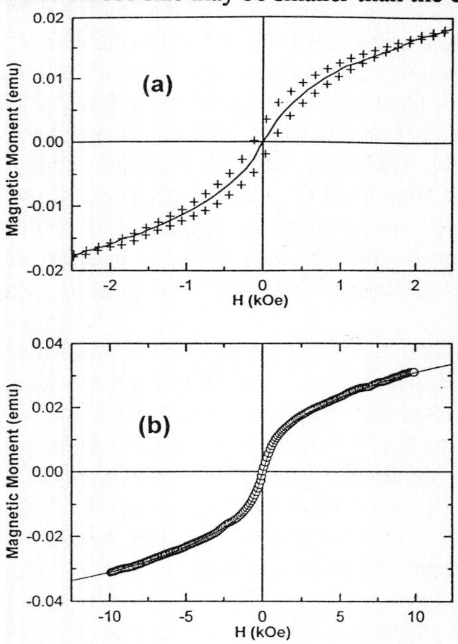

Fig. 2. (a) hysteresis loop and anhysteretic curve of $Cu_{90}Co_{10}$ (as quenched ribbon); (b) fit of the anhysteretic curve shown in (a) by means of four weighted Langevin curves

3) the measured hysteresis loops always look like broadened Langevin functions, and are rather poorly fitted by a superposition of a Langevin function and of a hysteretic magnetisation curve.

The experimental evidence leads us to introduce a different approach to study the size distribution of Co clusters in the considered systems. From this viewpoint, more information can be extracted from the overall shape of the magnetisation curves than from the coercivity values. As a consequence, the anhysteretic magnetisation curves, whose correspondence to the hysteresis loops is always apparent (see Fig.2a) may be proposed as the most convenient data sets to be analysed in terms of Langevin functions. It should be explicitly noted that the superparamagnetic character of an assembly of fine particles is just a matter of the time scale of the experiments performed to determine their magnetic behavior [12].In principle, therefore, the hysteresis loops measured on the considered samples should reduce to the corresponding anhysteretic curves by merely performing magnetic measurements on a much longer time scale.

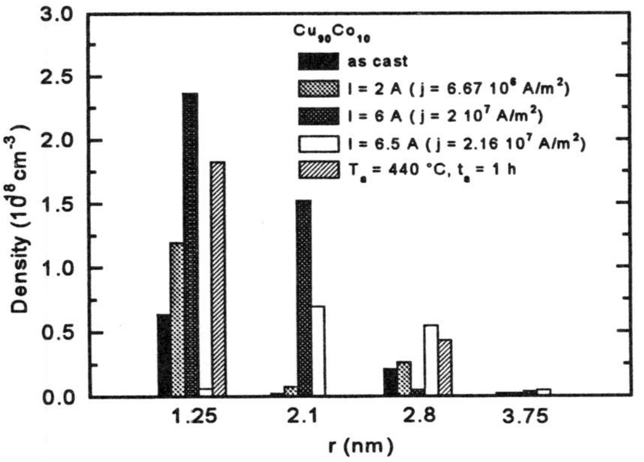

Fig. 3 Radius distribution of Co particles in selected melt-spun ribbons

In the present case, the Co grains dispersed in the Cu matrix are assumed to be distributed in size. The Langevin function depends on the grain size through the particle's magnetic moment. The experimental data are well described in terms of a weighted sum of few Langevin functions, each describing the contribution to the resulting magnetic moment of a set of clusters of given size. Shorthand information about the structure of the studied granular materials can be therefore obtained through an analysis of this type, completely based upon magnetic data.

The fitting function is therefore of the form:

$$M = M_0 \sum_i p_i \mathscr{L}_i (r_i) \tag{1}$$

where $\mathscr{L}_i$ is the Langevin function, M_0 and r_i are the alloy's magnetisation and particle radius, respectively, and the weights p_i, satisfy the conditions $\Sigma_i p_i = 1$ and $p_i \geq 0$.

In the present analysis, the radius of Co particles was assumed to be distributed in the range 1.25 ÷ 3.75 nm, in agreement with the results reported in the literature [1,6]. The fits appeared to be systematically worsened by using radii exceeding both limits. The best results were obtained using four Langevin functions in Eq. 1. In this case the r_i's (i = 1,4) take the values 1.25 nm, 2.1 nm, 2.9 nm, 3.75 nm. The fit obtained through Eq. 1 on an as-cast sample of $Cu_{90}Co_{10}$ is shown in Fig. 2b as an example of the validity of such a procedure. The agreement is excellent in all cases. The present analysis was exploited in samples submitted to different thermal treatments (either Joule heating or conventional annealing) in order to get information on the evolution of Co clusters. It should be noted that the observation of small Co particles through X-ray diffraction or electrical microscopy is difficult owing to the inherent difficulty in separating the contributions of Cu and Co, which are characterised by closely similar lattice parameters [12].

Typical results obtained by using this technique of analysis on $Cu_{90}Co_{10}$ are summarised in Fig. 3. The density of Co particles of a given size, N_i, was obtained by multiplying the weights p_i by the total number of particles per unit volume, N, obtained from the equation:

$$N\bar{\mu} = M_s^* , \tag{2}$$

where M_s^* is the alloy's saturation magnetisation, and $\bar{\mu}$ is the average magnetic moment:

$$\bar{\mu} = \sum_i p_i \left(\frac{4\pi}{3} r_i^3 \right) M_s^{(Co)} , \tag{3}$$

$M_s(Co)$ being the saturation magnetisation of bulk Co. As a consequence,

$$N_i = Np_i = \frac{NM_s^*}{\bar{\mu}} = \frac{p_i}{\frac{4\pi}{3}\sum_i p_i r_i^3} \frac{M_s^*}{M_s^{(Co)}} . \quad (4)$$

The value of M_s^* is determined by extrapolating the fitting function of Eq. 2b to $H \to \infty$.

The hystogram shown in Fig. 3 includes the as-quenched ribbon, three Joule-heated samples and the high-GMR sample obtained by conventional annealing (T_a = 440 °C for 1 hour). The Joule-heated samples were selected in order to show the changes in cluster-size distribution in the ribbon submitted to the optimum electrical current density for GMR (j_{opt}), and in two ribbons where j was respectively lower and higher than j_{opt}.

The heat treatments induce a moderate shift in the average cluster size towards higher values. The density of very small magnetic particles (r = 1.25 and 2.1 nm) is the highest in the sample displaying the maximum GMR (Joule-heated, I = 6 A). Comparing high-GMR systems obtained by furnace annealing and Joule heating, one finds that the latter is more effective in producing a high density of small clusters, and a slightly lower density of large particles. The reduction in GMR for electrical current density vales beyond j_{opt} appears to be related to a strong reduction in the number of the smallest grains (r = 1.25 nm), as observed in Fig.3 for the sample submitted to a current of 6.5 A. This material exhibits a significant fraction of magnetic particles having larger radii, possibly indicating a process of growth of pre-existing grains.

Finally, a comment about the grain sizes obtained through the present analysis is needed. In all cases, the anhysteretic curves are definitely not fitted by admitting the existence of particles having larger sizes ($r \geq 4$ nm). It should be noted that all reported size distributions rapidly tail off in their right sides, indicating that the particle radius is rather small in the majority of cases. This outcome is in line with other findings on similar alloy systems [1,6]. In the light of these results, the appearance of a non-zero coercivity may indicate either that the critical diameter for superparamagnetism is in the range 7 - 8 nm (as expected for elongated particles of fcc Co [12]), or that the Co clusters cannot be considered as independent. Further measurements of anhysteretic magnetisation curves as functions of temperature (typically, above RT) are however needed to check the validity of the approach proposed here to determine the cluster size distribution from purely magnetic data. In particular, the obtained fractions of particles of given radius will be confirmed to be actually significant only if anhysteretic magnetisation curves measured at different temperatures will be found to remain unchanged when plotted as functions of the ratio H/T.

Representative GMR curves obtained at room temperature on as-prepared samples [a thin film produced by sputtering under non-optimised preparing conditions (a) and two ribbon strips (b,c)] of the system $Cu_{90}Co_{10}$ are shown in Fig. 4. The existence of a measurable GMR effect in all these systems indicates that the as-prepared metastable materials are not ideal solid solutions of Co in Cu. A low fraction of particles of clusterized Co is responsible for the small GMR effect, the maximum GMR value never exceeding 1.5%. In particular, the thin film (prepared on a substrate whose temperature was maintained at T = 300 °C) displays the worst magnetoresistance behaviour. Note that the high-field tails of this GMR curve appear to saturate much more rapidly than in melt-spun ribbons. This feature point to the existence of rather large Co particles, and may explain why the maximum GMR value is so low. The comparison between melt spun ribbons of different thickness (prepared under different quenching rates from the melt) shows that the fraction of quenched-in Co particles (indicated by the maximum GMR value) is higher in the thicker material, obtained at a lower quenching rate, as expected.

In this work we perform a study of the magnetic properties and the GMR effect in granular $Cu_{90}Co_{10}$ alloys obtained both by rapid quenching in ribbon form and as heterogeneous sputtered thin films. Rapidly quenched samples were submitted to both conventional annealing and DC Joule heating, and sputtered thin films were furnace-annealed, in order to optimise the GMR response. The distribution of the magnetic particle radii in rapidly quenched samples was studied as a function of the heat treatment by analysing the room-temperature anhysteretic magnetisation curves. A comparison between magneto-transport properties of granular $Cu_{90}Co_{10}$ materials obtained by rapid solidification and by sputtering is performed. The sputtering parameters are shown to play a major role on the GMR of thin films. The maximum GMR values of melt-spun ribbon strips submitted to proper heat treatment are found to be about seven times larger than the ones measured on annealed thin films of the same composition.

Experimental

A continuous ribbon of $Cu_{90}Co_{10}$ was produced by planar flow casting in He atmosphere on a CuZr rotating drum. Heterogeneous thin films of the same nominal composition have been produced by sputtering on glass substrates using a single target of the master alloy.

Different ribbon strips (width 5 10^{-3} m, thickness 5 10^{-5} m) and thin films were submitted to conventional furnace annealing in Ar atmosphere in the interval 440÷460 °C for 1 hour to induce the precipitation of Co particles. In addition, several melt-spun ribbon strips were submitted to dc Joule-heating in vacuo by clipping the sample between two copper electrodes (length of the sample 0.1 m), and applying an electrical current (in the range $2\ A \leq I \leq 10\ A$) for a fixed time ($t = 60$ s) [9]

Magnetoresistance curves were obtained at room temperature on both ribbon strips and thin films in the as-prepared state and after thermal treatments. The magnetoresistance was measured up $|H| = 20$ kOe through a conventional four-contact technique. The magnetic field was applied in the ribbon's plane, perpendicular to the bias current (transverse configuration). The value of the Giant Magnetoresistance (GMR) for a given value of H was defined as $[(R(H)\text{-}R(H=0))/R(H=0)] \cdot 100$.

Magnetic hysteresis loops of melt-spun samples were measured using a vibrating-sample magnetometer VSM (LDJ, model 9500) with an applied field H varying between ±10kOe. In addition, the corresponding anhysteretic magnetisation curves were determined starting from the demagnetised state and gradually increasing the applied field in either direction. In this case, the field was again set to zero after each measurement.

Results and Discussion

Typical GMR curves are shown in Fig. 1 for four different $Cu_{90}Co_{10}$ samples [a) as cast, b) Joule-heated, I = 6 A t = 60s, c) Joule-heated, I = 6.5 A t = 60 s, d) furnace-annealed, T = 440 °C, t = 1 h]. The maximum GMR value at H = 20 kOe is observed to increase with respect to the as-prepared material for all thermally treated samples. The GMR curve of samples a) b) and d) is clearly non-saturating up to H = 20 kOe, indicating the presence of a large fraction of superparamagnetic Co clusters in the Cu matrix.

Figure 2a shows the hysteresis loop and the anhysteretic curve of an as-cast sample of $Cu_{90}Co_{10}$. This is an example of the close relationship existing in all studied samples between hysteretic and anhysteretic magnetisation curves.

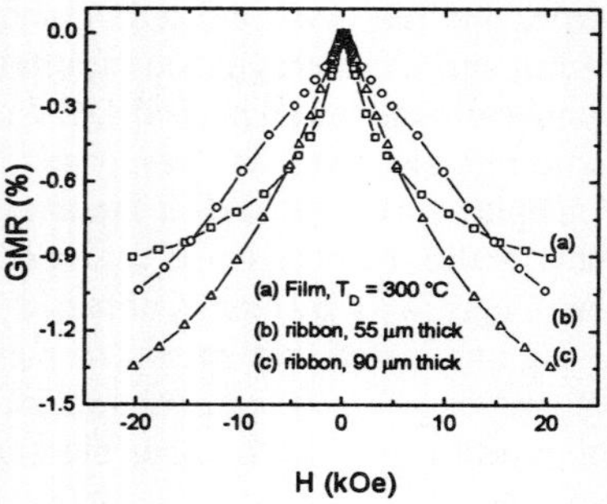

Fig. 4 GMR behavior for a thin film sample (curve a) and two melt spun strips (curves b,c).

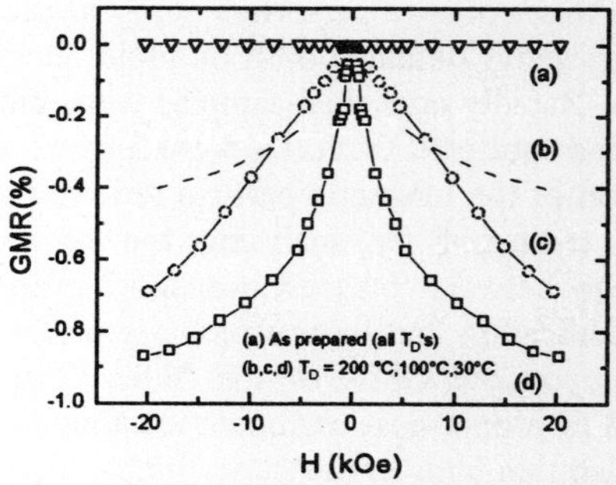

Fig. 5 GMR curves of thin films prepared with different substrate temperatures. b-d: annealed films (T = 460 °C, 1 h).

The preparing conditions play an even more interesting role in sputtered thin films, as shown by the preliminary results reported in Fig. 5. All films were prepared in this case by reducing the exposed surface of the glass substrate, and resulted to be always characterised by the absence of magnetoresistance effect in the as-prepared condition (curve a), indicating that nearly ideal solid solutions of Co in Cu were actually obtained, even starting from a single master-alloy target. In addition, the role of the glass-substrate temperature was elucidated by annealing all films in a furnace at the same temperature (T = 460 °C) for 1 hour. The resulting magnetoresistance (curves b,c,d) turns out to be strongly dependent on the deposition temperature T_D, and increases with reducing T_D. However, the maximum GMR values obtained after annealing are still far from the ones measured on annealed melt-spun ribbons of the same composition. This result could be related either to a reduction in the total Co content in the sputtered films with respect to the starting alloy, or to the annealing treatment, which is still to be optimised. Further measurements on similar films will be performed in order to check this point.

References

[1] A.E. Berkowitz, J.R. Mitchell, M.J. Carey, A.P. Young, S. Zhang, F.E. Spada, F.T. Parker, A. Hutten, and G. Thomas, Phys. Rev. Lett. **68**, 3745 (1992).

[2] J.Q. Xiao, J.S. Jiang, and C.L. Chien, Phys. Rev. Lett. **68**, 3749 (1992).

[3] M.N. Baibich, J.M. Broto, A. Fert, F. Nguyen van Dau, F. Petroff, P. Etienne, G. Creuzet, A. Friederich, and J. Chazeles, Phys. Rev. Lett. **61**, 2472 (1988).

[4] A.R. Yavari, P.J. Desré, and T. Benameur, Phys. Rev. Lett. **68**, 2235 (1992).

[5] J. Wecker, R. von Helmolt, L. Schultz, and K. Samwer, Appl. Phys. Lett. **62**, 1985 (1993); IEEE Trans. Magn. **29**, 3087 (1993).

[6] T.A. Rabedeau, M.F. Toney, R.F. Marks, S.S.P. Parkin, R.F.C. Farrow, G.R. Harp, Phys. Rev. **B48**, 16810 (1993).

[7] M. Rubinstein, Phys. Rev. **B50**, 3830 (1994).

[8] D.J. Kubinski and H. Holloway, J. Appl. Phys. **77**, 2508 (1995).

[9] P. Allia, M. Baricco, P. Tiberto and F. Vinai, Rev. Sci. Instr. **64**, 1053 (1993)

[10] M.B. Stearns and Y. Cheng, J: Appl. Phys. **75**, 6894 (1994)

[11] B.D. Cullity : "Introduction to Magnetic Materials" (Addison-Wesley, Reading 1972), *passim*

[12] A. Hütten and G. Thomas, Ultramicroscopy **52**, 581 (1993)

[13] D.V. Berkov and S.V. Meshkov, IEEE Trans. Magn. **26**, 1804 (1990)

Materials Science Forum Vol. 195 (1995) pp. 209-214

Synthesis, Characterization and Magnetic Properties of Fine Fe-Cr-B Amorphous Particles

L. Suber[1], D. Fiorani[1], H. Romero[1], A. De Pamphilis[1], M. Vittori[2], A. Montone[2], J.L. Dormann[3] and J. Maknani[3]

[1] CNR, ICMAT, CP 10, I-00016 Monterotondo St., Roma, Italy

[2] ENEA, Nuovi Materiali, C.R.E. Casaccia, C.P.2400, I-00100 Roma, Italy

[3] CNRS, Laboratoire de Magnetisme, F-92195 Meudon Cedex, France

ABSTRACT

We report the preparation of amorphous $Fe_{80-x}Cr_xB_{20}$ ($5 \leq x \leq 70$) fine particles by chemical reduction, their morphological and structural characterization by SEM, TEM measurements and X-ray diffraction, and the study of their magnetic properties by means of Mössbauer and magnetization measurements.
The particles are spherical and their size ($10 < \phi < 200$ nm) is composition dependent, decreasing with increasing Cr-content. The magnetic properties, determined by chemical composition and by particle sizes distribution, move with increasing chromium content x, from a ferromagnetic character to a magnetically disordered state, characterized by the coexistence of ferro and antiferromagnetic interactions.

INTRODUCTION

Amorphous fine particles have attracted much attention in recent years because of their distinctive properties (with respect to those of ribbons and films of the same composition) and for potential applications (e.g. in powder metallurgy, catalysis, ferrofluids, magnetorecording...). Different synthesis methods have been developed for producing fine powders of amorphous alloys, e.g. mechanically alloying of component metals [1], hydrogenation of intermetallic compounds [2], thermal decomposition of iron pentacarbonyl [3], chemical reduction of transition metal ions by alcaline metal borohydrides [4-6].
The chemical reduction method has the advantage of allowing to obtain in great quantity, at low temperature, more homogeneous products in a larger composition range. In particular, this method, using $NaBH_4$ or KBH_4 as reducing agent, has proved to be successful in the synthesis of transition metals boron amorphous alloys (e.g. $M_{80}B_{20}$), leading to ultrafine powders suitable for compactation in samples of various shapes for different kinds of technological applications.
In this paper we report the synthesis of Fe-Cr-B particles by reducing metal aqueous solution with aqueous solution of $NaBH_4$, their morphological and structural characterization and the study of their magnetic properties.

EXPERIMENTAL

Preparation

Fine $Fe_{80-x}Cr_xB_{20}$ ($5 \leq x \leq 70$) particles were prepared by reduction of $FeSO_4$ and $CrCl_3$ with $NaBH_4$ (typical molecular ratio : $FeSO_4 + CrCl_3/ NaBH_4$: 8 / 100). The reactions were carried out in an atmosphere of high purity nitrogen. 50 ml 1M $NaBH_4$ aqueous solution was added dropwise to a 50 ml $8x10^{-2}$ M metal salts solution mantaining the reaction vessel in an alcohol bath at T = -10 °C. The black fine powder was washed thoroughly with water to remove residual ions from the reaction mixture and with acetone to remove water. The sample was then dried overnight under vacuum. Finally air was introduced in the vessel in a controlled way in order to passivate the fine particles, which are pyrophoric. $FeCl_3.6H_2O$ was also used, instead of $FeSO_4$ and the way of mixing liquids was changed, i.e. metal salts aqueous solution was added to the reducing solution. Moreover , with the aim of covering the particles with a SiO_2 layer, tetraethoxysilane was added to a suspension of $5x10^{-5}$M Fe-Cr-B in a mixture of 0.1M isopropylic alcohol , $3x10^{-2}$M water and $3x10^{-2}$M NH_3 at 50°C. The mixture was stirred for 5 days and finally the ultrafine powder was filtered, washed with isopropylic alcohol and dried under vacuum.

Structural and magnetic characterization

The morphology, the chemical composition and the structure of the powder specimens were examined by TEM, SEM and X-ray diffraction measurements. The TEM instrument was a Jeol 4000 FX, equipped with an Ultra Thin Window (UTW) X- ray detector and operated at 400 kV. To this purpose the powders were dispersed by an ultrasonic bath and a drop of the resulting suspension was allowed to dry on a microscope grid coated with a thin carbon film.

The magnetic properties were investigated by Mössbauer and magnetization measurements. Mössbauer spectra were recorded in the temperature range 4.2 < T < 600 K, using a conventional spectrometer operating in the constant acceleration mode with a source of ^{57}Co in Rh. Magnetization measurements were performed in the temperature range 4.2 < T < 300 K, using a commercial SQUID magnetometer (H_{max} = 5.5 T)

RESULTS AND DISCUSSION

A) Morphological and structural characterization

X- ray and electron diffraction show that the particles are amorphous. TEM measurements show that the series of samples with variable composition x consists of agglomerates or long chains of spherical particles of different sizes. The mean particle diameter and the width of size distribution depends on the composition x (e.g. for x = 40, the diameter ranges from few nanometers to 200 nm, with a mean value $\langle \phi \rangle$ = 110 nm and with width at half height in the distribution curve $\Gamma_{1/2}$ = 50 nm; for x = 60, ϕ_{max} = 90, $\langle \phi \rangle$ = 50 nm, $\Gamma_{1/2}$ = 40 nm). Within each sample there is a chemical composition distribution, related to the size distribution, the smaller particles being richer in chromium. This indicates that the nucleation process in this system is controlled by the chemical composition, an higher chromium content promoting an easier nucleation, whereas the subsequent growth occurs in an iron enriched solution .

B) Thermal behaviour

The thermal behaviour of the $Fe_{80-x}Cr_xB_{20}$ alloys has been studied by thermogravimetric and differential thermal analysis (DTA). For x ≥ 60 the material remains amorphous at least till 700°C, whereas for x ≤ 60 an increase of weight associated with exothermic processes denotes, as checked by X-ray powder diffraction measurements , the formation of iron and chromium borates.

C) Magnetic Properties

C1) Mössbauer spectra

At low temperature, the spectra consist, for all the samples, in an enlarged hyperfine pattern characteristic of amorphous structure, superposed to a central paramagnetic doublet. Increasing the temperature, the paramagnetic fraction increases as reported for x = 25 in fig. 1a. The corresponding hyperfine field distribution is shown in fig 1b. At low temperature a hump is observed for H_{hyp} at about 450 kOe, characteristic of Fe^{3+}. H_{hyp} has been correlated with the isomer shift δ with the relation $\delta = 1.16 \times 10^{-3} \times \Delta H_{hyp} - 0.132$.

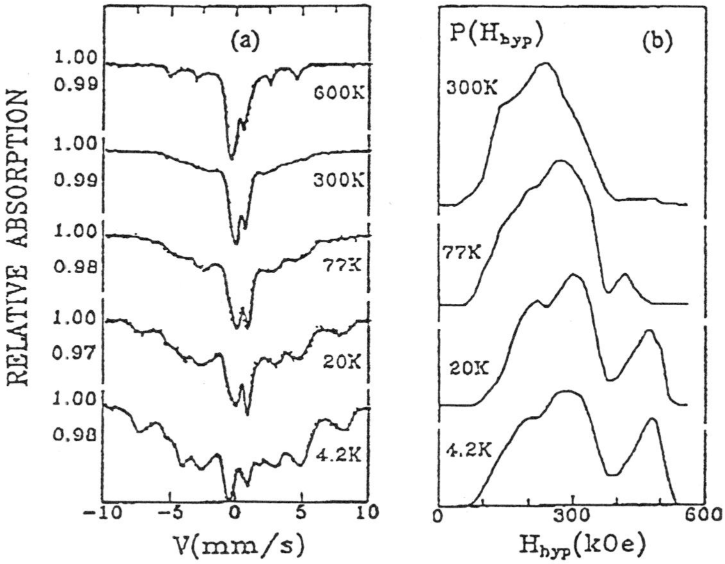

Fig. 1 (a) Mössbauer spectra at different temperatures for x = 25
(b) Corresponding hyperfine field distributions.

Four iron species can be distinguished, whose population is reported in table1;

x	Fe^{2+}(a)	Fe^{3+} (b)	Fe^{3+}(c)	Fe^{0}	$\langle H_{hyp}\rangle$ 4.2 K	(kOe) 300 K
5	0.03	0.02	0.13	0.82	251	222
16	0.06	0.05	0.16	0.73	238	200
25	0.06	0.05	0.22	0.67	235	200
40	0.06	0.11	0.18	0.65	238	200

Table 1. Populations of the different iron species (±0.02) (see text);and $\langle H_{hyp}\rangle$ for Fe^{0} (±5 kOe).

(a) Fe^{2+}ions, antiferromagnetically ordered with $20 < T_N < 60$ K ;
(b) paramagnetic Fe^{3+}ions (down to 4.2 K).They are probably paramagnetic at zero temperature too, because no variation of spectra occurs between 10 and 4.2 K (fig.2);
(c) Fe^{3+}ions, antiferromagnetically ordered, with $40 < T_N < 70$ K .
(d) metallic Fe^{0}, ferromagnetic, with a large T_c distribution, depending on x. The average T_c value is higher than that of the corresponding bulk composition. This indicates that the amorphous alloy particles are richer in Fe^{0} than the nominal value. An example of the temperature dependence of the Fe^{3+} (b) and Fe^{0} population is reported in fig.2.

Oxidized iron species (a), (b) and the fraction of (c), showing T_N values similar to species (a) are mainly included in the smallest particles. Passivation leads to oxidation on the particle surface which can extend to the particle core when the particle volume is small. The oxidation of the core may be partial (Fe^{2+}). Moreover, the smallest particles are also the richest in Cr, as determined by TEM measurements.

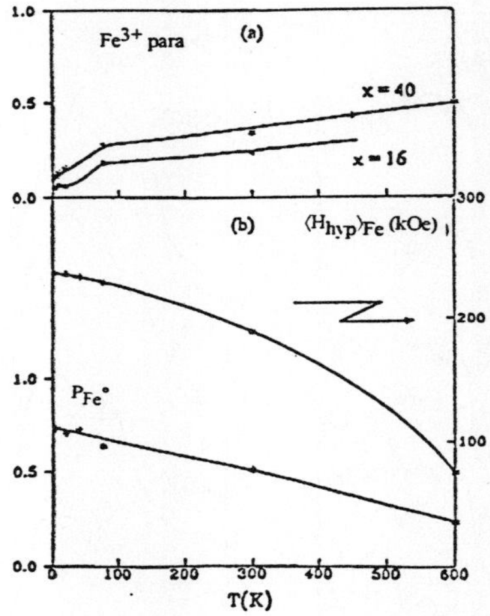

Fig 2 (a) Temperature dependence of paramagnetic Fe^{3+} populations for x = 16 and 40; (b) Thermal variation of the magnetic population $P_{Fe°}$ and of the mean hyperfine field $\langle H_{hyp}\rangle_{Fe°}$ for x = 16.

With increasing x, the transition temperature is lowered and above a certain amount of Cr, no magnetic state occurs. This is in agreement with the fact that for x = 5 the paramagnetic Fe^{2+} population is very low and increases with x (table1). A perturbed antiferromagnetic order should occur in those particles. The remaining fraction of species (c) corresponds to oxidized iron on the surface of the particles. The population of metallic iron decreases with increasing x, according to the decrease of the mean particle size .

Finally, crystallization is observed between 450 and 600 K for $x \leq 25$ and above 600 K for x = 40. At 600 K, the hyperfine parameters of the crystallized fraction are $\delta \approx -0.2$ mm/s and $H \approx 300$kOe, characteristic of Fe^0. The population is equal to about 0.25 for x = 16 and 25. This means that increasing the temperature, the biggest particles, poorer in Cr , crystallize first and that the smaller ones, richer in Cr, mantain the amorphous structure.

C.2)Magnetization Measurements

Magnetization measurements were performed as a function of the magnetic field (H_{max} = 5T) at T = 5K (fig.3). A progressive change with the composition x is observed, reflecting the change of the relative populations of the different iron species. For low x values ($x \leq 25$) a ferromagnetic behaviour is observed, with saturation and low coercive fields (e.g. for x = 5, H_c =14Oe),as expected for soft ferromagnetic $Fe_{80-x}Cr_xB_{20}$ particles. For x > 25, the magnetization does not saturate, showing a linear increase in the high field region. This reflects the increasing fraction of paramagnetic and antiferromagnetic oxidized iron species with increasing x. The saturated magnetization is close to that evaluated from the $\langle H_{hyp}\rangle$value of metallic iron (from tab.1, using the factor 100 kOe = 0.65 μ_B and by supposing that the oxidized particles are antiferromagnetic) [7,8].

The coercive field is found to increase with x (e.g. for x = 40, H_c = 70 Oe; for x = 70, H_c = 200 Oe). This should be mainly due to the decrease of the average particle size with x, leading to an increase of the anisotropy. Surface should give the dominant contribution to the anisotropy with decreasing particle size, being negligible the magnetocrystalline anisotropy of amorphous particles. This is

confirmed by the effect of covering particles with SiO_2. An increase of the coercive field is observed (e.g. for x = 70, H_c = 500 Oe) with respect to the uncovered particles (H_c = 200 Oe) indicating that the surface modification leads to an increase of surface anisotropy.
Magnetization measurements as a function of temperature were performed after cooling the sample in zero field (ZFC curve), as well as after cooling it in the measuring field (FC) (fig.4). A broad maximum is observed in the ZFC curves and an irreversible magnetic behaviour in all the investigated temperature range. The difference between the two curves should be mainly due to the temperature variation of H_c. The irreversible behaviour as a function of the temperature was also studied by measuring the temperature dependance of the so called thermoremanent magnetiazation TRM (the remanent magnetization measured after cooling the sample in a magnetic field). An exponential decrease with temperature is observed .

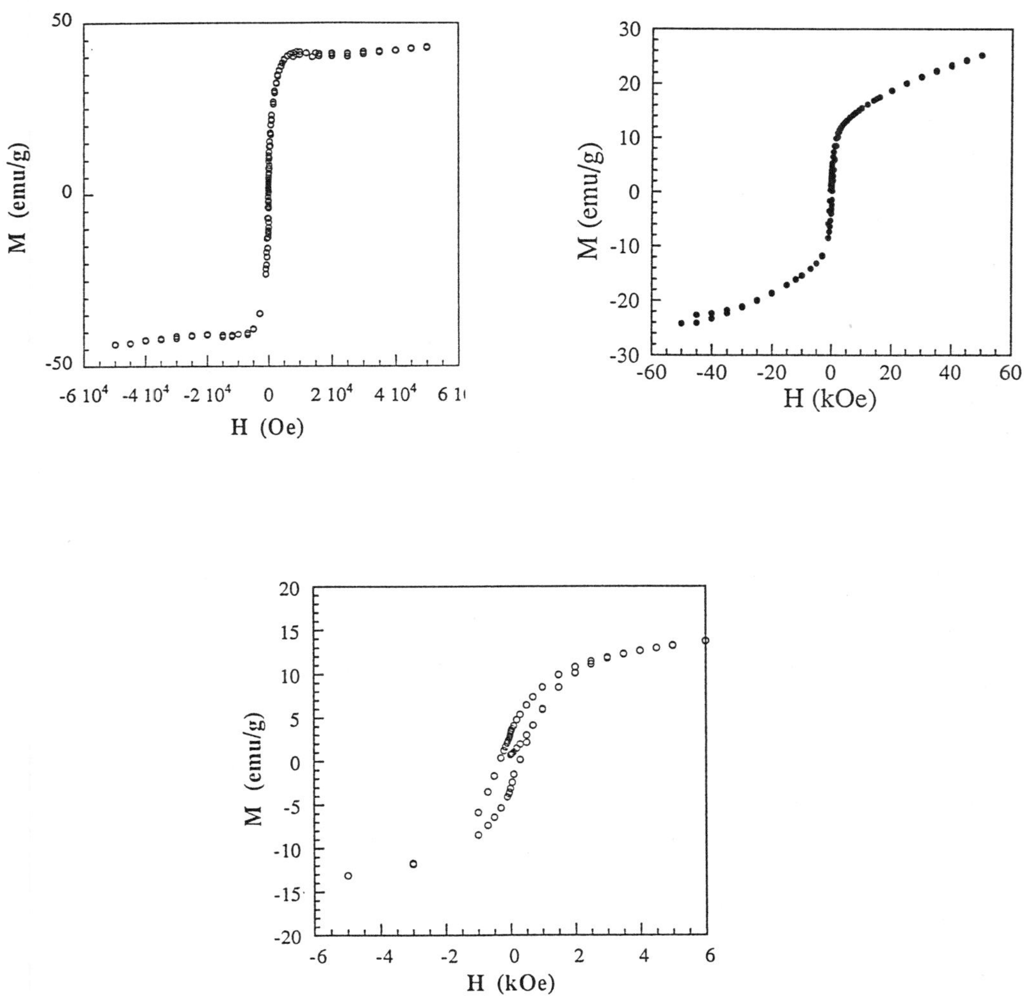

Fig3 (a) M vs H at T = 5 K for x = 25 ; (b) M vs H at T = 4.2 K for x = 70 ; (c) M vs H at T = 5 K for x = 60

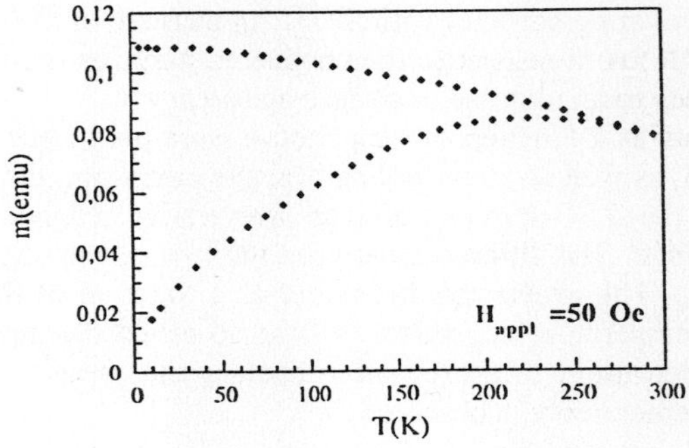

Fig. 4 M vs temperature for x = 40

CONCLUSIONS

The results indicate that Fe-Cr-B amorphous alloys powders, prepared by a chemical reduction method, present a quite large particle size aggregation state (chains or agglomerates) and the particle size distribution are found to depend on the preparation parameters (e.g. on the type of starting materials, on the way of mixing the solutions, pH...).The composition dependence of the magnetic properties, moving from a ferromagnetic to a complex disorder magnetic state (where ferromagnetic, antiferromagnetic and paramagnetic components coexist) with increasing x, is determined by the close relationship between chemical composition and particle size, and by the change of the relative fraction of the different iron species.

AKNOWLEDGMENTS

We thank for technical assistance Mr Gianni Chiozzini, Mr. Claudio Veroli, and Mr. Pasquale Filaci for thermal analysis, powder X-ray diffraction patterns, and magnetization measurements respectively.

REFERENCES

[1] E.Hellestern and L.Schultz, Appl.Phys.Lett.4,124(1986)
[2] X.L.Yeh, K.Samwer and W.L.Johnson, Appl.Phys.Lett.42,242 (1983)
[3] J.van Wonterghem, S.Morup, S.W.Charles,S.Wells and J.Villadsen, Phys.Rev.Lett. 55,410(1985)
[4] M.Takagi, Y.Kawamura, M. Araki, Y.Yuroyana and T.Imura, Mater.Scien.Eng.98, 457 (1988)
[5] S.Wells, S.W. Charles, S.Linderoth, J.Van Wonterghen, J.Larsen and M.B.Madsen, J.Phys.Cond.Mat. 1, 8199(1989)
[6] J.Rivas,M.A.Lopez Quintela,M.G.Bonome,R.J.Duro,J.M:BonomeJ.M.Greneche and J.Mag.Magn.Mat.122,1,(1993)
[7] L.Suber,H.Romero,D.Fiorani,A.Montone,M.Vittori Antisari,J.L.Dormann,J.Maknani and N.Sparvieri,Nanostruct.Mat,6,949,(1995)
[8] D.Fiorani,H.Romero,L.Suber,A.M.Testa,J.L.Dormann,J.Maknani and N.Sparvieri,JMMM 140-144,411(1995)

Materials Science Forum Vol. 195 (1995) pp. 215-220

Determination of the Magnetic Parameters of thin Ni Films by Brillouin Light Scattering

G. Gubbiotti, G. Carlotti and G. Socino

INFM, Dipartimento di Fisica dell'Università, Via Pascoli, I-06100 Perugia, Italy

Keywords: Brillouin Scattering, Magnetic Properties, Thin Films

Abstract

The Brillouin light scattering technique has been exploited in order to investigate the magnetic properties of Ni films about 20 nm thick, deposited by thermal evaporation onto sapphire, silicon and gallium arsenide substrates. Experiments were performed in the backscattering interaction geometry, using a Sandercock type, 3+3 pass tandem Fabry-Perot interferometer. Measurement of the frequency shift of the Damon-Eshbach (DE) surface spin wave versus the applied magnetic field H_o enabled us to determine the values of characteristic magnetic parameters of the Ni films, such as the spectroscopic separation factor and the surface saturation magnetization. The former parameter is in fairly good agreement with the literature value, while the latter shows to be lower than in bulk Ni and to depend appreciably on the substrate material. In addition, the exchange stiffness constant has been evaluated by studying the dependence of the DE wave on the angle of incidence of light. This proves that a systematic study of the DE wave can provide the amount of information necessary for a complete magnetic characterization of magnetic films, even in the case of thicknesses below 30 nm, where bulk standing spin waves are not experimentally accessible.

1. Introduction

In recent years great attention has been devoted to the study of the magnetic properties of thin magnetic films, for both fundamental and applicative reasons. The theoretical interest for layered systems is motivated by the fact that the reduced dimensionality strongly modifies their magnetic properties with respect to those of bulk materials [1]. As for the applicative aspect, magnetic thin films are finding steadily growing application in the tecnology of magnetic recording media [2,3].

A powerful non-destructive technique which has proved to be effective for the investigation of the magnetic properties of thin films and multilayers is Brillouin light scattering (BS), which relies upon detection of the small amount of light which is inelastically scattered by thermal spin waves. As far as layered structures are concerned, BS has many advantages over alternative methods for the investigation of spin waves, such as neutron scattering (NS) and microwave absorption (MA) [4]. In fact NS can probe a large wave vector range, but it is generally not sensitive enough for application to thin films, while MA can probe only excitations with wave vector close to zero.

In a BS experiment, a beam of monochromatic light is focused onto the surface of the specimen and the light scattered within a solid angle analysed by means of a multipass Fabry-Perot interferometer. The probe is therefore local, so that measurements performed on different points of the specimen can give information about its homogeneity. In addition, because of the conservation of momentum in the photon-magnon interaction, the wave vector of spin waves revealed in BS experiments is of about $10^5 cm^{-1}$, with frequencies in the range 1-100 GHz. In recent years the BS tecnique has been successfully applied to the study of magnetic properties of layered structures. In particular it has been shown that the simultaneous analysis of both surface and bulk standing modes in films with different thicknesses in submicron range permitted the magnetic parameter to be evaluated [1,4].

In this paper we present the results of a BS investigation of the magnetic properties of polycrystalline Nichel films, a few nanometers thick, deposited on different substrates. We show that a systematic study of the surface DE spin wave can supply the amount of information necessary for determining the characteristic magnetic parameters of the films. The g-factor and the surface saturation magnetization are obtained from the measured dispersion of the DE mode with the applied magnetic field, while the exchange stiffness constant is evaluated from the dispersion of the DE mode with the angle of incidence of light.

2. Experimental

The samples were produced by thermal evaporation of Ni at 250°C onto single-crystal [0001] sapphire, [001] silicon and [111] gallium arsenide. X-ray experiments with the scattering vector normal to the film plane (θ-2θ scan) have shown that these films have a polycrystalline structure consisting of a highly aligned columnar texture, with the crystallites oriented along the (111) crystallografic axis. The specimes analyzed are summarized in Table 1.

Table 1 *Characteristics of the specimens investigated*

Sample #	Substrate material	L (*nm*)
1	Si	20
2	GaAs	22
3	Al_2O_3	26

Brillouin spectra were taken in air, at room temperature, with typical acquisition time of two hours; the source was an Ar^+ laser operating on a single mode of the 514.5 nm line. An analyzer was placed at the entrance of the interferometer in order to select the light scattered by magnons which is cross-polarized with respect to the incident light. A p-polarized light beam, with power of 300 mW, was focused onto the surface of the specimens and the back-scattered light was analyzed using a Sandercock type, 3+3 tandem Fabry-Perot interfometer [5], shown in Fig. 1, characterized by a finesse of about 120 and a contrast ratio higher than 5×10^{10}.

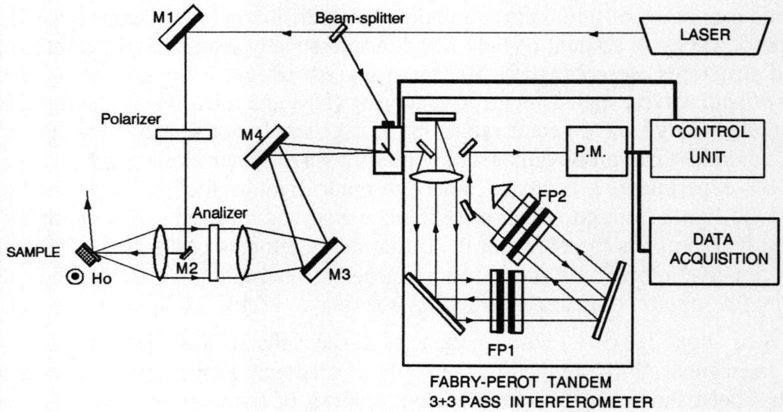

Fig.1 *Experimental apparatus used for Brillouin light scattering measurements.*

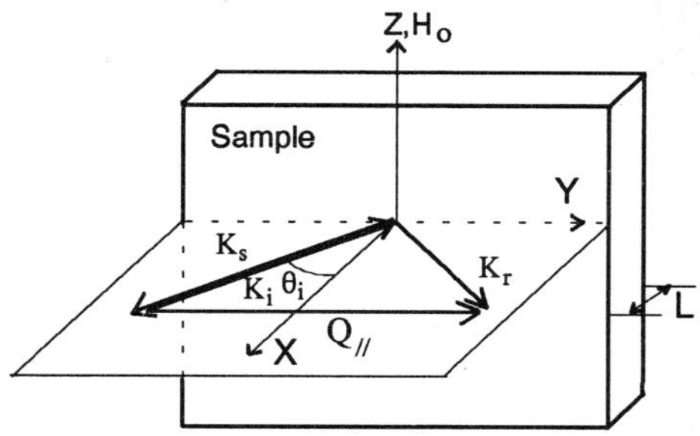

Fig.2 *Geometry of the Brillouin scattering measurements. K_i, K_r and K_s are the wave vectors of incident, reflected and scattered light, respectively, while $Q_{\|}$ is the magnon wave vector, θ_i the angle of incidence of light and H_o the applied magnetic field.*

The sample was placed between the poles of an electromagnet with the magnetic field H_o applied in the plane of the film, perpendicular to the scattering plane. In Fig.2 it is shown a schematic diagram of the wave vectors interaction geometry, from which it turns out that the wave vector $Q_{\|}$ of the surface magnon involved into the scattering process is given by:

$$Q_{\|} = 2K_i \sin\theta_i \qquad (1)$$

where K_i is the optical wavevector and θ_i the angle of incidence of light.

3. Brillouin scattering and magnetic parameters

There are two types of magnetic excitations in stratified ferromagnetic system: surface spin waves which are localized at the film interfaces and the bulk spin waves which are excitations of the whole magnetic system [1,4]. The former, known as Damon-Eshbach modes [6], are non-reciprocal waves; they can propagate along each surface of the film in only one direction across the magnetization. Referring to the coordinate system shown in Fig. 2, with the applied magnetic field along the Z direction, the DE waves propagate in the +Y direction at the surface X=0 and in the -Y direction at the surface X=-L. The two waves are degenerate in frequency and have the same penetration depth $1/Q_{\|}$. The angular frequency of the DE wave for propagation orthogonal to the magnetic field, taking into account both dipolar and exchange interaction, can be written as [7,8]:

$$\omega_s = \gamma\left[\left(H_o + 1.76DQ_{\|}^2 + 2\pi M_s\right)^2 - 4\pi^2 M_s^2\, exp\left(-2Q_{\|}L\right)\right]^{1/2} \qquad (2)$$

where γ is the gyromagnetic ratio, L the film thickness, D is the exchange stiffness constant and M_s the surface saturation magnetization, which is generally different from the bulk saturation magnetization [9]. Inspection of Eq.2 suggests that, in principle, the characteristic magnetic parameters γ, D and M_s can be obtained from a study of the dispersion of the angular frequency ω_s vs. the applied magnetic field H_o. However, since D is of the order of 10^{-9} Oe cm^2 and $Q_{\|}\cong10^5$ cm^{-1}, it follows that the exchange term $DQ_{\|}^2$ is negligible and the D constant cannot be obtained by a best fit procedure of the experimental data to the curve of Eq.2. On the other hand, one can make the fitting procedure sensitive to the $DQ_{||}^2$ term by fixing the value of H_o and varying $Q_{||}$, i.e. the angle of incidence of light. Therefore, a study of the angular dispersion of the DE mode permits the exchange constant D to be determined.

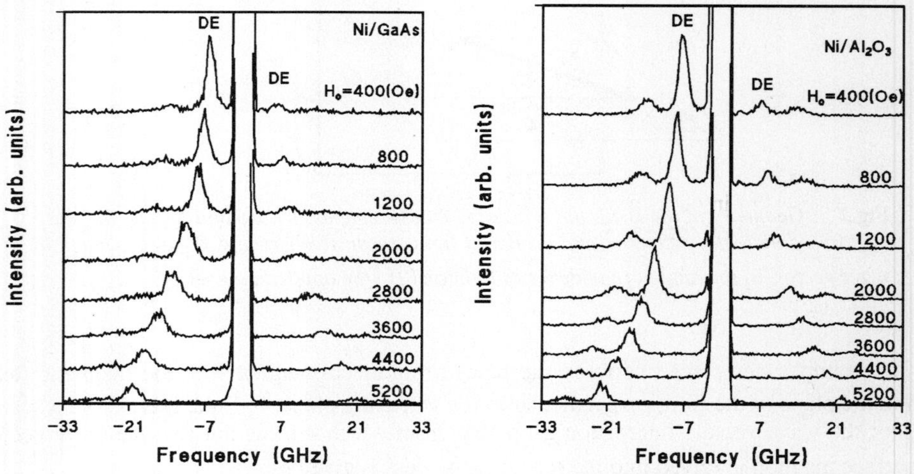

Fig.3 *BS spectra from specimens # 2 and 3 for an angle of incidence* θ_i*=50°*

4. Results and discussion

Fig.3 presents typical Brillouin spectra relative to the Ni films, supported by GaAs and Al_2O_3 substrates, for different values of the applied magnetic field. The DE wave peak is present in both the Stokes and the anti-Stokes side of the spectra, even if with a remarkable asymmetry of the intensity. For the films studied here, this is caused by the contribution of the off-diagonal spin-spin correlation function to the light scattering intensity, while the thermal population factor, the localization of the modes, and the characteristics of the magneto-optic tensor do not play a significant role in determining the Stokes anti-Stokes asymmetry [10]. We can observe a marked decrease in the intensity of the Brillouin peak associated with the DE wave when the applied magnetic field increases. This is mainly due to a decrease of the thermal amplitude of the DE wave [11]. The weak peak which appears at frequency higher than the DE wave, more clearly observable in the case of the sapphire substrate, corresponds to the first bulk standing mode.

In order to determine the characteristic magnetic parameters of the Ni films, we have first studied the frequency dispersion of the DE mode, varying the applied magnetic field between 400 and 5200 Oe, at a fixed angle of incidence of light $\theta_i = 50°$. A best fit procedure of the experimental data to Eq.2, in the dipolar limit, enabled us to obtain the values of the surface saturation magnetization M_S and of the g-factor ($g=\hbar\gamma/\mu_B$), which are summarized in Table 2. The values obtained for g-factor are in fairly good agreement with those already measured in Ni films by other authors (g=2.2) [4,12], while the value obtained for M_S can only be compared with those of the bulk saturation magnetization (M_B=480 Oe) [12]. To this respect, it turns out that the surface saturation magnetization M_S is smaller than M_B and the deviation from this value is more marked in the case of silicon and gallium arsenide substrates, because of the formation of amorphous compounds at the interface [13,14]. The presence of such a modified layer is confirmed by the fact that, different from the case of sapphire substrate, for Ni/Si and Ni/GaAs films with thicknesses below 10 nm no magnetic excitation could be observed; in addition for thicker Ni/Si films the analysis of the bulk standing spin waves gives evidence of a marked surface anisotropy [15].

Table 2 *Characteristic magnetic parameters of the specimens investigated*

Substrate material	M_S (*Oe*)	g	D (*Oe cm*^{2})
Al_2O_3	405±25	2.15±0.03	3.0±0.3
Si	266±15	2.15±0.04	3.1±0.4
GaAs	301±15	2.14±0.04	3.0±0.2

The second step in our analysis deals with the determination of the exchange constant D. To this aim, we have followed the approach outlined in the previous paragraph, i.e. for a fixed value of the applied magnetic field H_o, we have studied the dispersion of the DE wave with the angle of incidence of light θ_i which was changed between 35° and 75°. As shown in Fig.4, when θ_i increases the DE peak exhibits a slight shift to higher frequency, because $Q_{\parallel}$ depends on θ_i according to Eq.1. Accurate measurement of the frequency shift enabled us to determine the value of the exchange stiffness constant D (last column of Table 2), using for g and M_S the values obtained before and given in the first two columns of Table 2. The experimental values of D do not exhibit any appreciable dependence on the substrate material and are in fairly good agreement with literature values obtained from a study of the bulk standing modes in thicker films [12]. In addition to the moderate frequency shift, Fig. 4 also shows that the intensity of the DE peak undergoes a marked decrease when the incidence angle increases. This is mainly due to the presence of the term $\cos^2(\theta_i)$ in the expression of the light scattering intensity, while the decrease of the thermal amplitude of the mode plays a negligible role [11].

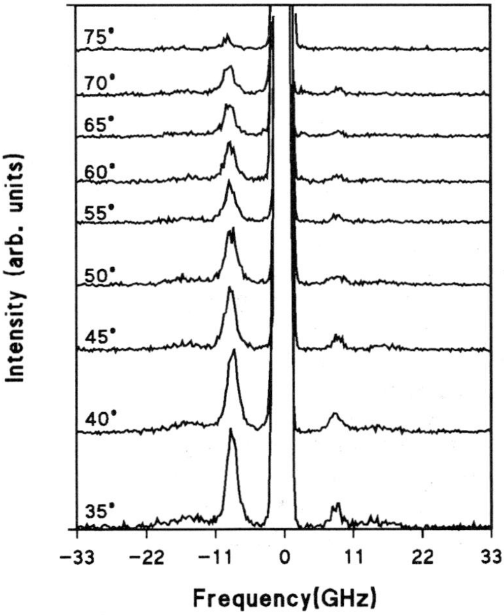

Fig.4 *Angular dispersion for the Damon-Eshbach wave for fixed applied field.*

5. Conclusions

The Brillouin light scattering technique has been used in order to determine the magnetic parameters of thin polycrysttaline Ni films, about 20 nm thick, deposited on three different substrates. We have shown that a systematic study of the frequency dispersion of the DE wave can supply the amount of information necessary for determining the characteristic magnetic parameters of the films. In particular, the g-factor and the surface saturation magnetization can be obtained taking measurements at different applied fields, while analysis of the frequency change caused by a variation of the incidence angle leads to the determination of the exchange stiffness constant D. We notice that to our knowledge this is the first experimental determination of the D constant in magnetic films as thick as a few nanometers. This is because the only method used in the past for determining the value of D relied upon the study of the ω vs. H_0 dispersion of the bulk standing modes, which are only accessible in films of thickness higher than about 30 nm.

Acknowledgements

The authors gratefully acknowledge Dr. E. Verona and Dr. R.Felici for preparing the Ni films and Dr C. Petrillo for the x-ray experiments.

References

[1] S. Demokritov, E. Tsymbal, J.Phys. **C6**, 7145 (1994)
[2] For a recent overview, see Physics Today, Vol.**48**, Iss.4, April 1995; MRS Bullettin, Vol.**XV** Iss.3 (1990)
[3] E. Terenzio, P. Visigalli, in *Magnetic properties of matter* ed. by G. Asti, D. Fiorani and F. Lucari, (World Scientific, Singapore, 1991) p.576; P.L. Cavallotti, A Cecchetti and R.M. White, in *Magnetic properties of matter* ed. by V. Tognetti and F.Borsa, (World Scientific, Singapore, 1988) p.511
[4] P. Grunberg, in *Light scattering in solids* V, Ed. M. Cardona e G. Guntherodt
[5] F. Nizzoli and J.R. Sandercock, in *Dynamical Properties of Solids*, vol. 6, ed. by G.K. Horton and A.A. Maradudin (North Holland, Amsterdam, 1990) p.307; for the experimental apparatus used here, see also G. Carlotti, D. Fioretto, L. Palmieri, G. Socino, L. Verdini and E. Verona., IEEE Trans. Ultrason. Ferroelectr. Frequency Control, **38**, 56 (1991)
[6] R.W. Damon, J.R. Eshbach, J.Chem. Sol. Vol.19, 308 (1961)
[7] T. Wolfram, R.E. De Wames, J.Appl.Phys.**41**, 987 (1970)
[8] R.E. Camley, L.Mills, Phys.Rev.**B18**, 4821 (1978)
[9] J.R. Sandercock, W.Wettling, J.Appl.Phys. **50(11)**, 7784 (1979)
[10] R.E. Camley, L.Mills, Phys.Rev.**B26**, 2609 (1982)
[11] M.G. Cottam, J.Phys.C16, 1573 (1983)
[12] P. Grunberg, C.M. Mayr, W. Vach, M. Grimsditch, J. Magn. Magn. Mat. **28**, 319 (1982)
[13] D.M. Vanderwalker, Appl.Phys. Lett.**48(11)**, 707 (1986)
[14] A.J. Freeman, A. Continenza and Chun Li, MRS Bullettin, Vol.**XV**, Iss.3, 30 (1990)
[15] G. Gubbiotti, Tesi di Laurea, Università di Perugia (1995)

AUTHOR INDEX

KEYWORD INDEX